GROUP THEORY IN PHYSICS

Group Theory in Physics

Editor:

Maria Emilova Velinova, Ph.D.

ArclerPress

www.arclerpress.com

Group Theory in Physics
Editor: Maria Emilova Velinova, Ph.D.

708 3rd Avenue, 6th Floor
New York
NY 10017
United States of America
www.arclerpress.com

ISBN: 978-1-68094-438-9

Library of Congress Control Number: 2016951129

About the Editor

Maria Emilova Velinova, Ph.D.

Maria Velinova is Ph.D. holder in Quantum chemistry at the University of Sofia since April 2012. Her major research experience is in the field of Computational Chemistry, especially in statistical mechanics methods applied to different sorts of biomolecules. Member of the Laboratory of Quantum and Computational Chemistry at the University of Sofia.

Contents

List of Contributors

Rudolf Schmid
Department of Mathematics, Emory University, Atlanta, GA 30032, USA

M. Nadjafikhah
School of Mathematics, Iran University of Science and Technology, Narmak, Tehran 1684613114, Iran
A. Mokhtary
Department of Complementary Education, Payame Noor University, Tehran 19395-3697, Iran

Muhammad Awais
Department of Mathematics, COMSATS Institute of Information Technology, Attock, 43600, Pakistan

Tasawar Hayat
Department of Mathematics, Quaid-I-Azam University 45320, Islamabad, 44000, Pakistan Abdul Wali Khan University Mardan, PAKISTAN

Sania Irum
Department of Mathematics, COMSATS Institute of Information Technology, Attock, 43600, Pakistan

Salman Saleem
Department of Mathematics, COMSATS Institute of Information Technology, Attock, 43600, Pakistan

Tetsuya Misawa
Faculty of Economics, Nagoya City University, Nagoya 467-8501, Japan

W. A. Khan
Department of Engineering Sciences, PN Engineering College, National University of Sciences & Technology, Karachi 75350, Pakistan

Md. Jashim Uddin
School of Mathematical Sciences, Universiti Sains Malaysia, Penang 11800, Malaysia
American International University-Bangladesh, Banani, Dhaka 1213, Bangladesh

A. I. Md. Ismail
School of Mathematical Sciences, Universiti Sains Malaysia, Penang 11800, Malaysia

Gözde Sarı
Department of Mechanical Engineering, Celal Bayar University, Muradiye, 45140 Manisa, Turkey
Applied Mathematics and Computation Center, Celal Bayar University, Muradiye, 45140 Manisa, Turkey

Mehmet Pakdemirli
Department of Mechanical Engineering, Celal Bayar University, Muradiye, 45140 Manisa, Turkey
Applied Mathematics and Computation Center, Celal Bayar University, Muradiye, 45140 Manisa, Turkey

Tasawar Hayat
Department of Mathematics, Quaid-i-Azam University 45320, Islamabad 44000, Pakistan
Department of Mathematics, King Saud University, P.O. Box 2455, Riyadh 11451, Saudi Arabia

Yiğit Aksoy
Department of Mechanical Engineering, Celal Bayar University, Muradiye, 45140 Manisa, Turkey

M. B. Akgül
Department of Mechanical Engineering, Celal Bayar University, Muradiye, 45140 Manisa, Turkey

G. Sarı
Applied Mathematics and Computation Center, Celal Bayar University, Muradiye, 45140 Manisa, Turkey

M. Pakdemirli
Applied Mathematics and Computation Center, Celal Bayar University, Muradiye, 45140 Manisa, Turkey

Jian-Qiang Sun
Department of Mathematics, College of Information Science and Technology, Hainan University, Haikou 570228, China

Rong-Fang Huang
Department of Mathematics, College of Information Science and Technology, Hainan University, Haikou 570228, China

Xiao-Yan Gu
Department of Physics, East China University of Science and Technology, Shanghai 200237, China

Ling Yu
Department of Mathematics, College of Information Science and Technology, Hainan University, Haikou 570228, China

Wenjie Yu
College of Electromechanical Engineering, Qingdao University of Science and Technology, Qingdao, P.R. China
College of Information Engineering, Qingdao University, Qingdao, P.R. China

Zhenkuan Pan
College of Information Engineering, Qingdao University, Qingdao, P.R. China

V V Kornyak
Laboratory of Information Technologies, Joint Institute for Nuclear Research, 141980 Dubna, Russia

Vita Solomko
Department of General and Theoretical Physics, Sumy State University, 2 Rimsky-Korsakov Street, Sumy 40007, Ukraine

Petro Kondratenko
Department of Theoretical Physics, Aerospace Institute, National Aviation University, 1 Kosmonavta Komarova Prospekt, Kiev 03680, Ukraine

Yuriy Lopatkin
Department of General and Theoretical Physics, Sumy State University, 2 Rimsky-Korsakov Street, Sumy 40007, Ukraine

Giuseppe Nistic`o
Dipartimento di Matematica e Informatica, Universit`a della Calabria, Italy, and INFN gr. Collegato di Cosenza, Italy

Huaqing Huang
Department of Physics and State Key Laboratory of Low-Dimensional Quantum Physics, Tsinghua University, Beijing 100084, People's Republic of China

Fawei Zhen
LCP, Institute of Applied Physics and Computational Mathematics, Beijing 100088, People's Republic of ChinaBeijing Computational Science Research Center, Beijing 100084, People's Republic of China

Ping Zhang
LCP, Institute of Applied Physics and Computational Mathematics, Beijing 100088, People's Republic of China
Beijing Computational Science Research Center, Beijing 100084, People's Republic of China

Jian Wu
Department of Physics and State Key Laboratory of Low-Dimensional Quantum Physics, Tsinghua University, Beijing 100084, People's Republic of China

Bing-Lin Gu
Institute for Advanced Study, Tsinghua University, and Beijing 100084, People's Republic of-China
Collaborative Innovation Center of Quantum Matter, Tsinghua University, Beijing 100084, Peoples Republic of China

Wenhui Duan
Department of Physics and State Key Laboratory of Low-Dimensional Quantum Physics, Tsinghua University, Beijing 100084, People's Republic of China
Institute for Advanced Study, Tsinghua University, and Beijing 100084, People's Republic of-China

Collaborative Innovation Center of Quantum Matter, Tsinghua University, Beijing 100084, Peoples Republic of China

Eugene Kogan
Department of Physics, Bar-Ilan University, Ramat-Gan, Israel

Guillaume Vasseur
Institut Jean Lamour, UMR 7198, Université de Lorraine, B.P. 70239, 54506 Vandœuvre-lès-Nancy, France

Yannick Fagot-Revurat
Institut Jean Lamour, UMR 7198, Université de Lorraine, B.P. 70239, 54506 Vandœuvre-lès-Nancy, France

Bertrand Kierren
Institut Jean Lamour, UMR 7198, Université de Lorraine, B.P. 70239, 54506 Vandœuvre-lès-Nancy, France

Muriel Sicot
Institut Jean Lamour, UMR 7198, Université de Lorraine, B.P. 70239, 54506 Vandœuvre-lès-Nancy, France

Daniel Malterre
Institut Jean Lamour, UMR 7198, Université de Lorraine, B.P. 70239, 54506 Vandœuvre-lès-Nancy, France

Preface

Group theory studies that has been considered by physicists as a very valuable tool for the clarification of the symmetry aspects of physical problems. The book intents to describe in detail the most important fundamental ideas of the group theory, its significant developments and various applications in: Hamiltonian systems, nonlinear systems, thermo-fluid dynamics, quantum mechanics and solid-state physics. In particular, different applications of Lie's group theory to the above said fields are shown.

The examination of the exact solutions of nonlinear equations takes an important place in physics. One of the noteworthy and efficient methods for gaining solutions of systems of nonlinear differential equations is the classical symmetries method, also called Lie's group analysis. This method is employed for the constructions of solutions for the magnetohydrodynamic (MHD) flow of an upper-convected Maxwell (UCM) fluid over a porous shrinking wall, for the boundary layer equations for the Sisko fluid, and for a two-dimensional, unsteady flow and heat transfer of a viscous fluid over a surface in the presence of variable suction/injection. Another interesting application about the design of Lie's group integrators of multibody system dynamics is presented. The quantum behavior of a physical system is a natural consequence of its symmetries. Hence, it is a fundamental to study the invariants of symmetry groups of them. In particular, invariant bilinear forms are very important for quantum physics, because these forms provide the link between mathematical description and experimental observations. The group theoretical analysis of the electronic and vibrational structure of the trimethine cyanine dye molecules is described. Other example of application of the group theory in the quantum mechanics is the establishment of a method for the description of an interacting spin-0 particle. The electronic energy band structure is a basic theory in condensed matter physics and can be used to study many physical properties of crystal materials. Here are presented a general method to unfold energy bands of supercell calculations to a primitive Brillouin zone and the results of the symmetry classification of the electron energy bands in graphene and silicene. The band degeneracy at high symmetry points or the existence of energy gaps, usually reflect the symmetry of the crystal, and this property is analyzed by considering two-dimensional (2D)-hexagonal lattices.

Editor

Maria Emilova Velinova, Ph.D.

INTRODUCTION

Quantum mechanics showed that the elementary systems that matter is made of, such as electrons and protons, are truly identical, not just very similar, so that symmetry in their arrangement is exact, not approximate as in the macroscopic world. Systems were also seen to be described by functions of position that are subject to the usual symmetry operations of rotation and reflection, as well as to others not so easily described in concrete terms, such as the exchange of identical particles. Elementary particles were observed to reflect symmetry properties in more esoteric spaces. In all these cases, symmetry can be expressed by certain operations on the systems concerned, which have properties revealed by Group Theory, a rather obscure branch of mathematics that had previously been mainly a curiosity without practical application.

Physics uses that part of Group Theory known as the theory of representations, in which matrices acting on the members of a vector space is the central theme. It allows certain members of the space to be created that are symmetrical, and which can be classified by their symmetry. It is found that all the observed spectroscopic states of atoms and molecules correspond to such symmetrical functions, and can be classified accordingly. Among other things, it gives *selection rules* that specify which transitions are observed, and which are not. These matters are so commonplace in spectroscopy that the fact that they are extraordinary and wonderful is hardly realized.

In introducing a subject, especially one as abstract as Group Theory, it is important to begin with concrete, explicit examples and not with general principles. The lofty and general approach may look elegant to its author who already has a good mental appreciation of the subject, but is totally opaque to the beginner. Let us start, then, with a function f(x). If f(-x) = f(x) we have an *even* function. Graphing it, its symmetry is evident. A function such that f(-x) = -f(x) is also symmetric; it is an *odd* function. We can express any function as the sum of an even and an odd part: f(x) = [f(x)+f(-x)]/2 + [f(x)-f(-x)]/2 = f(even) + f(odd). We are doing pretty well here without any Group Theory, and this is not unusual. Anything that Group Theory does can also be done without it, and in many places physicists and chemists have gone ahead algebraically instead of learning Group Theory, often proudly. However, not using Group Theory is like not using a map--you never see the big picture and may go down many blind paths. Group theory can give you a lot of information with very little input.

Introduction to Groups and Representations

Well, then, into Group Theory. Let the symbol σ stand for the transformation x = -x. In three dimensions, this would be a reflection in the yz-plane. We can use σ as an *operator*: $\sigma f(x) = f(-x)$. Then, $\sigma f(-x) = f(x)$ as well. Indeed, $\sigma\sigma f(x) = f(x)$ for any f(x), so it looks like $\sigma^2 = E$, where E is the *identity operator*, such that $Ef(x) = f(x)$ for any f(x). E stands for *Einheit*, and the letter I is also used. The *product* of two operations simply means to do the two transformations one after the other. It has nothing to do with arithmetic multiplication in this case. Now, we see that if we make any products we like from the pair E, σ we never get anything other than E or σ. This means that the set $\{E,\sigma\}$ is *closed* with respect to their products in the current sense.

If you look at transformations in general, you notice that there is always the unit transformation that changes nothing, like E above. Also, if you make a transformation, there should be a way to reverse it. Otherwise, you will lose information that you cannot get back, and this is not good. We see that σ reverses itself; it is its own *inverse*, or $\sigma^{-1} = \sigma$. The exponent -1 is used to denote an inverse, by analogy with the usual rules for exponents. Finally, and here is something you might overlook, the effect of a series of transformations is the same however they are grouped, just so long as the order is not changed. This may be expressed by (AB)C = A(BC) = ABC. This is rather obvious for transformations, but might not be with other kinds of operations. Therefore, the group product must be *associative*. This happens automatically when we begin with a concrete representation of a group, but is very restrictive if we are just building up groups abstractly.

A *group* is a set of members {R} that is (1) closed under the group operation, so that the product of any two members is again a member of the group; (2) the group operation is associative; (3) {R} includes the unit element E; and (4) includes the inverse R^{-1} to any element R, such that $RR^{-1} = R^{-1}R = E$. Closure is the main property that is used. The others just ensure that the corners are tucked in. The number of member's g is the *order* of the group. Groups may be finite, if g is finite, or infinite. Infinite groups are not very different from finite groups, but have some extra interesting properties. The set of operations $\{E, \sigma\}$ that we looked at form a group of order 2 that we can call C_i. This group is too small to show some of the more interesting aspects of groupp structure, but it is indeed a group, and an important one.

Now consider some other sets, such as the signed integers {1, -1} under ordinary multiplication, or the inversion transformation I f(x,y,z) - f(-x, -y, -z), or the exchange operation X f(x,y) = f(y,x), together with the identity E in the latter two cases. If we call the elements E and A, we see that all obey the same *multiplication table*. This table defines the *abstract group* C_i, which has many *representations*, or concrete realizations. The members of the group are listed in the first row and first column. We can agree that the result of any product is found at the intersection of the row corresponding to the first factor and the column of the second factor. As operators, we consider them applied from right to left. That is, EA means we do A first, then E, and the result is found at the intersection of the E row and the A column in the multiplication table. I am sorry for this complication, but things have to be kept straight, and many authors do not make clear what conventions they observe.

The Group C_i

E	A
A	E

One group structure, as defined by the multiplication table, can have many representations, which are concrete groups *isomorphic* to each other. Isomorphism implies a one-to-one correspondence of distinct elements of the groups. The group properties of isomorphic groups are exactly the same.

As an exercise, investigate the set of numbers that arise when all powers of the imaginary unit i are considered, and show that they form a group of order 4 under ordinary multiplication, writing down the multiplication table. This is the abstract group C_4. If A and B are any two members of this group, then AB = BA, which means that the order of the members does not matter (this is, a fundamental property of multiplying numbers, after all). A and B are said to *commute*, and the group is said to be *commutative*, or *Abelian*. We noticed that {1, -1} was a representation of C_i as well as {E, σ}, and that these numbers appeared before an odd f(x) when operated on by E or σ. Also, {1, 1} obeys the multiplication table and appeared before an even f(x). All Abelian groups have representations by numbers in this way, with 1 corresponding to E. One of the representations consists only of 1's, and the number of distinct representations is equal to the order of the group. You already know two of the representations of C_4. Can you find the other two? One has +i and -i interchanged; the other has two -1's. Considered as g-dimensional vectors, these representations all have the squares of their lengths equal to g, and the scalar product (sum of products of the elements) of two different ones is zero. The representations containing i's of C_4 must be massaged a little to make them orthogonal to each other, but it can be done.

This may be mildly interesting so far, but the really interesting stuff happens when, for some members of a group, AB ≠ BA. That is, they do not *commute*, and the group is then called *non commutative.* As Schensted points out, the fact that the smallest non commutative group has g = 6 is something of an instructional disaster. However, we can make an important point simply from the fact of non-commutatively. We have seen that we can find numerical representations of Abelian groups, which is natural, because numbers commute. To find a representation of a non-commutative group that preserves the non commutativity, we have to go beyond numbers to something that can be non-commutative. Now, matrices have this property, so we require matrices to form some representations of non-commutative groups. If we consider operation on functions, this means that we operate on *sets* of functions that act like vectors, for matrices operate on vectors. This occurrence of functions in natural families is exciting: they represent degenerate states (states with the same energy) in quantum mechanics.

$$\begin{bmatrix} 0 & 1 \\ 1 & 0 \end{bmatrix} \begin{bmatrix} 1 & 0 \\ 0 & -1 \end{bmatrix} = \begin{bmatrix} 0 & -1 \\ 1 & 0 \end{bmatrix}$$

$$\begin{bmatrix} 1 & 0 \\ 0 & -1 \end{bmatrix} \begin{bmatrix} 0 & 1 \\ 1 & 0 \end{bmatrix} = \begin{bmatrix} 0 & 1 \\ -1 & 0 \end{bmatrix}$$

An example of non-commuting of matrices is shown at the left. If you have forgotten how to multiply matrices, the i_j element in the product is the sum of the products of elements in the i_{th} row of the first matrix with elements in the j_{th} column of the second matrix, each by each. For example the 11 element in the top product (row 1, column 1) is 0 x 1 + 1 x 0 = 0, the 12 element 0 x 0 + 1 x -1 = -1, and so on. These matrices are members of a group of order 8, and you can find all the other members by multiplying matrices until you get no new ones. Each of the matrices multiplied in the Figure is its own inverse, which can easily be seen by multiplication. The fourth powers of the product matrices are the identity, and their product is the identity, so they are inverses of each other. The group of these matrices is called C_{4v}, and we will come back to it later.

Symmetry was first studied mathematically with reference to crystal symmetry. If you rotate a crystal by certain angles about certain axes, or reflect it in certain planes, you find equivalent faces in equivalent places. It took genius to realize that under the imperfect symmetry of actual crystals there was an exact symmetry reflected in the orientations of crystal faces, not their apparent shapes, which are greatly affected by accidents of growth. A symmetry operation took a face of a certain orientation into an equivalent face at a symmetrical orientation. Crystals are an interesting, but somewhat difficult case to study, so we will introduce spatial symmetry by operations that keep one point fixed, which make up the *point* groups. The symmetry operations are rotations about an axis, and reflections in a plane (both of which pass through the one point). C_n will be the symbol for a rotation of $2\pi/n$ radians about a certain axis, σ_h for a reflection in a plane perpendicular to an axis, and σ_v for a reflection in a plane containing the axis. A symmetry operation takes a symmetrical figure into itself; the result is indistinguishable from the original. The symmetry operations of a given figure form a group, since the operations satisfy all four group postulates.

Let us suppose that a rotation C_3 around a certain axis, and a σ_v containing this axis, are symmetry elements of a certain figure, which may be, for example, an ammonia molecule, NH_3. By multiplying these elements (performing them successively) we find that the set of six operations E, C_3, C_3^2, σ, σ', σ" is closed, and forms a finite group of order 6, which we shall call C_{3v}. It is easier to appreciate the group operations using the diagram on the right. We are looking directly down the 3-fold axis, and the original symmetry plane is denoted by σ. Rotations carry this plane into equivalent symmetry planes σ' and σ". A general point P is carried into the other two solid points by rotations, and into the open points by reflections in the three symmetry planes. From this diagram, it should be easy to write down the multiplication table for the group, and to verify that it is non-commutative, since rotations and reflections will be found not to commute.

There is a curious interaction between the rotations and reflections in this group. Suppose we reflect P in σ, then rotate clockwise by C_3, and finally reflect again in σ. The result is the same as rotating anticlockwise by C_3^2. Algebraically, we can write $\sigma^{-1} C_3\sigma = C_3^2$, since σ is its own inverse. If we do the same thing with a C instead of a σ, we always get a C back, never a σ That is, if we consider the products $A^{-1}CA$, where C is either of the rotations, and A is any member of the group, we get a C. In an Abelian group, this result is not interesting, since we always get back the same individual element: $A^{-1}BA = A^{-1}AB = B$. Here, however, we put in C_3 and got back C_3^2. All the members of a group that are associated in this way form a class. The members of a class are all of a similar nature, simply in a different orientation. For our group, now consider the products $C^{-1}\sigma C$, and show that the three reflections also fall into a single class. E is obviously always in a class by itself, since it commutes with every member. C3v, therefore, has three classes, E, 2C, and 3σ. It turns out to be very important to group these similar elements.

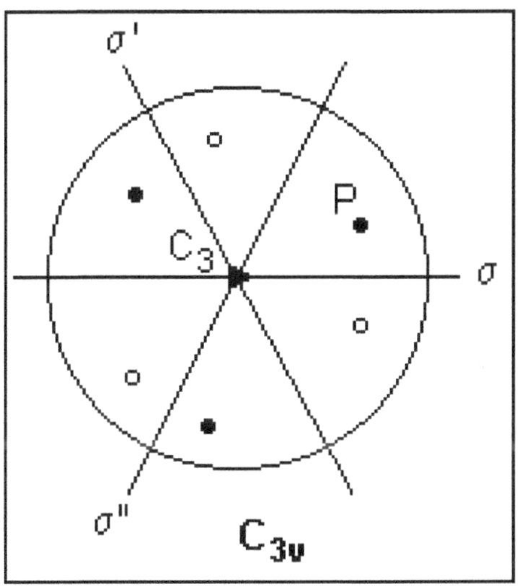

Let us now generate some representations of the group. We can set up rectangular coordinates x,y,z, perhaps with z along the symmetry axis and x along one of the symmetry planes, with y making a right-handed system. Then point P can be expressed as (x,y,z). Any symmetry operation will take these coordinates into (x',y',z') which are linear functions of the original coordinates. Explicitly, the transformation can be expressed as 3 x 3 matrices acting on 3-dimensional column vectors. There will be six matrices, and E will correspond to the indentity matrix. The matrices are easily found from the formulas for rotation of coordinates. We notice that z is not affected. It is multiplied by 1 in each transformation and the values of x,y are never mixed in. Therefore, z is the basis of a representation all by itself, the unit representation where each member corresponds to 1. The values of x and y do get mixed under the transformations, and it is easy to see that no choice of coordinates could ever change this. They form the basis of a two-dimensional matrix representation. The representation in terms of (x,y,z) is said to be *reducible* to the two smaller representations, and each of these is *irreducible* to smaller representations. It is the irreducibe representations that give the essential information; reducible representations can be formed ad libitum, and are a dime a dozen. A 1-dimensional representation is naturally irreducible, since there is nowhere lower to go.

It was easy to find the reducibility of the vector representation because we chose the coordinates wisely. If we had taken the coordinate axes in some arbitrary position, we would have obtained six rather full matrices and the reducibility with a fortunate selection of axes would not be at all obvious. Any representation changes its explicit form if we make a different choice of basis, but it is still the same at heart. How can we find out what it is really like? The answer is in the realization that choosing a different basis is a linear transformation y = Ax, under which the matrices R of a representation change into R› = A^{-1}RA. This is called a *similarity* transformation. Is there something invariant under such a transformation? Yes, indeed: the diagonal sum or *trace* of the matrix R. That is, Tr R› = Tr R, and this trace is called the *character* of the member in these *equivalent* representations, often denoted by $\chi(R)$. Two equivalent representations differ only in the choice of basis functions use to express them explicitly; their basic nature is the same. The converse of the theorem we have just established is the important result: if the characters of two representations are the same, then the representations are equivalent, and we can look for a matrix A connecting them with hope of success. This result, and others that we will quote, apply to *unitary* matrices, and are certainly true for rotation and reflection matrices.

If a representation is reducible, it can be expressed with the matrices for irreducible representations along the diagonal by a suitable choice of coordinates. Therefore, the character of the reducible representation is the sum of the characters for the irreducible representations that comprise it.

The characters of matrices representing the members of a class will be the same, as we know that the members of a class are related by similarity transformations. The number of irreducible representations of a group is equal to the number of classes. This is one of the remarkable theorems about group representations whose proof we will leave to the references. We can construct a table showing the characters of its irreducible representations that turns out to be very useful. For C_{3v}, the table is shown at the left. The letter A usually designates a one-dimensional representation with characters 1 for the rotations, and E a two-dimensional representation. Note that the sum of the squares of the characters, summed over all elements, is equal to the order of the group. For representation E, this is 2x2 + 1x1 + 1x1 + 0x0 + 0x0 + 0x0 = 6. Also, considered as group vectors with 6 components, the characters are orthogonal to each other. These surprising relations are the reason why there is a finite number of irreducible representations, and why there is one for each class.

We can now show one of the real uses of group theory. Consider the vector representation of C_{3v} that we constructed above. Its characters were 3, 0, 1. The sum of squares is 3x3 + 2x0x0 + 3x1x1 = 12, greater than g = 6. Therefore, it is reducible. If we take the scalar product of this group vector with one of the vectors of an irreducible representation, we will find the number of times this representation is contained in the reducible representation, times the order of the group. For A_1, we get 3x1 + 2x0x1 + 3x1x1 = 6. For A_2, we have 3x1 + 2x0x1 + 3x1x(-1) = 0. For E, we find 3x2 + 2x(-1)x0 + 3x1x0 = 6. Therefore, our reducible vector representation is reducible to A_1 + E. We could have determined this even with a bad choice of coordinates, since the characters would be unaffected. It is often easier to find the characters of a representation than the explicit matrices.

C_{3v}	E	2C	3σ
A_1	1	1	1
A_2	1	1	-1
E	2	-1	0

Let›s now return to the group C_{4v}, which we mentioned briefly above. To generate this group, start with a 4-fold axis and one reflection plane. Draw a diagram like the one for C_{3v}. As you consider the symmetrical positions into which a general point P is carried by the members of the group, you will see diagonal symmetry planes appear as if by magic, arising from products like $C_4\sigma$. The rotations now fall into two classes, one containing $C_4^2 = C_2$ alone, and one the other two rotations. The reflections are also divided into two classes, one class in the right angles between members of the other class. There are five classes, and so five irreducible representations. The new one-dimensional representations B_1 and B_2 appear. The character table is shown at the left.

C_{4v}	E	$2C_4$	C_2	2σ	2σ'
A_1	1	1	1	1	1
A_2	1	1	1	-1	-1
B_1	1	-1	1	1	-1
B_2	1	-1	1	-1	1
E	2	0	-2	0	0

Find the vector representation for this group, and show that it reduces to A_1 + E, as for C_{3v}, but the matrices are completely different. In fact the matrix elements are only 0, +1 and -1, and it is not difficult to find matrices for the E representation if the coordinates are chosen wisely. We have already seen some of them above. Consider the function $x^2 - y^2$. Show that it is transformed into itself by all group operations, so that it is a basis for a one-dimensional representation. This can be done without algebra, merely by considering the effects of the operations on x and y. To which representation does it belong? Do the same for the functions xy and $x^2 + y^2$.

A subset of the members of the group may form a group by themselves. This subset must contain the identity E, of course, and the inverses of each member. It is called a *subgroup*. A subgroup made up of whole classes is called an *invariant subgroup*. The order of a subgroup is a factor of the order of the group (Lagrange's Theorem). Improper subgroups are the identity E alone and the whole group; all others are *proper* subgroups. The members of a group that commute with all members of the group form an Abelian subgroup called the *center* of the group. There are many algebraic results such as these that aid an understanding of the structure of groups, but are not directly applicable to representations. Any representation of a group also gives a representation of a subgroup, but an irreducible representation of the group may give a reducible representation of a subgroup.

The n! permutations of n identical objects also form a group, the *symmetric group* P_n. You should be able to convince yourself that P_3 is isomorphic to C_{3v} by setting up a one-to-one correspondence of the members of each group. The diagrams that were presented above may help, if the point P and its images are numbered. Subgroups of P_n are also called permutation groups. For example, the even permutations in P_n (those that can be accomplished by an even number of interchanges) form a group of order n!/2 called the *alternating* group A_n. Any finite group is isomorphic to a certain permutation group (Cayley's Theorem). In fact, permutation groups were the first ones studied, and there is considerable lore on their representations.

The Four Major Theorems

Most of the results of matrix representation theory that are useful in physics are derived from the following four theorems.

- A matrix that commutes with all matrices of an irreducible representation can only be a multiple of the identity matrix.
- If A and B are the matrices of two irreducible representations of a group, of dimension n and m respectively, then if there is a matrix M with m rows and n columns such that MA = BM for all members of the group, it is the null matrix if $m \neq n$, and if m = n, it is either a null matrix or a matrix with non-vanishing determinant. In the latter case, the inverse M^{-1} exists, and the representations A, B are equivalent.
- If A and B are two in equivalent, irreducible representations of a group, then $\Sigma A^*_{ij}B_{kl} = 0$, and for a single unitary irreducible representation A we have $\Sigma A^*_{ij}A_{kl} = (g/n)\delta_{ik}\delta_{jl}$, where g is the order of the group, and n the dimension of the representation. The sum is over all the members of the group. The deltas are zero when their indices are different, unity when their indices are equal.

The scalar product of two vectors is the sum of the products of the individual elements, symbolically written (**a**,**b**), and is a number, not a vector. A *unitary* transformation or matrix U is one that preserves this scalar product, (U**a**,U**b**) = (**a**,**b**). This will be so if $U^{-1}_{ij} = U^{*}_{ji}$, if the inverse equals the hermitian conjugate (complex conjugate transposed) of the matrix. The concept can be extended to function spaces where the scalar product is an integral.

Applications to Quantum Mechanics

The Hamiltonian operator H is a function of the coordinates and momenta of a system. In the Schrödinger representation, solutions of the eigenvalue equation $H\psi = E\psi$ exist for certain values of the number E that can be cast into an orthonormal set of functions of the coordinates $\{\psi\}$, the Eigen functions. If more than one state ψ corresponds to the same energy E, this energy level is said to be degenerate (not the states, since there is nothing degenerate about the states involved--they simply have the same energy). In general, this does not occur except when there is some reason forcing the states to have the same energy, as we shall see. When it occurs without a compelling reason, it is called accidental degeneracy.

Multiply the eigenvalue equation by some operator Q from the left, and insert $Q^{-1}Q$ between H and ψ, to get $(QHQ^{-1})Q\psi = QE\psi = EQ\psi$. If $QHQ^{-1} = H$, or what is the same thing, QH = HQ, then $HQ\psi = EQ\psi$, or the states ψ and $Q\psi$ belong to the same value of the energy E; that is, the level is degenerate. If Q represents some change of basis functions or coordinates, QHQ^{-1} is the operator H in the new frame of reference. Should Q be a symmetry operation, then we must have $QHQ^{-1} = H$. The set of operators commuting with H is a group, called the symmetry group of the Hamiltonian.

Every state can be classified by the irreducible representation to which it belongs, and any coincidence in energy is regarded as accidental, unless the representation is multidimensional, in which case the coincidence in energy is necessary. Spectroscopic classifications are essentially identifications of the irreducible representation to which the state belongs.

It is often useful to express H as the sum of H_0, an approximate Hamiltonian that is simple and corresponds to most of the energy, and a perturbation Hamiltonian H' that includes finer details. If the symmetry group of H' is smaller than the symmetry group of H_0, then the greater symmetry is *broken* by the perturbation. In this case, previously degenerate levels may split into distinct levels, some of which may still be degenerate. Group theory will provide suitable functions for this calculation that can greatly reduce the effort involved.

As an example, the states of an isolated atom are classified by the total angular momentum J, and belong to irreducible representations of the rotation group in three dimensions with dimension 2J + 1. As long as there is spherical symmetry, these states all have exactly the same energy. If you apply a magnetic field, the spherical symmetry is broken, and there is now only symmetry about an axis in the direction of the magnetic field. The 2J + 1 levels now acquire different energies, and the non-degenerate states can be classified by the magnetic quantum number M, M = J, J - 1, ..., -J. Group theory can determine these states in advance, so that the splitting is given by a simple diagonal matrix element.

As a second example, the energy levels of the hydrogen atom depend only on the principal quantum number N. For any N, there are states with J = 0, 1, N-1. As far as the rotation group is concerned, this degeneracy is accidental. However, the Coulomb 1/r potential gives the Hamiltonian more symmetry than just rotation in 3-space. In fact, the Hamiltonian is invariant under the four-dimensional rotation group (actually, a group isomorphic to the four-dimensional rotation group), and its irreducible representations explain the added degeneracy, which is really not accidental at all. The degeneracy is lifted when the potential has a different radial dependence in more complex atoms, although the spherical symmetry is still there.

As still another example, it was noted in the late 1960›s that the nucleons (neutron and proton) and the & Lambda were not greatly different in mass. It was conjectured that they were members of an irreducible representation of dimension 3 of a group called SU_3, and that the differences in the observed masses was a splitting due to a perturbation related to the property called strangeness. There was some success in arranging the known particles in irreducible representations of SU_3, but later a theory of the internal structure of heavy elementary particles superseded this idea. However, group theory is still quite important in the field.

Group theory is good for more than classification, however. The probability amplitude for a transition between two states ψ and φ is given by some integral like ∫ψ*Pφ dτ, where P is an operator characteristic of the mechanism involved in the transition (electric dipole, etc), and dτ is the differential volume element. If ψ, φ and P are all classified according to their irreducible representations, then their product belongs to a representation whose characters are the product of the characters of each of the representations involved. This representation is almost always reducible, and the component irreducible representations can be found by character analysis. Unless the unit representation (all matrices +1) is there, the integral must vanish, since otherwise it changes under some symmetry operation, which by definition cannot change the integral. This gives *selection rules* for the non-vanishing of certain transitions or of certain matrix elements.

An example from spectroscopy might show how this operates. The electric dipole moment is a vector, and corresponds to the irreducible representation J = 1 of the rotation group. Operating on a state of angular momentum J, it gives representations corresponding to J + 1, J, and J - 1. The other state involved must have one of these values of J, or the result will not contain the unit representation J = 0. Therefore, we have the selection rule that J changes only by ±1 or 0 in an electric dipole transition.

1

CHAPTER

INFINITE-DIMENSIONAL LIE GROUPS AND ALGEBRAS IN MATHEMATICAL PHYSICS

Rudolf Schmid

INTRODUCTION

Lie groups play an important role in physical systems both as phase spaces and as symmetry groups. Infinite-dimensional Lie groups occur in the study of dynamical systems with an infinite number of degrees of freedom such as PDEs and in field theories. For such infinite-dimensional dynamical systems, diffeomorphism groups and various extensions and variations thereof, such as gauge groups, loop groups, and groups of Fourier integral operators, occur as symmetry groups and phase spaces. Symmetries are fundamental for Hamiltonian systems. They provide conservation laws (Noether currents) and reduce the number of degrees of freedom, that is, the dimension of the phase space.

The topics selected for review aim to illustrate some of the ways infinite-dimensional geometry and global analysis can be used in mathematical problems of physical interest. The topics selected are the following.(1)Infinite-Dimensional Lie Groups.(2)Lie Groups as Symmetry Groups of Hamiltonian Systems.(3)Applications.(4)Gauge Theories, the Standard Model, and Gravity.(5)SUSY (supersymmetry).

INFINITE-DIMENSIONAL LIE GROUPS

Basic Definitions

A general theory of infinite-dimensional Lie groups is hardly developed. Even Bourbaki [1] only develops a theory of infinite-dimensional manifolds, but all of the important theorems about Lie groups are stated for finite-dimensional ones.

An infinite-dimensional Lie group G is a group and an infinite-dimensional manifold with smooth group operations

Citation: Rudolf Schmid, "Infinite-Dimensional Lie Groups and Algebras in Mathematical Physics," Advances in Mathematical Physics, vol. 2010, Article ID 280362, 35 pages, 2010. doi:10.1155/2010/280362

$$m : \mathcal{G} \times \mathcal{G} \longrightarrow \mathcal{G}, \quad m(g,h) = g \cdot h, \quad C^{\infty}, \tag{2.1}$$

$$i : \mathcal{G} \longrightarrow \mathcal{G}, \quad i(g) = g^{-1}, \quad C^{\infty}. \tag{2.2}$$

Such a Lie group G is locally diffeomorphic to an infinite-dimensional vector space. This can be a Banach space whose topology is given by a norm $\|\cdot\|$, a Hilbert space whose topology is given by an inner product $\langle\cdot,\cdot\rangle$, or a Frechet space whose topology is given by a metric but not by a norm. Depending on the choice of the topology on G, we talk about Banach, Hilbert, or Frechet Lie groups, respectively

The Lie algebra g of a Lie group G is defined as g = {left invariant vector fields on G} ≃ *Te*G tangent space at the identity e). The isomorphism is given (as in finite dimensions) by

$$\xi \in T_e\mathcal{G} \longmapsto X_{\xi} \in \mathfrak{g}, \quad X_{\xi}(g) := T_e L_g(\xi), \tag{2.3}$$

and the Lie bracket on g is induced by the Lie bracket of left invariant vector fields $[\xi,\eta] = [X_{\xi}, X_{\eta}](e), \xi, \eta \in T_e\mathcal{G}$.

These definitions in infinite dimensions are identical with the definitions in finite dimensions. The big difference although is that infinite-dimensional manifolds, hence Lie groups, are *not* locally compact. For Frechet Lie groups, we have the additional nontrivial difficulty of the question how to define differentiability of functions defined on a Frechet space; see the study by Keller in 2 . Hence the very definition of a Frechet manifold is not canonical. This problem does not arise for Banach- and Hilbert-Lie groups; the differential calculus extends in a straightforward manner from R*n* to Banach and Hilbert spaces, but not to Frechet spaces

Finite- versus Infinite-Dimensional Lie Groups

Infinite-dimensional Lie groups are NOT locally compact. This causes some deficiencies of the Lie theory in infinite dimensions. We summarize some classical results in finite dimensions which are NOT true in general in infinite dimensions as follows.

(1) There is NO Implicit Function Theorem or Inverse Function Theorem in infinite dimensions! (Except Nash-Moser-type theorems).

(2) If G is a finite-dimensional Lie group, the exponential map exp :g→G is defined as follows. To each ξ∈g, we assign the corresponding left invariant vector field Xξ defined by (2.3). We take the flow φξ(t) of Xξ and define exp(ξ)=φξ(1). The exponential map is a local diffeomorphism from a neighborhood of zero in g onto a neighborhood of the identity in G; hence exp defines canonical coordinates on the Lie group G. This is not true in infinite dimensions.

(3) If f1,f2:G1→G2 are smooth Lie group homomorphisms (i.e., fi(g·h)=fi(g)·fi(h), i=1,2) with Tef1=Tef2, then locally f1=f2. This is not true in infinite dimensions.

(4) If f:G→H is a continuous group homomorphism between finite-dimensional Lie groups, then f is smooth. This is not true in infinite dimensions.

(5) If $\mathfrak{g}$ is any finite-dimensional Lie algebra, then there exists a connected finite-dimensional Lie group G with $\mathfrak{g}$ as its Lie algebra; that is, $\mathfrak{g} \simeq TeG$. This is not true in infinite dimensions.

(6) If G is a finite-dimensional Lie group and H⊂G is a closed subgroup, then H is a Lie subgroup (i.e., Lie group and submanifold). This is not true in infinite dimensions.

(7) If G is a finite-dimensional Lie group with Lie algebra $\mathfrak{g}$ and $\mathfrak{h} \subset \mathfrak{g}$ is a subalgebra, then there exists a unique connected Lie subgroup H⊂G with $\mathfrak{h}$ as its Lie algebra; that is, $\mathfrak{h} \simeq T_eH$. This is not true in infinite dimensions.

Some classical examples of finite-dimensional Lie groups are the matrix groups GL(n), SL(n), O(n), SO(n), U(n), SU(n), and Sp(n) with smooth group operations given by matrix multiplication and matrix inversion. The Lie algebra bracket is the commutator [A,B]=AB-BA with exponential map given by exp (A)= $\sum_{i=0}^{\infty}(1/i!)A^i = e^A$.

Examples of Infinite-Dimensional Lie Groups

The Vector Groups G = (V, +)

Let V be a Banach space and take $\mathcal{G}$=V with m(x,y)=x+y, i(x)=-x, and e=0, which makes $\mathcal{G}$ into an Abelian Lie group; that is, m(x,y)=m(y,x). For the Lie algebra we have $\mathfrak{g} \simeq TeV \simeq V$. For u∈TeV the corresponding left invariant vector field X_u is given by $X_u(v)=u$, ∀v∈V; that is, X_u=const. Hence the Lie algebra $\mathfrak{g}$=V with the trivial Lie bracket [u,v]=0 is Abelian. For the exponential map we get exp :=V→$\mathcal{G}$=V, exp =idV.

The General Linear Group G = GL(V), ∘0)

Let V be a Banach space and L(V,V) the space of bounded linear operators A:V→V. Then L(V,V) is a Banach space with the operator norm $\|A\|=\sup_{\|x\|\leq 1}\|A(x)\|$, and the group $\mathcal{G}$=GL(V) of all invertible elements is open in L(V,V). So GL(V) is a smooth Lie group with m(f,g)=f∘g, $i(f)=f^{-1}$, and $e=id_V$. Its Lie algebra is $\mathfrak{g}$=L(V,V) with the commutator bracket [A,B]=AB-BA and exponential map exp $A=e^A$.

The Abelian Gauge Group G = (C^∞M), +)

Let M be a finite-dimensional manifold and let $\mathcal{G}=C^\infty(M)$ (smooth functions on M). With group operation being addition, that is, m(f,g)=f+g, i(f)=-f, and e=0. $\mathcal{G}$ is an Abelian C^∞ (addition is smooth) Frechet Lie group with Lie algebra $\mathfrak{g}=T_eC^\infty(M)\simeq C^\infty(M)$, with trivial bracket [ξ,η]=0, and exp =id. If we complete these spaces in the C^k-norm, $k<\infty$ (denoted by $\mathcal{G}^k$), then $\mathcal{G}^k$ is a Banach-Lie group, and if we complete in the Hs-Sobolev norm with s>(1/2)dim M then $\mathcal{G}^s$ is a Hilbert-Lie group.

The Abelian Gauge Group $\mathcal{G} = (C^\infty(M, \mathbb{R} - \{0\}), \cdot)$

Let M be a finite-dimensional manifold and let $\mathcal{G}=C^\infty(M,\mathbb{R}-\{0\})$, with group operation being multiplication; that is, $m(f,g)=f\cdot g$, $i(f)=f^{-1}$, and $e=1$. For $k<\infty$, $C^k(M,\mathbb{R}-\{0\})$ is open in $C^\infty(M,\mathbb{R})$, and if M is compact, then $C^k(M,\mathbb{R}-\{0\})$ is a Banach-Lie group. If $s>(1/2)\dim M$, then $H^s(M,\mathbb{R}-\{0\})$ is closed under multiplication, and if M is compact, then $Hs(M,\mathbb{R}-\{0\})$ is a Hilbert-Lie group.

Loop Group $\mathcal{G} = (C^k(M, G), \cdot)$

We generalize the Abelian example (see Section 2.3.4) by replacing $\mathbb{R}-\{0\}$ with any finite-dimensional (non-Abelian) Lie group G. Let $\mathcal{G}=C^k(M,G)$ with pointwise group operations $m(f,g)(x)=f(x)\cdot g(x)$, $x\in M$, and $i(f)(x)=(f(x))^{-1}$, where "$\cdot$" and "$(\cdot)^{-1}$" are the operations in G. If $k<\infty$ then $C^k(M,G)$ is a Banach-Lie group. Let g denote the Lie algebra of G, then the Lie algebra of $\mathcal{G}=C^k(M,G)$ is $\mathfrak{g}=Ck(M,\mathfrak{g})$, with pointwise Lie bracket $[\xi,\eta](x)=[\xi(x),\eta(x)]$, $x\in M$, the latter bracket being the Lie bracket in g. The exponential map $\exp : \mathfrak{g}\to G$ defines the exponential map $EXP:\mathfrak{g}=C^k(M,\mathfrak{g})\to\mathcal{G}=C^k(M,G)$, $EXP(\xi)=\exp\circ\xi$, which is a local diffeomorphism. The same holds for $H^s(M,G)$ if $s>(1/2)\dim M$.

Applications of these infinite-dimensional Lie groups are in gauge theories and quantum field theory, where they appear as groups of gauge transformations. We will discuss these in Section 5.

Special Case: $\mathcal{G} = (C^k(S^1, G), \cdot)$

As a special case of example mentioned in Section 2.3.5 we take $M=S^1$, the circle. Then $\mathcal{G}=C^k(S^1,G)=\mathcal{L}^k(G)$ is called a loop group and $\mathfrak{g}=C^k(S^1,\mathfrak{g})=l^k(\mathfrak{g})$ is its loop algebra. They find applications in the theory of affine Lie algebras, Kac-Moody Lie algebras (central extensions), completely integrable systems, soliton equations (Toda, KdV, KP), and quantum field theory; see, for example, [3] and Section 5. Central extensions of loop algebras are examples of infinite-dimensional Lie algebras which need not have a corresponding Lie group.

Certain subgroups of loop groups play an important role in quantum field theory as groups of gauge transformations. We will discuss these in Section 2.4.4.

Diffeomorphism Groups

Among the most important "classical" infinite-dimensional Lie groups are the diffeomorphism groups of manifolds. Their differential structure is not the one of a Banach Lie group as defined above. Nevertheless they have important applications.

Let M be a compact manifold (the noncompact case is technically much more complicated but similar results are true; see the study by Eichhorn and Schmid in [4]) and let $\mathcal{G}=\mathrm{Diff}^\infty(M)$ be the group of all smooth diffeomorphisms on M, with group operation being composition; that is, $m(f,g)=f\circ g$, $i(f)=f^{-1}$, and $e=\mathrm{id}M$. For C^∞ diffeomorphisms, $\mathrm{Diff}^\infty(M)$ is a Frechet manifold and there are nontrivial problems with the notion of smooth maps between Frechet spaces. There is no canonical extension of the differential calculus from Banach spaces (which is the same as for $\mathbb{R}^n$) to Frechet spaces; see the study by Keller in [2]. One possibility is to generalize the notion of differentiability. For example, if we use the so-called C^∞_Γ differentiability, then $\mathcal{G}$=Diff$^\infty$(M) becomes a C^∞_Γ Lie group with C^∞_Γ differentiable group operations. These notions of differentiability are difficult to apply to concrete examples. Another possibility is to complete Diff$^\infty$(M) in the Banach C^k-norm, $0\leq k<\infty$, or in the Sobolev Hs-norm, s>(1/2)dim M. Then Diff^k(M) and Diff^s(M) become Banach and Hilbert manifolds, respectively. Then we consider the inverse limits of these Banach- and Hilbert-Lie groups, respectively:

$$\mathrm{Diff}^\infty(M) = \varprojlim \mathrm{Diff}^k(M) \tag{2.4}$$

becomes a so-called ILB- (Inverse Limit of Banach) Lie group, or with the Sobolev topologies

$$\mathrm{Diff}^\infty(M) = \varprojlim \mathrm{Diff}^s(M) \tag{2.5}$$

becomes a so-called ILH- (Inverse Limit of Hilbert) Lie group. See the study by Omori in [5] for details. Nevertheless, the group operations are not smooth, but have the following differentiability properties. If we equip the diffeomorphism group with the Sobolev Hs-topology, then Diff^s(M) becomes a C^∞ Hilbert manifold if s>(1/2)dim M and the group multiplication

$$m : \mathrm{Diff}^{s+k}(M) \times \mathrm{Diff}^s(M) \longrightarrow \mathrm{Diff}^s(M) \tag{2.6}$$

is C^k differentiable; hence for k=0, m is only continuous on Diff^s(M). The inversion

$$i : \mathrm{Diff}^{s+k}(M) \longrightarrow \mathrm{Diff}^s(M) \tag{2.7}$$

is C^k differentiable; hence for k=0, i is only continuous on Diffs(M). The same differentiability properties of m and i hold in the C^k topology.

The Lie algebra of Diff$^\infty$(M) is given by $g=T_e\mathrm{Diff}^\infty(M)\simeq\mathrm{Vec}^\infty(M)$ being the space of smooth vector fields on M. Note that the space Vec (M) of all vector fields is a Lie algebra only for C^∞ vector fields, but not for C^k or H^s vector fields if $k<\infty$, $s<\infty$, because one loses derivatives by taking brackets.

The exponential map on the diffeomorphism group is given as follows. For any vector field $X\in\mathrm{Vec}\infty(M)$, take its flow $\varphi_t\in\mathrm{Diff}\infty(M)$, then define $\mathrm{EXP}:\mathrm{Vec}^\infty(M)\rightarrow\mathrm{Diff}^\infty(M):X\mapsto\varphi_1$, the flow at time t=1. The exponential map EXP is NOT a local diffeomorphism; it is not even locally surjective.

We see that the diffeomorphism groups are not Lie groups in the classical sense, but what we call nested Lie groups. Nevertheless they have important applications as we will see.

Subgroups of Diff$^\infty$(M)

Several subgroups of Diff$^\infty$(M) have important applications.

Group of Volume-Preserving Diffeomorphisms

Let μ be a volume on M and

$$\mathcal{G} = \mathrm{Diff}_{\mu}^{\infty}(M) = \{f \in \mathrm{Diff}^{\infty}(M) \mid f^{*}\mu = \mu\} \tag{2.8}$$

the group of volume-preserving diffeomorphisms. $\mathrm{Diff}_{\mu}^{\infty}(M)$ is a closed subgroup of $\mathrm{Diff}^{\infty}(M)$ with Lie algebra

$$\mathfrak{g} = \mathrm{Vec}_{\mu}^{\infty}(M) = \{X \in \mathrm{Vec}^{\infty}(M) \mid \mathrm{div}_{\mu}X = 0\} \tag{2.9}$$

being the space of divergence-free vector fields on M. $\mathrm{Vec}_{\mu}^{\infty}(M)$ is a Lie subalgebra of $\mathrm{Vec}^{\infty}(M)$.

We cannot apply the finite-dimensional theorem that if $\mathrm{Vec}_{\mu}^{\infty}(M)$ is Lie algebra then there exists a Lie group whose Lie algebra it is; nor the one that if $\mathrm{Diff}_{\mu}^{\infty}(M) \subset \mathrm{Diff}(M)$ is a closed subgroup then it is an Lie subgroup.

Nevertheless $\mathrm{Diff}_{\mu}^{\infty}(M)$ is an ILH-Lie group.

Symplectomorphism Group

Let ω be a symplectic 2-form on M and

$$\mathcal{G} = \mathrm{Diff}_{\omega}^{\infty}(M) = \{f \in \mathrm{xDiff}^{\infty}(M) \mid f^{*}\omega = \omega\} \tag{2.10}$$

the group of canonical transformations (or symplectomorphisms). $\mathrm{Diff}_{\omega}^{\infty}(M)$ is a closed subgroup of $\mathrm{Diff}^{\infty}(M)$ with Lie algebra

$$\mathfrak{g} = \mathrm{Vec}_{\omega}^{\infty}(M) = \{X \in \mathrm{Vec}^{\infty}(M) \mid L_{X}\omega = 0\} \tag{2.11}$$

being the space of locally Hamiltonian vector fields on M. $\mathrm{Vec}_{\omega}^{\infty}(M)$ is a Lie subalgebra of $\mathrm{Vec}^{\infty}(M)$. Again $\mathrm{Diff}_{\omega}^{\infty}(M)$ is an ILH-Lie group.

Group of Gauge Transformations

The diffeomorphism subgroups that arise in gauge theories as gauge groups behave nicely because they are isomorphic to subgroups of loop groups which are not only ILH-Lie groups but actually Hilbert-Lie groups.

Let $\pi: P \to M$ be a principal G bundle with G being a finite-dimensional Lie group (structure group) acting on P from the right $p \in P$, $g \in G$, and $p \cdot g \in P$.

The Gauge group $\mathcal{G}$ is the group of gauge transformations defined by

$$\mathcal{G} = \{\phi \in \mathrm{Diff}^{\infty}(P);\ \phi(p \cdot g) = \phi(p) \cdot g,\ \pi(\phi(p)) = \pi(p)\}. \tag{2.12}$$

$\mathcal{G}$ is a group under composition, hence a subgroup of the diffeomorphism group $\mathrm{Diff}^{\infty}(P)$. Since a gauge transformation $\phi \in \mathcal{G}$ preserves fibers, we can realize each such $\phi \in \mathcal{G}$ via $\phi(p) = p \cdot \tau(p)$, where $\tau: P \to G$ satisfies $\tau(p \cdot g) = g^{-1}\tau(p)g$, for $p \in P$, $g \in G$. Let

$$\mathrm{Gau}(P) = \left\{\tau \in C^{\infty}(P,G);\ \tau(p \cdot g) = g^{-1}\tau(p)g\right\}. \tag{2.13}$$

Gau(P) is a group under pointwise multiplication, hence a subgroup of the loop group $C^{\infty}(P,G)$ (see Section 2.4.3), which extends to a Hilbert-Lie group if equipped with the Hs-Sobolev topology. We give Gau(P) the induced topology and extend it to a Hilbert-Lie group denoted by $\mathrm{Gau}^s(P)$. Another interpretation is that Gau(P) is isomorphic to $C^{\infty}(\mathrm{AdP})$ the space of sections of the associated vector bundle $\mathrm{Ad}(P)=P\times_G G$. Completed in the H^s Sobolev topology, we get $\mathrm{Gau}^s(P)\simeq H^s(\mathrm{Ad}\,P)$.

Let g denote the Lie algebra of G. Then the Lie algebra $\mathfrak{gau}(P)$ of Gau(P) is a subalgebra of the loop algebra $H^s(P,g)$ under pointwise bracket in g, the finite-dimensional Lie algebra of G; that is, for any $\xi,\eta\in H^s(P,g)$ the bracket is defined by $[\xi,\eta]_{\mathfrak{gau}}(P)(p)=[\xi(p),\eta(p)]_g$, $p\in P$. Then $\mathfrak{gau}^s(P)$ is the subalgebra of Ad -invariant g-valued functions on P; that is

$$\mathfrak{gau}(P) = \{\xi \in C^{\infty}(P,\mathbf{g});\ \xi(p\cdot g) = Ad_{g^{-1}}\xi(p)\}. \tag{2.14}$$

The Lie algebra $\mathfrak{lie}\ \mathcal{G}$ (running out of symbols) of the gauge group $\mathcal{G}$ is the Lie subalgebra of $\mathfrak{X}^{\infty}(P)$ consisting of all G-invariant vertical vector fields X on P; that is,

$$\mathfrak{lie}\,\mathcal{G} = \left\{X \in \mathfrak{X}^{\infty}(P);\ R_g^{*}X = X,\ X(p) \in \mathbf{g},\ g \in G,\ p \in P\right\} \tag{2.15}$$

with commutator bracket $[X_1, X_2] = X_1X_2 - X_2X_1 \in \mathfrak{lie}\ \mathcal{G}$.

On the other hand, the Lie algebra of $C^{\infty}(\mathrm{Ad\,P})$ is $C^{\infty}(\mathrm{ad}(P))$ being the space of sections of the associated vector bundle $\mathrm{ad}(P)\equiv(P\times G_g)\to M$ with pointwise bracket.

We have three versions of gauge groups: $\mathcal{G}$,Gau(P), and $C^{\infty}(\mathrm{Ad\,P})$. They are all group isomorphic. There is a natural group isomorphism $\mathrm{Gau}(P)\to\mathcal{G}:\tau\mapsto\phi$ defined by $\phi(p)=p\cdot\tau(p)$, $p\in P$, which preserves the product $\tau_1\cdot\tau_2\mapsto\phi_1\circ\phi_2$. Identifying $\mathcal{G}$ with Gau(P), we can avoid the troubles with diffeomorphism groups and we can extend $\mathcal{G}$ to a Hilbert-Lie group $\mathcal{G}^s$. So $\mathcal{G}^s$ is actually a Hilbert-Lie group in the classical sense; that is, the group operations are C^{∞}. Also the three Lie algebras $\mathfrak{lie}\,\mathcal{G}$, $\mathfrak{g}(P)$, and $C^{\infty}(\mathrm{ad\,P})$ are canonically isomorphic. Indeed, for $s\in C^{\infty}(\mathrm{ad\,P})$ define $\xi\in\mathfrak{gau}(P)\xi:P\to g$ by $\xi(p\cdot a):=Ad^{a-1}\xi(p)$; and for $\xi\in\mathfrak{gau}(P)$ define $s\in C^{\infty}(\mathrm{adP})$ by $s(\pi(p)):=[p,\xi(p)]$.

On the other hand, for $\xi\in\mathfrak{g}(P)$ define $Z_\xi\in\mathfrak{lie}\,\mathcal{G}$ by

$$Z_\xi(p) = \frac{d}{dt}|_{t=0}R(p,\exp t\xi(p)) \quad (= \xi(p)^{*}(p)), \tag{2.16}$$

that is, Z_ξ is the fundamental vector field on P, generated by $\xi\in g$. Z_ξ is invariant if and only if $\xi(p\cdot g)=Ad_{g-1\xi}(p)$.

To topologize $\mathfrak{lie}\ \mathcal{G}$, we complete $C^{\infty}(\mathrm{ad\,P})$ in the H^s-Sobolev norm. If $s>(1/2)$ dim M, then $\mathfrak{lie}\,\mathcal{G}^s\simeq H^s(\mathrm{ad\,P})\simeq\mathfrak{gau}^s(P)$ are isomorphic Hilbert-Lie algebras.

There is a natural exponential map $\mathrm{Exp}:u(P)\to\mathrm{Gau}(P)$, which is a local diffeomorphism. Let $\exp:g\to G$ be the finite-dimensional exponential map. Then define

$$\mathrm{Exp} : \mathfrak{gau}^s(P) \to \mathrm{Gau}^s(P) : (\mathrm{Exp}\,\xi)(p) = \exp(\xi(p)), \quad \xi \in \mathfrak{gau}^s(P). \tag{2.17}$$

Or in terms of $\mathcal{G}$, Exp:e $\mathcal{G}^s \to \mathcal{G}^s$:(Exp ξ)(p)=p·exp (ξ_p).

We have the following theorem (Schmid [6]).

Theorem 2.2. For s>(1/2)dim M

$$\mathcal{G}^s \simeq \mathrm{Gau}^s(P) \simeq H^s(Ad\,P) \tag{2.18}$$

is a smooth Hilbert-Lie group with Lie algebra

$$\mathfrak{lie}\,\mathcal{G}^s \simeq \mathfrak{gau}^s(P) \simeq H^s(ad\,P) \tag{2.19}$$

and smooth exponential map, which is a local diffeomorphism

$$\mathrm{EXP} : \mathfrak{lie}\,\mathcal{G}^s \longrightarrow \mathcal{G}^s : (\mathrm{EXP}\,\xi)(p) = p \cdot \exp(\xi(p)). \tag{2.20}$$

See [1–5, 7–19].

LIE GROUPS AS SYMMETRY GROUPS OF HAMILTONIAN SYSTEMS

A short introduction and "crash course" to geometric mechanics can be found in the studies by Abraham and Marsden [20], Marsden [21], as well as Marsden and Ratiu [22]. For the general theory of infinite-dimensional manifolds and global analysis, see, for example, the studies by Bourbaki [9], Lang [14], as well as Palais [18].

Hamilton's Equations on Poisson Manifolds

A Poisson manifold is a manifold P (in general infinite-dimensional) equipped with a bilinear operation {·,·}, called Poisson bracket, on the space $C^\infty(P)$ of smooth functions on P satisfying the following.

(i) $(C\infty(P),\{\cdot,\cdot\})$ is a Lie algebra; that is, $\{\cdot,\cdot\}:C^\infty(P)\times C^\infty(P)\to C^\infty(P)$ is bilinear, skew symmetric and satisfies the Jacobi identity $\{\{F,G\},H\}+\{\{H,F\},G\}+\{\{G,H\},F\}=0$ for all $F,G,H\in C^\infty(P)$.

(ii) {·,·} satisfies the Leibniz rule; that is, {·,·} is a derivation in each factor: $\{F\cdot G,H\}=F\cdot\{G,H\}+G\cdot\{F,H\}$, for all $F,G,H\in C\infty(P)$.

The notion of Poisson manifolds was rediscovered many times under different names, starting with Lie, Dirac, Pauli, and others. The name Poisson manifold was coined by Lichnerowicz.

For any $H\in C^\infty(P)$ we define the Hamiltonian vector field X_H by

$$X_H(F) = \{F, H\}, \quad F \in C^\infty(P) \tag{3.1}$$

It follows from (ii) that indeed XH defines a derivation on $C\infty(P)$, hence a vector field on P. Hamilton's equations of motion for a function $F\in C\infty(P)$ with Hamiltonian

$H \in C\infty(P)$ (energy function) are then defined by the flow (integral curves) of the vector field XH; that is,

$$\dot{F} = X_H(F) = \{F, H\}, \quad \text{where } \dot{} = \frac{d}{dt}. \tag{3.2}$$

We then call F a Hamiltonian system on P with energy (Hamiltonian function) H.

Examples of Poisson Manifolds and Hamilton's Equations

Poisson manifolds are a generalization of symplectic manifolds on which Hamilton's equations have a canonical formulated.

Finite-Dimensional Classical Mechanics

For finite-dimensional classical mechanics we take $P=\mathbb{R}^{2n}$ with coordinates $(q^1,\ldots,q^n,p_1,\ldots,p_n)$ with the standard Poisson bracket for any two functions $F(q^i,p_i)$, $H(q^i,p_i)$ given by

$$\{F, H\} = \sum_{i=1}^{n} \frac{\partial F}{\partial p_i}\frac{\partial H}{\partial q^i} - \frac{\partial H}{\partial p_i}\frac{\partial F}{\partial q^i}. \tag{3.3}$$

Then the classical Hamilton's equations are

$$\dot{q}^i = \left\{q^i, H\right\} = \frac{\partial H}{\partial p_i}, \quad \dot{p}_i = \{p_i, H\} = -\frac{\partial H}{\partial q^i}, \tag{3.4}$$

Where $i=1,\ldots,n$. This finite-dimensional Hamiltonian system is a system of ordinary differential equations for which there are well-known existence and uniqueness theorems; that is, it has locally unique smooth solutions, depending smoothly on the initial conditions.

Example 3.1 (Harmonic Oscillator). As a concrete example we consider the harmonic oscillator. Here $P=\mathbb{R}^2$ and the Hamiltonian (energy) is $H(q,p)=(1/2)(q^2+p^2)$. Then Hamilton's equations are

$$\dot{q} = p, \qquad \dot{p} = -q. \tag{3.5}$$

Infinite-Dimensional Classical Field Theory

Let V be a Banach space and V^* its dual space with respect to a pairing $\langle\cdot,\cdot\rangle : V\times V^* \to \mathbb{R}$ (i.e., $\langle\cdot,\cdot\rangle$ is a symmetric, bilinear, nondegenerate function). On $P=V\times V^*$ we have the canonical Poisson bracket for F, $H\in C^\infty(P)$, $\varphi \in V$, and $\pi \in V^*$, given by

$$\{F, H\} = \left\langle \frac{\delta F}{\delta \pi}, \frac{\delta H}{\delta \varphi} \right\rangle - \left\langle \frac{\delta H}{\delta \pi}, \frac{\delta F}{\delta \varphi} \right\rangle,$$

(3.6)

where the functional derivatives $\delta F/\delta\pi \in V$, $\delta F/\delta\varphi \in V^*$ are the "duals" under the pairing $\langle\cdot,\cdot\rangle$ of the partial gradients $D1F(\pi)\in V^*, D2F(\varphi)\in V^{**}\simeq V$. The corresponding Hamilton's equations are

$$\dot{\varphi} = \{\varphi, H\} = \frac{\delta H}{\delta \pi}, \quad \dot{\pi} = \{\pi, H\} = -\frac{\delta H}{\delta \varphi}.$$

(3.7)

As a special case in finite dimensions, if $V\simeq\mathbb{R}^n$, so that $V^*\simeq\mathbb{R}n$ and $P=V\times V^*\simeq\mathbb{R}^{2n}$, and the pairing is the standard inner product in $\mathbb{R}^n$, then the Poisson bracket (3.6) and Hamilton's equations (3.7) are identical with (3.3) and (3.4), respectively.

Example 3.2 (Wave Equations). As a concrete example we consider the wave equations. Let $V=C^\infty(\mathbb{R}3)$ and $V^*=\mathrm{Den}(\mathbb{R}3)$ (densities) and the L^2 pairing $\langle\varphi,\pi\rangle=\int\varphi(x)\pi(x)dx$. We take the Hamiltonian to be $H(\varphi,\pi)=\int((1/2)\pi^2+(1/2)|\nabla\varphi|^2+F(\varphi))dx$, where F is some function on V. Then Hamilton's (3.7) become

$$\dot{\varphi} = \pi, \qquad \dot{\pi} = \nabla^2\varphi - F'(\varphi), \quad \text{where } ' = \frac{d}{d\varphi},$$

(3.8)

which imply the wave equation $\partial^2\varphi/\partial t^2=\nabla^2\varphi-F'(\varphi)$. Different choices of F give different wave equations; for example, for F=0 we get the linear wave equation $\partial^2\varphi/\partial t^2=\nabla^2\varphi$. For $F=(1/2)m\varphi$ we get the Klein-Gordon equation $\nabla^2\varphi-\partial^2\varphi/\partial t^2=m\varphi$. So these wave equations and the Klein-Gordon equation are infinite-dimensional Hamiltonian systems on $P=C^\infty(\mathbb{R}^3)\times\mathrm{Den}(\mathbb{R}^3)$.

Cotangent Bundles

The finite-dimensional examples of Poisson brackets (3.3) and Hamilton's (3.4) and the infinite-dimensional examples (3.6) and (3.7) are the local versions of the general case where $P=T^*Q$ is the cotangent bundle (phase space) of a manifold Q (configuration space). If Q is an n-dimensional manifold, then T^*Q is a 2n-Poisson manifold locally isomorphic to $\mathbb{R}^{2n}$ whose Poisson bracket is locally given by (3.3) and Hamilton's equations are locally given by (3.4). If Q is an infinite-dimensional Banach manifold, then T^*Q is a Poisson manifold locally isomorphic to $V\times V^*$ whose Poisson bracket is given by (3.6) and Hamilton's equations are locally given by (3.7).

Symplectic Manifolds

All the examples above are special cases of symplectic manifolds (P,ω). That means that P is equipped with a symplectic structure ω which is a closed $(d_\omega=0)$, (weakly) nondegenerate 2-form on the manifold P. Then for any $H\in C^\infty(P)$ the corresponding Hamiltonian vector field X_H is defined by $dH=\omega(XH,\cdot)$ and the canonical Poisson bracket is given by

$$\{F, H\} = \omega(X_F, X_H), \quad F, H \in C^\infty(P). \tag{3.9}$$

For example, on $\mathbb{R}^{2n}$ the canonical symplectic structure ω is given by $\omega = \sum_{i=1}^{n} dp_i \wedge dq^i$ = =dθ, where $\theta = \sum_{i=1}^{n} p_i \wedge dq^i$.. The same formula for ω holds locally in T^*Q for any finite-dimensional Q (Darboux's Lemma). For the infinite-dimensional example $P=V\times V^*$, the symplectic form ω is given by $\omega((\varphi_1,\pi_1),(\varphi_2,\pi_2))=\langle\varphi_1,\pi_2\rangle-\langle\varphi_2,\pi_1\rangle$. Again these two formulas for ω are identical if $V=\mathbb{R}^n$.

Remark 3.3 (A) If P is a finite-dimensional symplectic manifold, then P is even dimensional.

(B) If the Poisson bracket $\{\cdot,\cdot\}$ is nondegenerate, then $\{\cdot,\cdot\}$ comes from a symplectic form ω; that is, $\{\cdot,\cdot\}$ is given by (3.9).

The Lie-Poisson Bracket

Not all Poisson brackets are of the form given in the above examples (3.3), (3.6), and (3.9); that is, not all Poisson manifolds are symplectic manifolds. An important class of Poisson bracket is the so-called Lie-Poisson bracket. It is defined on the dual of any Lie algebra. Let G be a Lie group with Lie algebra $\mathfrak{g}=T_eG\simeq$ {left invariant vector fields on G}, and let $[\cdot,\cdot]$ denote the Lie bracket (commutator) on $\mathfrak{g}$. Let $\mathfrak{g}^*$ be the dual of a $\mathfrak{g}$ with respect to a pairing $\langle\cdot,\cdot\rangle:\mathfrak{g}^*\times\mathfrak{g}\to\mathbb{R}$. Then for any $F,H\in C^\infty(\mathfrak{g}^*)$ and $\mu\in\mathfrak{g}^*$, the Lie-Poisson bracket is defined by

$$\{F, H\}(\mu) = \pm\left\langle \mu, \left[\frac{\delta F}{\delta \mu}, \frac{\delta H}{\delta \mu}\right]\right\rangle, \tag{3.10}$$

where $\delta F/\delta\mu, \delta H/\delta\mu\in\mathfrak{g}$ are the "duals" of the gradients $DF(\mu), DH(\mu)\in\mathfrak{g}^{**}\simeq\mathfrak{g}$ under the pairing $\langle\cdot,\cdot\rangle$. Note that the Lie-Poisson bracket is degenerate in general; for example, for G=SO(3) the vector space $\mathfrak{g}^*$ is 3 dimensional, so the Poisson bracket (3.10) cannot come from a symplectic structure. This Lie-Poisson bracket can also be obtained in a different way by taking the canonical Poisson bracket on T*G (locally given by (3.3) and (3.6)) and then restricting it to the fiber at the identity $T_e^*G = \mathfrak{g}^*$In this sense the Lie-Poisson bracket (3.10) is induced from the canonical Poisson bracket on T*G. It is induced by the symmetry of left multiplication as we will discuss in Section 3.3.

Example (3.4 Rigid Body). A concrete example of the Lie-Poisson bracket is given by the rigid body. Here G=SO(3) is the configuration space of a free rigid body. Identifying the Lie algebra $((3),[\cdot,\cdot])$ with $(\mathbb{R}^3,\times)$, where × is the vector product on $\mathbb{R}^3$, and $\mathfrak{g}^*=\mathfrak{so}(3)^*\simeq\mathbb{R}^3$, the Lie-Poisson bracket translates into

$$\{F, H\}(m) = -m \cdot (\nabla F \times \nabla H). \tag{3.11}$$

For any $F \in C^\infty(\mathfrak{so}(3)^*)$, we have $(dF/dt)(m)=\nabla F \cdot \dot{m}=\{F,H\}(m)=-m\cdot(\nabla F\times\nabla H)=\nabla F\cdot(m\times\nabla H)$; hence $\dot{m}=m\times\nabla H$. With the Hamiltonian $H = (1/2)(m_1^2/I_1^2 + m_2^2/I_2^2 + m_3^2/I_3^2)$ we get Hamilton's equation as

$$\dot{m}_1 = \frac{I_2 - I_3}{I_2 I_3} m_2 m_3, \qquad \dot{m}_2 = \frac{I_3 - I_1}{I_3 I_1} m_3 m_1, \qquad \dot{m}_3 = \frac{I_1 - I_2}{I_1 I_2} m_1 m_2. \tag{3.12}$$

These are Euler's equations for the free rigid body.

Reduction by Symmetries

The examples we have discussed so far are all canonical examples of Poisson brackets, defined either on a symplectic manifold (P,ω) or T^*Q, or on the dual of a Lie algebra $\mathfrak{g}^*$. Different, noncanonical Poisson brackets can arise from symmetries. Assume that a Lie group G is acting in a Hamiltonian way on the Poisson manifold $(P,\{\cdot,\cdot\})$. That means that we have a smooth map $\varphi:G\times P\to P:\varphi(g,p)=g\cdot p$ such that the induced maps $\varphi g=\varphi(g,\cdot):P\to P$ are canonical transformations, for each $g\in G$. In terms of Poisson manifolds, a canonical transformation is a smooth map that preserves the Poisson bracket. So the action of G on P is a Hamiltonian action if $\varphi_g^*\{F,H\} = \{\varphi_g^* F, \varphi_g^* H\}$, for all $F,H \in C^\infty(P), g \in G$. For any $\xi\in\mathfrak{g}$ the canonical transformations $\varphi\exp(t\xi)$ generate a Hamiltonian vector field ξF on P and a momentum map $J:P\to\mathfrak{g}^*$ given by $J(x)(\xi)=F(x)$, which is Ad* equivariant.

If a Hamiltonian system XH is invariant under a Lie group action, that is, $H(_{\varphi g}(x))=H(x)$, then we obtain a reduced Hamiltonian system on a reduced phase space (reduced Poisson manifold). We recall the following Marsden-Weinstein reduction theorem [23].

Theorem 3.5 (Reduction Theorem). For a Hamiltonian action of a Lie group G on a Poisson manifold $(P,\{\cdot,\cdot\})$, there is an equivariant momentum map $J:P\to\mathfrak{g}^*$ and for every regular $\mu\in\mathfrak{g}^*$ the reduced phase space $P\mu\equiv J^{-1}(\mu)/G\mu$ carries an induced Poisson structure $\{\cdot,\cdot\}_\mu$ (G_μ being the isotropy group). Any G-invariant Hamiltonian H on P defines a Hamiltonian H_μ on the reduced phase space P_μ, and the integral curves of the vector field X_H project onto integral curves of the induced vector field $\hat{X}_{H_\mu}$ on the reduced space P^μ.

Example 3.6 (Rigid Body). The rigid body discussed above can be viewed as an example of this reduction theorem. If $P=T^*G$ and G is acting on T^*G by the cotangent lift of the left translation $l_g:G\to G, l_g(h)=gh$, then the momentum map $J:T^*G\to\mathfrak{g}^*$ is given by $J(\alpha_g) = T_e^* R_g(\alpha_g)$ and the reduced phase space $(T^*G)\mu=J\text{-}1(\mu)/G_\mu$ is isomorphic to the coadjoint orbit $\mathcal{O}_\mu$ through $\mu\in\mathfrak{g}^*$. Each coadjoint orbit $\mathcal{O}_\mu$ carries a natural symplectic structure ω_μ, and in this case, the reduced Lie-Poisson bracket $\{\cdot,\cdot\}\mu$ on the coadjoint orbit $\mathcal{O}_\mu$ is induced by the symplectic form ω_μ on $\mathcal{O}_\mu$ as in (3.9). Furthermore $T^*G/G\simeq\mathfrak{g}^*$ and the induced Poisson bracket $\{\cdot,\cdot\}_\mu$ on $\mathcal{O}_\mu$ are identical with the Lie-Poisson bracket restricted to the coadjoint orbit $\mathcal{O}_\mu\subset\mathfrak{g}^*$. For the rigid body we apply this construction to G=SO(3).

See [1, 8, 10, 17, 19–31].

APPLICATIONS

We now discuss some infinite-dimensional examples of reduced Hamiltonian systems.

Maxwell's Equations

Maxwell's equations of electromagnetism are a reduced Hamiltonian system with the Lie group $\mathcal{G}=(C^\infty(M),+)$ discussed in Section 2.3.3 as symmetry group.

Let E,B be the electric and magnetic fields on $\mathbb{R}^3$, then Maxwell's equations for a charge density ρ are

$$\dot{E} = \operatorname{curl} B, \qquad \dot{B} = -\operatorname{curl} E, \tag{4.1}$$

$$\operatorname{div} B = 0, \qquad \operatorname{div} E = \rho. \tag{4.2}$$

Let A be the magnetic potential such that B=-curl A. As configuration space we take $V=\mathrm{Vec}(\mathbb{R}^3)$, vector fields (potentials) on $\mathbb{R}^3$, so $A\in V$, and as phase space we have $P=T^*V\simeq V\times V^*\ni(A,E)$, with the standard L^2 pairing $\langle A,E\rangle=\int A(x)E(x)dx$, and canonical Poisson bracket given by (3.6), which becomes

$$\{F,H\}(A,E) = \int\left(\frac{\delta F}{\delta A}\frac{\delta H}{\delta E} - \frac{\delta H}{\delta A}\frac{\delta F}{\delta E}\right)dx. \tag{4.3}$$

As Hamiltonian we take the total electromagnetic energy

$$H(A,E) = \frac{1}{2}\int\left(|\operatorname{curl} A|^2 + |E|^2\right)dx. \tag{4.4}$$

Then Hamilton's equations in the canonical variables A and E are $\dot{A}=\delta H/\delta E=E\Rightarrow \dot{B}=-\operatorname{curl} E$ and $\dot{E}=-\delta H/\delta A=-\operatorname{curl}\operatorname{curl} A=\operatorname{curl} B$. So the first two equations of Maxwell's equations (4.1) are Hamilton's equations; we get the third one automatically from the potential $\operatorname{div} B=-\operatorname{div}\operatorname{curl} A=0$ and we obtain the 4th equation $\operatorname{div} E=\rho$ through the following symmetry (gauge invariance). The Lie group $\mathcal{G}=(C^\infty(\mathbb{R}^3),+)$ acts on V by $\varphi\cdot A=A+\nabla\varphi$, $\varphi\in\mathcal{G}, A\in V$. The lifted action to $V\times V^*$ becomes $\varphi\cdot(A,E)=(A+\nabla\varphi,E)$, and has the momentum map $J:V\times V^*\to g^*\simeq\{\text{charge densities}\}$:

$$J(A,E) = \operatorname{div} E. \tag{4.5}$$

With $g=C\infty(\mathbb{R}^3)$ and $g^*=\mathrm{Den}(\mathbb{R}^3)$, we identify elements of g^* with charge densities. The Hamiltonian H is $\mathcal{G}$ invariant; that is, $H(\varphi\cdot(A,E))=H(A+\nabla_\varphi,E)=H(A,E)$. Then the reduced phase space for $\rho\in g^*$ is $(V\times V^*)\rho=J^{-1}(\rho)/G=\{(E,B)|\operatorname{div} E=\rho,\operatorname{div} B=0\}$ and the reduced Hamiltonian is

$$H_\rho(E,B) = \frac{1}{2}\int\left(|E|^2 + |B|^2\right)dx. \tag{4.6}$$

The reduced Poisson bracket becomes for any functions F,H on $(V\times V^*)_\rho$

$$\{F, H\}_\rho(E, B) = \int \left(\frac{\delta F}{\delta E} \cdot \text{curl}\, \frac{\delta H}{\delta B} - \frac{\delta H}{\delta E} \cdot \text{curl}\, \frac{\delta F}{\delta B} \right) dx, \tag{4.7}$$

and a straightforward computation shows that

$$\dot{F} = \{F, H_\rho\}_\rho \Longleftrightarrow \begin{cases} \dot{E} = \text{curl}\, B, & \dot{B} = -\,\text{curl}\, E, \\ \text{div}\, B = 0, & \text{div}\, E = \rho. \end{cases} \tag{4.8}$$

So Maxwell's equations (4.1), (4.2) are an infinite-dimensional Hamiltonian system on this reduced phase space with respect to the reduced Poisson bracket.

Fluid Dynamics

Euler's equations for an incompressible fluid

$$\frac{\partial u}{\partial t} + u \cdot \nabla u = -\nabla p, \quad \text{div}\, u = 0 \tag{4.9}$$

are equivalent to the equations of geodesics on Diffμ∞(M). See the study by Marsden et al. in [15] for details.

Plasma Physics

The Maxwell-Vlasov's equations are a reduced Hamiltonian system on a more complicated reduced space. See the study by Marsden et al. in [32] for details.

Maxwell-Vlasov's equations for a plasma density f(x,v,t) generating the electric and magnetic fields E and B are the following set of equations

$$\begin{aligned} &\frac{\partial f}{\partial t} + v \cdot \frac{\partial f}{\partial x} + (E + v \times B)\frac{\partial f}{\partial v} = 0, \\ &\frac{\partial B}{\partial t} = -\,\text{curl}\, E, \\ &\frac{\partial E}{\partial t} = \text{curl}\, B - J_f, \quad J_f = \text{current density}, \\ &\text{div}\, E = \rho_f, \quad \rho_f = \text{charge density}, \\ &\text{div}\, B = 0. \end{aligned} \tag{4.10}$$

This coupled nonlinear system of evolution equations is an infinite-dimensional Hamiltonian system of the form $\dot{F} = \{F, H\}_{\rho_f}$ on the reduced phase space

$$\mathcal{MV} = \left(T^*\text{Diff}^\infty_\omega\left(\mathbb{R}^6\right) \times T^*V\right)/C^\infty\left(\mathbb{R}^6\right) \tag{4.11}$$

(V being the same space as in the example of Maxwell's equations) with respect to the following reduced Poisson bracket, which is induced via gauge symmetry from the canonical Poisson bracket on $^{T^*\mathrm{Diff}^\infty_\omega(\mathbb{R}^6)\times T^*V}$:

$$\begin{aligned}\{F,G\}_{\rho_f}(f,E,B) &= \int f\left\{\frac{\delta F}{\delta f},\frac{\delta G}{\delta f}\right\}dx\,dv \\ &+ \int\left(\frac{\delta F}{\delta E}\cdot \operatorname{curl}\frac{\delta G}{\delta B} - \frac{\delta G}{\delta E}\cdot \operatorname{curl}\frac{\delta F}{\delta B}\right)dx\,dv \\ &+ \int\left(\frac{\delta F}{\delta E}\cdot\frac{\partial f}{\partial v}\frac{\delta G}{\delta f} - \frac{\delta G}{\delta E}\cdot\frac{\partial f}{\partial v}\frac{\delta F}{\delta f}\right)dx\,dv \\ &+ \int fB\cdot\left(\frac{\partial}{\partial v}\frac{\delta F}{\delta f}\times\frac{\partial}{\partial v}\frac{\delta G}{\delta f}\right)dx\,dv,\end{aligned} \tag{4.12}$$

and with Hamiltonian

$$H(f,E,B) = \frac{1}{2}\int v^2 f(x,v,t)dv + \frac{1}{2}\int\left(|E|^2 + |B|^2\right)dx. \tag{4.13}$$

More complicated plasma models are formulated as Hamiltonian systems. For example, for the two-fluid model the phase space is a coadjoint orbit of the semidirect product (⋉) of the group $\mathcal{G}=\mathrm{Diff}^\infty(\mathbb{R}^6)\ltimes(C^\infty(\mathbb{R}^6)\times C^\infty(\mathbb{R}^6))$. For the MHD model, $\mathcal{G}=\mathrm{Diff}^\infty(\mathbb{R}^6)\ltimes(C^\infty(\mathbb{R}^6)\times\Omega 2(\mathbb{R}^3))$.

The KdV Equation and Fourier Integral Operators

There are many known examples of PDEs which are infinite-dimensional Hamiltonian systems, such as the Benjamin-Ono, Boussinesq, Harry Dym, KdV, KP equations, and others. In many cases the Poisson structures and Hamiltonians are given ad hoc on a formal level. We illustrate this with the KdV equation, where at least one of the three known Hamiltonian structures is well understood [33].

The Korteweg-deVries (KdV) equation

$$u_t + 6uu_x + u_{xxx} = 0 \tag{4.14}$$

is an infinite-dimensional Hamiltonian system with the Lie group of invertible Fourier integral operators as symmetry group. Gardner found that with the bracket

$$\{F,G\} = \int_0^{2\pi}\frac{\delta F}{\delta u}\frac{\partial}{\partial x}\frac{\delta G}{\delta u}dx \tag{4.15}$$

and Hamiltonian

$$H(u) = \int_0^{2\pi}\left(u^3 + \frac{1}{2}u_x^3\right)dx \tag{4.16}$$

u satisfies the KdV equation (4.14) if and only if

$$\dot{u} = \{u, H\}. \tag{4.17}$$

The question is where this Poisson bracket (4.15) and Hamiltonian (4.16) come from? We showed [33–35] that this bracket is the Lie-Poisson bracket on a coadjoint orbit of Lie group $\mathcal{G}$=FIO of invertible Fourier integral operators on the circle S1. We briefly summarize the following.

A Fourier integral operators on a compact manifold M is an operator

$$A : C^{\infty}(M) \longrightarrow C^{\infty}(M) \tag{4.18}$$

locally given by

$$A(u)(x) = (2\pi)^{-n} \iint e^{i\varphi(x,y,\xi)} a(x,\xi) u(y) dy\, d\xi, \tag{4.19}$$

where φ(x,y,ξ) is a phase function with certain properties and the symbol a(x,ξ) belongs to a certain symbol class. A pseudodifferential operator is a special kind of Fourier integral operators, locally of the form

$$P(u)(x) = (2\pi)^{-n} \iint e^{i(x-y)\cdot\xi} p(x,\xi) u(y) dy\, d\xi. \tag{4.20}$$

Denote by FIO and ΨDO the groups under composition (operator product) of invertible Fourier integral operators and invertible pseudodifferential operators on M, respectively. We have the following results.

Both groups ΨDO and FIO are smooth infinite-dimensional ILH-Lie groups. The smoothness properties of the group operations (operator multiplication and inversion) are similar to the case of diffeomorphism groups (2.6), (2.7). The Lie algebras of both ILH-Lie groups ΨDO and FIO are the Lie algebras of all pseudodifferential operators under the commutator bracket. Moreover, FIO is a smooth infinite-dimensional principal fiber bundle over the diffeomorphism group of canonical transformations $\mathrm{Diff}^{\infty}_{\omega}(T^{*}\dot{M}-\{0\})$ with structure group (gauge group) ΨDO.

For the KdV equation we take the special case where M=S^1. Then the Gardner bracket (4.15) is the Lie-Poisson bracket on the coadjoint orbit of FIO through the Schrodinger operator P∈ΨDO. Complete integrability of the KdV equation follows from the infinite system of conserved integral in involution given by Hk=Trace(Pk); in particular the Hamiltonian (4.16) equals H=H_2.

See the study by Adams et al. in [34, 35] for details.

See [10, 15, 31–40].

GAUGE THEORIES, THE STANDARD MODEL, AND GRAVITY

Here we will encounter various infinite-dimensional Lie groups and algebras such as diffeomorphism groups, loop groups, groups of gauge transformations, and their cohomologies.

Gauge Theories: Yang-Mills, QED, and QCD

Consider a principal G-bundle $\pi:P\rightarrow M$, with M being a compact, orientable Riemannian manifold (e.g., $M=S^4,T^4$) and G a compact non-Abelian gauge group with Lie algebra g. Let $\mathcal{A}$ be the infinite-dimensional affine space of connection 1-forms on P. So each $A\in\mathcal{A}$ is a g-valued, equivariant 1-form on P (also called vector potential) and defines the covariant derivative of any field φ by $D_{A\varphi}=d\varphi+(1/2)[A,\varphi]$. The curvature 2-form FA (or field strength) is a g-valued 2-form and is defined as $FA=D_A A=dA+(1/2)[A,A]$. They are locally given by $A=A_\mu dx^\mu$ and $F=(1/2)F_{\mu\nu}dx^\mu \wedge dx^\nu$, where $F_{\mu\nu}=\partial_\mu A_\nu-\partial_\nu A_\mu+[A_\mu,A_\nu]$.

In pure Yang-Mills theory the action functional is given by

$$S(A)=\frac{1}{2}\|F_A\|^2=\frac{1}{2}\int_M \mathrm{Tr}(F_{\mu\nu}F^{\mu\nu}), \tag{5.1}$$

and the Yang-Mills equations become globally

$$d*F_A=0. \tag{5.2}$$

With added fermionic field ψ interaction, the action becomes

$$S(A,\psi)=\frac{1}{2}\|F_A\|^2+\langle \partial\!\!\!/_A\psi,\psi\rangle, \tag{5.3}$$

where ψ is a section of the spin bundle $\mathcal{S}pin^{\pm}(M)$ and, $\partial A:\mathcal{S}pin^{\pm}(M)\rightarrow\mathcal{S}pin^{\mp}(M)$ is the induced Dirac operator.

Gauge Invariance

In gauge theories the symmetry group is the group of gauge transformations. The diffeomorphism subgroups that arise in gauge theories as gauge groups behave nicely because they are isomorphic to subgroups of loop groups, as discussed in Section 2.4.4.

The group $\mathcal{G}$ of gauge transformations of the principal G-bundle $\pi:P\rightarrow M$ is given by

$$\begin{aligned}\mathcal{G}&=\{\phi\in \mathrm{Diff}^\infty(P);\ \phi(p\cdot g)=\phi(p)\cdot g,\ \pi\phi(p)=\pi(p)\}\\ &\cong\{\tau\in C^\infty(P,G);\ \tau(p\cdot g)=g^{-1}\tau(p)g\}=\mathrm{Gau}(P)\end{aligned} \tag{5.4}$$

$$\mathbf{YM}(A)=\int_\Omega \|F_A\|^2,\quad \Omega\subset M \text{ compact.} \tag{5.5}$$

which is a smooth Hilbert-Lie group with smooth group operations [6].

We only sketch here what role this infinite-dimensional gauge group $\mathcal{G}$ plays in these quantum field theories. A good reference for this topic is the study by Deligne et al. in [41, 42].

The gauge group $\mathcal{G}$ acts on $\mathcal{A}$ via pullback $\phi\in\mathcal{G}$, $A\in\mathcal{A}$, $\phi\cdot A=(\phi^{-1})^*A\in\mathcal{A}$, or under the isomorphism (see Section 2.4.4) $\mathcal{G}\cong\mathrm{Gau}(P)$, $\phi\Leftrightarrow\tau$ we have Gau(P) acting on $\mathcal{A}$ by $\tau\cdot A=\tau A\tau_{-1}+\tau d\tau^{-1}$. Hence the covariant derivative transforms as $D\tau\cdot A=\tau D_A\tau^{-1}$, and the action on the field is $\tau\cdot F_A:=F_{\tau\cdot A}=\tau F_{A\tau}^{-1}$.

The action functional (the Yang-Mills functional) is $S(A)=\|FA\|^2$, locally given by $\|FA\|2=(1/2)\int_M \mathrm{Tr}\,(F\mu\nu^{F\mu\nu})$. This action is gauge invariant $S(\phi\cdot A)=S(A)$, $\phi\in\mathcal{G}$, so the Yang-Mills functional is defined on the orbit space $\mathcal{M}=\mathcal{A}/\mathcal{G}$. The space $\mathcal{M}$ is in general not a manifold since the action of $\mathcal{G}$ on $\mathcal{A}$ is not free. If we restrict to irreducible connections, then $\mathcal{M}$ is a smooth infinite-dimensional manifold and $\mathcal{A}\rightarrow\mathcal{M}$ is an infinite-dimensional principal fiber bundle with structure group $\mathcal{G}$.

For self-dual connections $FA=*FA$ (instantons) on a compact 4-manifold, the moduli space $\mathcal{M}=\{A\in\mathcal{A};A \text{ self-dual}\}/\mathcal{G}$ is a smooth finite-dimensional manifold. Self-dual connections absolutely minimize the Yang-Mills action integral

$$W(f)=\int_{\mathcal{A}/\mathcal{G}} e^{-S(A)} f(A)\mathfrak{D}(A) \tag{5.6}$$

The Feynman path integral quantizes the action and we get the probability amplitude

$$\text{Yang-Mills: } S(\phi\cdot A)=S(A), \quad A\in\mathcal{A}, \phi\in\mathcal{G}, \tag{5.7}$$

for any gauge-invariant functional f(A).

Let $\mathcal{G}$ be the group of gauge transformations. So $\phi\in\mathcal{G}\Leftrightarrow\phi:P\rightarrow P$ is a diffeomorphism over idM; that is, $\phi(p\cdot g)=\phi(p)\cdot g$, $p\in P$, $g\in G$. Then $\mathcal{G}$ acts on $\mathcal{A}$ and $\mathcal{S}pin^{\pm}(M)$ by $\phi\cdot A=(\phi^{-1})^*A$ and $\phi\cdot\psi=(\phi^{-1})^*\psi$. The action functionals S are gauge invariant:

$$\text{QED: } S(\phi\cdot A,\phi\cdot\psi)=S(A,\psi), \quad A\in\mathcal{A}, \psi\in\mathcal{S}pin^{\pm}(M), \phi\in\mathcal{G}. \tag{5.8}$$

Quantum Electrodynamics (QED) and Quantum Chromodynamics (QCD)

In classical field theory, one considers a Lagrangian $\mathcal{L}(\phi_i,\partial_\mu\phi_i)$ of the fields $\varphi i:\mathbb{R}^n\rightarrow\mathbb{R}$, $i=1,\ldots,k$, and $\partial_\mu=\partial/\partial x_\mu$ and the corresponding action functional $S=\int\mathcal{L}(\phi_i,\partial_\mu\phi_i)d^nx$. The variational principle $\delta S=0$ then leads to the Euler-Lagrange equations of motion

$$\frac{\partial\mathcal{L}}{\partial\phi_i}-\partial_\mu\frac{\partial\mathcal{L}}{\partial(\partial_\mu\phi_i)}=0. \tag{5.9}$$

In QED and QCD the Lagrangian is more complicated of the form

$$\mathcal{L}(A,\psi,\varphi) = -\frac{1}{4g^2}\,\mathrm{Tr}\,F_{\mu\nu}F^{\mu\nu} - i\overline{\psi}\left[\gamma^{\mu}(\partial_\mu + ieA_\mu) + m\right]\psi + \left(D^{\mu}_{A}\varphi\right)^{\dagger}\left(D^{\mu}_{A}\varphi\right) - m^2\varphi^{\dagger}\varphi, \tag{5.10}$$

where $A_\mu(x)$ is a potential 1-form (boson), and the field strength F is given by $F_{\mu\nu} = \partial_\mu A_\nu - \partial_\nu A_\mu + [A_\mu, A_\nu]$. In QED the gauge group of the principal bundle is $G = U(1)$, and in QCD we have $G = SU(2)$. The Dirac γ-matrices are $\gamma^i = \begin{pmatrix} 0 & -\sigma_i \\ \sigma_i & 0 \end{pmatrix}$, where σ_i are the Pauli matrices (canonical basis of $\mathfrak{su}(2)$) and $\overline{\psi} = \psi^{\dagger}\gamma^{o}$ is the Pauli adjoint with $\gamma^{o} = \begin{pmatrix} 0 & 1 \\ 1 & 0 \end{pmatrix}$, m is the electron mass, e is the electron charge, and g is a coupling constant.

where $A_\mu(x)$ is a potential 1-form (boson), and the field strength F is given by $F_{\mu\nu}=\partial_\mu A_\nu - \partial_\nu A_\mu+[A_\mu,A_\nu]$. In QED the gauge group of the principal bundle is G=U(1), and in QCD we have G=SU(2). The Dirac γ-matrices are $\gamma^i = \begin{pmatrix} 0 & -\sigma_i \\ \sigma_i & 0 \end{pmatrix}$ where σi are the Pauli matrices (canonical basis of (2)) and $\overline{\psi} = \psi^{\dagger}\gamma^{o}$ is the Pauli adjoint with $\gamma^{o} = \begin{pmatrix} 0 & 1 \\ 1 & 0 \end{pmatrix}$, m is the electron mass, e is the electron charge, and g is a coupling constant.

The Equations of Motion

The variational principle of the Lagrangian (5.10) with respect to the fields A,ψ, and φ gives the corresponding Euler-Lagrange equations of motion. They describe, for instance, the motion of an electron ψ(x) (fermion, spinor) in an electromagnetic field F, interacting with a bosonic field φ. We get, from the variational principle, $\delta S/\delta A_\mu = 0 \Rightarrow \partial_\mu F^{\mu\nu} = e\overline{\psi}\gamma^{\nu}\psi$, which are Maxwell's equations for G=U(1).

In the free case, that is, when ψ=0, we get $\partial_\mu F^{\mu\nu}=0$, the vacuum Maxwell equations.

For G=SU(2) these equations become $D_\mu F^{\mu\nu}=0$, the Yang-Mills equations. Moreover, $\delta S/\delta\psi=0 \Rightarrow i(\not{\partial}A-m)\psi=0$, which are Dirac's equations, where $\partial A=\gamma_\mu(\partial_\mu+ieA_\mu)=\gamma^\mu D_A^{\ \mu}$. In the free case, that is, when A=0, we get $i(\not{\partial}A-m)\psi=0$, the classical Dirac equation.

Chiral Symmetry

The chiral symmetry is the symmetry that leads to anomalies and the BRST invariance. In QCD the chiral symmetry of the Fermi field ψ is given by $\psi \mapsto e^{i\beta\gamma_5}\psi$, where β is a constant and $\gamma_5=i\gamma^o\gamma^1\gamma^2\gamma^3$. The classical Noether current of this symmetry is given by $J_\mu = \overline{\psi}\gamma_\mu\gamma_5\psi$ which is conserved; that is, $\partial^\mu J_\mu=0$.

This conservation law breaks down after quantization; one gets

$$\partial^\mu J_\mu = 2\mathrm{im}\,\overline{\psi}\gamma_5\psi - \frac{g^2}{8\pi^2}\,\mathrm{Tr}\,F_{\mu\nu}F^{\mu\nu} \equiv \omega \neq 0. \tag{5.11}$$

This value ω is called the chiral anomaly.

Quantization

The quantization is given by the Feynman path integral:

$$\int_{\mathcal{A}/\mathcal{G}} \int_{\times Spin} e^{iS(A,\psi)} \mathcal{F}(A,\psi) \mathfrak{D}A \mathfrak{D}\psi = \langle \mathcal{F}(A,\psi) \rangle \tag{5.12}$$

which computes the expectation value $\langle \mathcal{F}(A,\psi) \rangle$ of the function $\mathcal{F}(A,\psi)$. This is an integral over two infinite-dimensional spaces: the gauge orbit space $\mathcal{A}/\mathcal{G}$ and the fermionic Berezin integral over the spin space $\mathcal{S}pin\pm(M)$. These integrals are mathematically not defined but physicists compute them by gauge fixing; that is, fixing a section $\sigma:/\mathcal{G}\to\mathcal{A}$, (e.g., $\sigma(A)=\partial_\mu A^\mu=0$, the Lorentz gauge) and then integrating over the section σ. Such a section does not exist globally, but only locally (Gribov ambiguity!). The effect of such a gauge fixing is that one gets extra terms in the Lagrangian (gauge-fixing terms) and one has to introduce new fields, so-called ghost fields η via the Faddeev-Popov procedure. The such obtained effective Lagrangian is no longer gauge invariant. This effective Lagrangian has the form in QCD:(

$$\begin{aligned} \mathcal{L}_{\text{eff}}(A,\psi,\eta) &= \frac{1}{2}\operatorname{Tr}(F_{\mu\nu}F^{\mu\nu}) && \text{kinetic energy} \\ &+ \frac{1}{2\alpha}\operatorname{Tr}(\partial_\mu A^\mu)^2 && \text{gauge-fixing term} \\ &- g\partial_\mu\bar{\eta}D^\mu_A\eta && \text{ghost term} \\ &+\cdots && \text{interaction terms.} \end{aligned} \tag{5.13}$$

We can write this globally as

$$\mathcal{L}_{\text{eff}} = \frac{1}{2}\|F_A\|^2 + \frac{1}{2}\|\sigma(A)\|^2 + \bar{\eta}\mathcal{M}\eta + \cdots, \tag{5.14}$$

where $\mathcal{M}=(\delta/\delta\phi)(\sigma(\phi\cdot A))$ is the Faddeev-Popov determinant, acting like the Jacobian of the global gauge variation $\delta/\delta\phi$ over the section σ. Writing this term in the exponent of the action functional like a "fermionic Gaussian integral" leads to the Faddeev-Popov ghost fields $\eta,\bar{\eta}$ in the form $\det \mathcal{M} = \int e^{-\bar{\eta}\mathcal{M}\eta} d\bar{\eta}\, d\eta$.

BRST Symmetry

The effective Lagrangian $\mathcal{L}$eff is NOT gauge invariant but has a new symmetry, called BRST symmetry.

$$\begin{aligned} \mathfrak{s}A &= d\eta + [A,\eta] \\ \mathfrak{s}\eta &= -\frac{1}{2}[\eta,\eta] \end{aligned} \qquad \mathcal{L}_{\text{eff}} \text{ is BRST invariant.} \tag{5.15}$$

Note that the BRST operator $\mathfrak{s}$ mixes bosons (A) and fermions (η). This is an example of supersymmetry which we will discuss in Section 6. Also, the BRST operator $\mathfrak{s}$ is nilpotent; that is, $\mathfrak{s}^2=0$. The question arises whether this operator $\mathfrak{s}$ is the coboundary operator of some kind of cohomology. The affirmative answer is given by the following theorem (Schmid [6, 44]).

Theorem 5.1. Let $\mathcal{C}^{q,p}(\mathrm{lie}\,\mathcal{G},\Omega_{\mathrm{loc}})$ be the Chevalley-Eilenberg complex of the Lie algebra $\mathrm{lie}\,\mathcal{G}$ of infinitesimal gauge transformations, with respect to the induced adjoint representation on local forms $\Omega\mathrm{loc}$, with boundary operator

$$\delta_{\mathrm{loc}} : \mathcal{C}^{q,p}(\mathrm{lie}\,\mathcal{G},\Omega_{\mathrm{loc}}) \longrightarrow \mathcal{C}^{q+1,p}(\mathrm{lie}\,\mathcal{G},\Omega_{\mathrm{loc}}), \qquad \delta^2_{\mathrm{loc}} = 0. \tag{5.16}$$

Then with $:=(-1)^{p+1}/(q+1)\delta_{\mathrm{loc}}$, one has $\mathfrak{s}^2=0$ and the following.

(1) For q=0, p=1, $A\in\mathcal{A}\subset\mathcal{C}^{0,1}$, then $\mathfrak{s}A=d\eta+[A,\eta]$.

(2) For q=1, p=0, $\eta\in\mathcal{C}^{1,0}$, then $\mathfrak{s}\eta=-(1/2)[\eta,\eta]$, the Maurer-Cartan form.

(3) The chiral anomaly ω (given by (5.11)) is represented as cohomology class of this complex $[\omega] \in \mathcal{H}^{1,0}_{\mathrm{BRST}}(\mathrm{lie}\,\mathcal{G},\Omega_{\mathrm{loc}})$.

The Chevalley-Eilenberg Cohomology

We are now going to explain the previous theorem, in particular the general definition of the Chevalley-Eilenberg [45] complex and the corresponding cohomology.

Let G be a Lie group with Lie algebra $\mathfrak{g}$ and let σ be a representation of $\mathfrak{g}$ on the vector space W. Denote by $\mathcal{C}^q(\mathfrak{g},W)$ the space of W-valued q-cochains on $\mathfrak{g}$ and define the coboundary operator $\delta:\mathcal{C}^q(\mathfrak{g},W)\to\mathcal{C}^{q+1}(\mathfrak{g},W)$ by

$$\delta\Phi(\xi_o,\ldots,\xi_q) = \sum_{i=0}^{q}(-1)^i\sigma(\xi_i)\Phi\left(\xi_o,\ldots,\widehat{\xi_i},\ldots,\xi_q\right) + \sum_{i<j}(-1)^{i+j}\Phi\left(\sigma(\xi_i)\xi_j,\ldots,\widehat{\xi_i},\ldots,\widehat{\xi_j},\ldots,\xi_q\right). \tag{5.17}$$

We have $\delta^2=0$, and define the Lie algebra cohomology of $\mathfrak{g}$ with respect to (σ,W) as $\mathcal{H}^*(\mathfrak{g},W)=\ker\delta/\mathrm{im}\,\delta$. This is called the Chevalley-Eilenberg cohomology [45] of the Lie algebra $\mathfrak{g}$ with respect to the representation σ.

Anomalies

The Noether current induced by the chiral symmetry (after quantization) for the free case (ψ=0), that is, for pure Yang-Mills becomes

$$\begin{aligned}\partial^\mu J_\mu &= -\frac{g^2}{8\pi^2}\varepsilon^{\mu\nu\rho\tau}\,\mathrm{Tr}\,F_{\mu\nu}F_{\rho\tau} \\ &= -\frac{1}{4}\pi^2\varepsilon^{\mu\nu\rho\tau}\,\mathrm{Tr}\,\partial_\mu\left(A_\nu\partial_\rho A_\tau + \frac{2}{3}A_\mu A_\rho A_\tau\right) \\ &= \omega \neq 0 \quad \text{anomaly.}\end{aligned} \tag{5.18}$$

See (5.11).

Note the similarity with the Chern-Simon Lagrangian

$$\mathcal{L}(A) = \int_M \mathrm{Tr}\left(Ad\,A + \frac{2}{3}A^3\right). \tag{5.19}$$

We are going to derive a representation of the chiral anomaly ω in the BRST cohomology that is $[\omega] \in \mathcal{H}^{1,0}_{\mathrm{BRST}}(\mathrm{lie}\,\mathcal{G}, \Omega_{\mathrm{loc}})$.

The question is "if $\mathfrak{s}\omega=0$, does there exist a local functional F(A), such that $\omega=\mathfrak{s}(F(A))$? That is, is ω BRST $\mathfrak{s}$-exact? The answer in general is NO; that is, ω represents a nontrivial cohomology class. This class is given by the Chern-Weil homotopy.

Let $\tilde{A}=A+\eta\in\mathcal{C}^{0,1}\times\mathcal{C}^{1,0}$ and $\tilde{F}\equiv s\tilde{A}+\tilde{A}^2=F_{\tilde{A}}$. For $t\in[0,1]$, let $\tilde{F}t=t\tilde{F}+(t2-t)\tilde{A}^2$ and define the Chern-Simons form

$$\omega_{2q-1} \equiv q\int_0^1 \mathrm{Tr}\left(\tilde{A}\tilde{F}_t^{q-1}\right)dt, \tag{5.20}$$

we get

$$\mathfrak{s}\omega_{2q-1} = \mathrm{Tr}\,\tilde{F}^q. \tag{5.21}$$

We write ω_{2q-1} as sum of homogeneous terms in ghost number (upper index) and degree (lower index) $\omega_{2q-1} = \omega^0_{2q-1} + \omega^1_{2q-2} + \omega^2_{2q-3} + \cdots + \omega_0^{2q-1}$. Let $\omega(X,A) = \int_M \omega^1_{2q-2}(X)$.

Theorem 5.2(see Schmid [46]).

The form $\omega(X,A) = \int_M \int_0^1 \tilde{A}\tilde{F}_t^{q-1}(X)dt$ satisfies the Wess-Zumino consistency condition $(\mathfrak{s}\omega)(X_0,X_1,A)=0$ and represents the chiral anomaly $\mathcal{H}^{1,0}_{\mathrm{BRST}}(\mathrm{lie}\,\mathcal{G}, \Omega_{\mathrm{loc}})$.

We have an explicit form of the anomaly in (2q – 2) dimensions:

$$\omega^1_{2q-2} = q(q-1)\int_0^1 (1-t)\,\mathrm{Tr}\left(\eta\delta_{\mathrm{loc}}\left(\tilde{A}\tilde{F}_t^{q-2}\right)\right)dt. \tag{5.22}$$

So for q=2 the non-Abelian anomaly in 2 dimensions becomes $\omega 21=\mathrm{Tr}\,(\eta\delta_{\mathrm{loc}}\tilde{A})$, and for q=3 the non-Abelian anomaly in 4 dimensions becomes

$$\omega^1_4 = \mathrm{Tr}\left(\eta\delta_{\mathrm{loc}}\left(\tilde{A}\delta_{\mathrm{loc}}\tilde{A} + \frac{1}{2}\tilde{A}^3\right)\right). \tag{5.23}$$

The Standard Model

The standard model is a Yang-Mills gauge theory. Recall that the free Yang-Mills equations are $D_A{}^*F=0$, where A is a connection 1-form (vector potential), and F is the associated curvature 2-form (field) on the principal bundle P. The connection A defines the covariant derivative D_A and the curvature F given by $F=D_AA=dA+(1/2)[A,A]$, or locally $F_{\mu\nu}=\partial_\mu A_\nu-\partial_\nu A_\mu+[A_\mu,A_\nu]$, and $D^\mu F_{\mu\nu}=0$. Again the connection A is the fundamental object.

For different choices of the gauge Lie group G, we obtain the 3 theories that make up the standard model. For G=U(1) on a trivial bundle (i.e., global symmetry, which gives charge conservation) the curvature 2-form F is simply the electromagnetic field, and the Yang-Mills equations $D_A{}^*F=0$ are Maxwell's equations dF=0, locally $\partial_\mu F^{\mu\nu}=0$. For G=U(1) as local gauge group we get the quantum mechanical symmetry and the equations of motion are Dirac's equations. Combing the two, we get QED as a U(1) gauge theory. For G=SU(N) we get the full non-Abelian Yang-Mills equations DA*F=0. For weak interactions with G=SU(2) and combining the two (spontaneous symmetry breaking, Higgs), we get the Glashow-Weinberg-Salam model as SU(2)×U(1) Yang-Mills theory of electroweak interactions. For G=SU(3) we obtain the Yang-Mills equations $D_A{}^*F=0$ for strong interactions and the equations of motion for QCD. Finally that standard model is a SU(3)×SU(2)×U(1) gauge theory governed by the corresponding Yang-Mills equations $D_A{}^*F=0$. Recall that F is the curvature in the corresponding principal bundle determined by the connection A.

For interactions, all the relevant fields involved can be considered as sections of corresponding associated vector bundles induced by representations of the gauge groups, for example, the Dirac operator on the associated spin bundle (induced by the spin representation of SU(2)) acting on spinors (sections of this bundle). The vector potentials are the corresponding connection 1-forms and the Yang-Mills fields are the corresponding curvature 2-forms on these bundles over spacetime.

Again we do not need the metric and the curvature is determined by the potential, so the potential is the fundamental object.

Gravity

Stop Looking for Gravitons

Stop looking for the graviton, not because it had been found but because it does not exist. The graviton is supposed to be the particle that communicates the gravitational force. But the gravitational force is not a fundamental force. Gravity is geometry. One might as well search for the Corioliston for the coriolis force or the Centrifugiton for the centrifugal force.

Since Einstein in the 1920s, physicists have tried to unify what are considered the four fundamental forces, namely, electromagnetism, weak and strong nuclear forces, and the gravitational force. In the 1970s, the three nongravitational forces were unified in the standard model. At high enough energy (about 10^{15} GeV) they become the same force.

Since then, with all the string theory, SUSY, branes, and extra dimensions, the gravitational force could not be incorporated into GUT that includes all 4 forces and no graviton has been found experimentally. The reason is simple: not many people, including Einstein himself, take/took the general theory of relativity seriously enough, according to which we know that the gravitational force does not exist as fundamental force but as geometry! We do not feel it. What we feel is the resistance of the solid ground on which we stand. In general relativity, free-falling objects follow geodesics of spacetime, and what we perceive as the force of gravity is instead a result of our being unable to follow those geodesics because of the mechanical resistance of matter. Newton's apple falls downward because the spacetime in which we exist is curved. The "gravitational force" is not a force but it is the geometry of spacetime as Einstein observed in [47, page 137]:

"Die Koeffizienten (gμv) dieser Mertik beschreiben in Bezug auf das gewählte Koordinatensystem zugleich das Gravitationsfeld."

("The coefficients (gμv) of this metric with respect to the chosen coordinate system describe at the same time the gravitational field") [47, page 146]:

"Aus pysikalischen Gründen bestand die Überzeugung, dass das metrische Feld zugleich das Gravitationsfeld sei."

("For physical reasons there was the conviction that the metric field was at the same time the gravitational field").

Therefore GUT, the grand unified theory had been completed since the 1970s with the standard model. Since the gravitational force does not exist as a fundamental force, there is nothing more to unify as forces. If we want to unify all four theories, then it has to be done in a geometric way. The equations governing gravity as well as the standard model are all curvature equations, Einstein's equation, and the Yang-Mills equations.

Einstein's Vacuum Field Equations

Let (M,g) be spacetime with Lorentzian metric g. Then Einstein's vacuum field equations are

$$\mathrm{Ric} = 0, \tag{5.24}$$

where Ric is the Ricci curvature of the Lorentz metric g. These are the Euler-Lagrange equations for the Lagrangian $\mathcal{L}(g)=\int R(g)\mu(g)$, where $\mu(g)=-\ \sqrt{-\det g d^4x}$ and R(g) is the scalar curvature of g.

Or in general, locally, in terms of the stress-energy tensor Tμv, Einstein's equations are $G_{\mu\nu}=\kappa T_{\mu\nu}$ with the Einstein tensor $G_{\mu\nu}=R_{\mu\nu}-(1/2)g_{\mu\nu}R$. The stress-energy tensor $T_{\mu\nu}$ is the conserved Noether current corresponding to spacetime translation invariance.

The Levi-Civita connection Γ of the Riemannian metric g is given by

$$\Gamma^{\lambda}_{\mu\nu} = \frac{1}{2}g^{\lambda\sigma}\left(\frac{\partial g_{\sigma\mu}}{\partial x^{\nu}} + \frac{\partial g_{\sigma\nu}}{\partial x^{\mu}} - \frac{\partial g_{\mu\nu}}{\partial x^{\sigma}}\right). \tag{5.25}$$

The curvature tensor R and the Ricci curvature Ric in Einstein's field equations are completely determined by the connection Γ.

First the curvature tensor R is locally given by

$$R^{\lambda}_{\mu\nu\kappa} = \left(\frac{\partial\Gamma^{\lambda}_{\mu\nu}}{\partial x^{\kappa}} - \frac{\partial\Gamma^{\lambda}_{\mu\kappa}}{\partial x^{\nu}}\right) + \left(\Gamma^{\eta}_{\mu\nu}\Gamma^{\lambda}_{\kappa\eta} - \Gamma^{\eta}_{\mu\kappa}\Gamma^{\lambda}_{\nu\eta}\right). \tag{5.26}$$

Taking its trace, we get the Rici tensor Ric as. $(\mathrm{Ric})^{\mu\nu} = R^{\lambda}_{\mu\lambda\nu}.$

So we can express Einstein's equations completely in terms of the connection (potential) Γ; we do not need the metric g; also the curvature R is determined by the potential Γ. So the potential Γ is the fundamental object.

The free motion in spacetime is along geodesic curves γ(t) which again are expressed in terms of the connection by

$$\ddot{\gamma}^{\alpha} + \Gamma^{\alpha}_{\beta\nu}\dot{\gamma}^{\beta}\dot{\gamma}^{\nu} = 0. \tag{5.27}$$

Symmetry

In general relativity the diffeomorphism group plays the role of a symmetry group of coordinate transformations. Then the vacuum Einstein's field equations Ric(g)=0 are invariant under coordinate transformations, that is, under the action of $\mathrm{Diff}^{\infty}(M)$. Denote by $\mathcal{M}$ the space of all metrics g on M. Then, Einstein's field equations Ric(g)=0 are a Hamiltonian system on the reduced space P=M/Diff∞(M); see the study by Marsden et al. in [15] for details.

Conclusions

The relation between the connection Γ occurring in Einstein's equations and the connection A in Yang-Mills equations is as follows. A is a Lie algebra-valued 1-form on a principal G-bundle (P,π,M) over spacetime (M,g) (or any associated vector bundle given by representations of G). The Levi-Civita connection Γ is a connection 1-form in this sense on the tangent bundle TM (frame bundle) with G=GL(n). So in this sense general relativity and the standard model are Yang-Mills gauge theories.

Therefore all four theories, electromagnetism, weak interaction, strong interaction, and gravity, are unified as curvature equations in vector bundles over spacetime. Different interactions require different bundles.

There is no hierarchy problem because there is no fundamental gravitational force. The question why gravitational interaction is so much weaker than electroweak and strong interactions is meaningless, comparing apples with oranges. Why are so many physicists still talking about gravitational force? It is like as if we are still talking about "sun rise" and "sun set", 500 years after Copernicus! only worse; these are serious scientists trying to unify all four "forces" to a TOE.

I am not saying that there are no open problems in physics. Of course there is still the problem of unifying quantum mechanics and general relativity on a geometric level (not as forces). The question is "how does spacetime look at the Planck scale? Do we have to modify spacetime to incorporate quantum mechanics or quantum mechanics to accommodate spacetime, or both? We need a theory of quantum gravity. There are several theories in the developing stage that promise to accomplish this.

(i) Superstring theory by E. Witten et al.

(ii) Discrete spacetime at Planck length by R. Loll in "Causal dynamical triangulation" and by J. Ambjorn, J. Jurkiewicz, and R. Loll in "The Universe from Scratch" [arXiv: hep-th/0509010].

(iii) Spacetime quantization: loop quantum gravity by L. Smolin in "Three Roads to Quantum Gravity" (London: Weidenfeld and Nicholson, 2000) and by S. O. Bilson-Thompson, F. Markopoulou, and L. Smolin in "Quantum Gravity and the Standard Model", preprint 2006.

(iv) Geometric formulation of quantum mechanics by A. Ashtekar and T. A. Schilling [arXiv: gr-qc/9706069].

(v) Deterministic quantum mechanics at Planck scale by G. t'Hooft in "Quantum Gravity as a Dissipative Deterministic System."

(vi) Branes and new dimensions: parallel universes by L. Randell et al., D. Deutsch, PS. In the brane world, gravity is again singled out as the only force not confined to one brane.

(vii) Noncommutative Geometry. A. Connes describes the standard model form general relativity.

(viii) The most recent new development is by Verlinde [48]. He agrees that gravity is not a fundamental force, but explains it as an emergent force (entropic force) caused by a change in the amount of information (entropy) associated with the positions of bodies of matter.

See [6, 11, 27, 36, 41–59].

SUSY (SUPERSYMMETRY)

Supersymmetry (SUSY) is an important idea in quantum filed theory and string theories. The BRST symmetry we described in Section 5.3 is an example of SUSY. Now we give a summary of a mathematical description of super Hamiltonian systems on supersymplectic supermanifolds, state a generalization of the Marsden-Weinstein reduction theorem in this context, and illustrate the method with examples. This is a very technical topic, so we only give a brief sketch; for details see the studies Glimm in [60] and Tuynman in [61].

The classical Marsden-Weinstein reduction theorem is a geometrical result stating that if a Lie group G acts on a symplectic manifold P by symplectomorphisms and admits an equivariant momentum map $J:P\to g^*$, then, for any regular value $\mu \in g$ of J, the quotient $P_\mu = J\text{-}1(_\mu)/G_\mu$ of the preimage $J^{-1}(\mu)$ by the isotropy group $G\mu$ of μ has a natural symplectic structure. A dynamical interpretation of the Marsden-Weinstein reduction theorem gives the following. If a given Hamiltonian $H \in C^\infty(P)$ is invariant under the action of the group G, then it projects to a reduced Hamiltonian H_μ on the reduced space P_μ. The integral curves of the Hamiltonian vector field XH project to the integral curves of $X_{H\mu}$. In this sense, one has reduced the system by symmetries. This reduction procedure unifies many methods and results concerning the use of symmetries in classical mechanics, some dating back to the time of Euler and Lagrange.

In [60], Glimm generalizes this result to the setting of supermanifolds, using the analytic construction of supercalculus and supergeometry due to DeWitt [62] and Tuynman [61]. The general idea of supermanifolds and superanalysis is to do geometry and analysis over a graded algebra of even supernumbers rather than $\mathbb{R}$. In "supermathematics," we have even and odd variables. Two variables a,b are called even, or commuting, or bosonic if $a\cdot b=b\cdot a$, whereas ξ,χ are called odd, or anticommuting, or fermionic if $\xi\cdot\chi=-\chi\cdot\xi$. The problem is doing analysis with even and odd variables.

Many classes of differential equations have extensions that involve odd variables. These are called superized versions, or supersymmetric extensions. An active area of research is in particular the construction of supersymmetric integrable systems.

As an example, consider the Korteweg-de Vries equation $ut=-u_{xxx}+6uu_x$. A possible supersymmetric extension is the following system of an even variable u(t,x) and an odd variable $\xi(t,x)$:

$$u_t = -u_{xxx} + 6uu_x - 3\xi\xi_{xx},$$

$$\xi_t = -\xi_{xxx} + 3u\xi_x + 3\xi u_x. \tag{6.1}$$

There is a different way of writing system (6.1). For this, one considers the so-called 2|1-dimensional superspace. This is a space with coordinates (x,t,ϑ), where x and t are even numbers as before and ϑ is an odd number. We can now gather the components u(x,t) and $\xi(x,t)$ into a superfield $\Phi(t,x,\vartheta)$ defined on superspace via

$$\Phi(t,x,\vartheta) = \xi(t,x) + \vartheta \cdot u(t,x). \tag{6.2}$$

The function $\Phi(t,x,\vartheta)$ takes as values odd numbers; it is thus called an odd function. Note how the right-hand side can be seen as a Taylor series in ϑ; indeed, all higher powers of ϑ are zero because ϑ is anticommuting. One defines the odd differential operator $D=\partial/\partial\vartheta+\vartheta\cdot(\partial/\partial x)$, acting on superfields. A computation yields that system (6.1) is equivalent to

$$\Phi_t = -D^6\Phi + 3D^2\Phi D\Phi. \tag{6.3}$$

(Note that $D^2=\partial_x$.) System (6.1) is called the component formulation; (6.3) is called the superspace formulation. In our example at hand, a justification for calling the system an "extension" of KdV would be that if one takes (6.3) and writes it in component form (6.1), one recovers the "usual" KdV by setting ξ to zero. Also, it can be shown that (6.3) is invariant under transformations $\Phi(t,x,\vartheta)\mapsto\Phi(t,x-\eta\vartheta,\vartheta+\eta)$, where η is an odd parameter. The infinitesimal version of this transformation is $\delta\Phi=\eta(\partial\vartheta-\vartheta\partial_x)\Phi$. This transformation is called a "supersymmetry" in the present context. In components, it reads

$$\delta u = \eta\xi_x, \qquad \delta\xi = \eta u. \tag{6.4}$$

These equations illustrate a characterization of supersymmetries: supersymmetries (as opposed to regular symmetries) "mix" even and odd variables.

One needs some concept of supermanifold even if one only works with the component formulation. For example, one has implicitly in system (6.1) the space of all $(u(t,x),\xi(t,x))$ on which the equations are defined; this is some kind of superspace itself. Also, there is some kind of supersubmanifold of those (u,ξ) which solve the equations.

In [60], Glimm proves a comprehensive result on supersymplectic reduction. He uses an analytic-geometric approach to the theory of supermanifolds, and not the Kostant theory of graded manifolds [49]. The Poisson bracket induced by odd supersymplectic forms is not a super Lie bracket on the space of supersmooth functions. This stands of course in contrast to both the usual ungraded case and the super case with even supersymplectic forms. While this makes the algebraic approach conceptually more difficult, no such problems arise in the analytic approach. Also, we do not require that the action be free and proper, but have the weaker requirement that the quotient space only has a manifold structure.

There are different approaches to supermanifolds. The Algebraic approach (Kostant [49], Berezin-Leĭtes, late 1970s) takes "superfunctions" as fundamental object. A graded manifold is a pair (M,A), where M is a conventional manifold and $U \mapsto A(U)$ is the following sheaf over M:

$$A(U) \simeq C^\infty(U) \otimes \bigwedge \mathbb{R}^n, \quad U \subseteq M. \tag{6.5}$$

So a superfunction $f \in A(U)$ can be written as

$$f = f_\emptyset + \sum_i f_{\{i\}} \vartheta^i + \sum_{i<j} f_{\{i,j\}} \vartheta^i \vartheta^j + \cdots. \tag{6.6}$$

The Geometric approach (DeWitt [62]) takes points as fundamental objects. Let $\mathcal{A}$ be the ring generated by L generators $\theta^1,\theta^2,\ldots,\theta^L$ (Grassmann numbers) with relations $\theta^i\theta^j=-\theta^j\theta^i$.

An element is written as

$$a = \underbrace{a_\emptyset}_{=Ba \text{ ("Body")}} + \underbrace{\sum_i a_{\{i\}}\theta^i + \sum_{i<j} a_{\{i,j\}}\theta^i\theta^j + \cdots}_{\text{nilpotent part}} \quad (a_\omega \in \mathbb{R}),$$

$$a \in \mathcal{A}_0 \Longleftrightarrow a = a_\emptyset + \sum a_{\{i,j\}}\theta^i\theta^j + \text{terms with even \# of } \theta\text{s},$$

$$a \in \mathcal{A}_1 \Longleftrightarrow a = \sum a_{\{i\}}\theta^i + \sum a_{\{i,j,k\}}\theta^i\theta^j\theta^k + \text{terms with odd \# of } \theta\text{s}. \tag{6.7}$$

For calculus we "replace" reals $\mathbb{R}$ by Grassmann numbers $\mathcal{A}$. A DeWitt supermanifold is a topological space M which is locally superdiffeomorphic to $\mathcal{A}^{m|n} \stackrel{\text{def}}{=} (\mathcal{A}^{m\oplus n})_0 = \mathcal{A}_0^m \times \mathcal{A}_1^n$.

The two approaches are equivalent. There is a one-to-one correspondence between isomorphism classes of m|n-dimensional DeWitt supermanifolds whose body is a fixed smooth m-dimensional manifold X and isomorphism classes of m|n-dimensional graded manifolds over X.

The DeWitt topology of $\mathcal{A}^{m|n}$ is defined as follows.

$U \subseteq \mathcal{A}^{m|n}$ is open if and only if B(U) is open in $B\mathcal{A}^{m\oplus n} = \mathbb{R}^m$, $U = B^{-1}(B(U))$.

Smooth functions $\mathcal{A}^{m|n} \to \mathcal{A}$ are defined as follows. Let $U \subseteq \mathcal{A}^{m|n}$ be an open set. A function $f: U \to \mathcal{A}$ is called smooth if there is a collection of smooth real functions defined on $BU \subseteq \mathbb{R}^m$:(

$$f_{i_1 \dots i_n} : BU \to \mathbb{R}, \quad \text{for } i_1, \dots, i_n = 0, 1 \tag{6.8}$$

such that

$$f(x_1, \dots, x_m, \xi_1, \dots, \xi_n) = \sum_{i_1, \dots, i_n = 0}^{1} \xi_1^{i_1} \cdots \xi_n^{i_n} \cdot Z f_{i_1 \dots i_n}(x_1, \dots, x_m), \tag{6.9}$$

where Zg is defined as follows. If $x \in \mathcal{A}^{m|0}$ has the decomposition $x = Bx + n$, then

$$Zg(\mathbf{x}) = \sum_{k=1}^{\infty} \frac{1}{k!} D^k g(B\mathbf{x})[\mathbf{n}, \dots, \mathbf{n}]. \tag{6.10}$$

Note. Smooth functions map the body to the body

Superdifferential Geometry

Versions of the inverse function theorem and implicit function theorem are still valid. Concepts of tangent space, vector fields, flows, and Lie groups can be developed as in the ungraded case.

Lie Supergroups

A Lie supergroup is a supermanifold G that is a group and for which the group operations of multiplication and inversion are smooth. If a Lie supergroup G acts freely and properly on a supermanifold M, then the quotient M/G can be given the structure of a supermanifold such that the projection $\pi: M \to M/G$ is a surjective submersion. The structure of T(M/G) is given by the following.

Theorem 6.1. Let $\varphi: M \times G \to M$ be an action. Suppose that M/G has a supermanifold structure such that $\pi: M \to M/G$ is a surjective submersion; that is, $T\pi(p)$ is onto for every $p \in M$. For any $p \in M$, one has

$$\ker T\pi(p) = \{\mathfrak{X}_M(p) \mid \mathfrak{X} \in \mathfrak{g}\} \tag{6.11}$$

and this is a proper subspace of T_pM. In particular,

$$T_{[p]}(M/G) \simeq T_pM / \{\mathfrak{X}_M(p) \mid \mathfrak{X} \in \mathfrak{g}\}. \tag{6.12}$$

Remark 6.2. There are examples where M/G does not have a supermanifold structure and $\{\mathfrak{X}M(p)|\mathfrak{X}\in g\}$ fails to be a free submodule.

Supersymplectic Structures

A supersymplectic supermanifold (M,ω) is a supermanifold M together with a closed (i.e., dω=0) nondegenerate homogeneous left 2-form $\omega \in \Omega_L^2(M)$.

Examples 6.3. (1) $\mathcal{A}2^{m|n}$ with coordinates (q^i,p^i,ξ^j),

$$\omega = \sum dq^i \wedge dp^i + \frac{1}{2}\sum d\xi^j \wedge d\xi^j \tag{6.13}$$

Defines an even supersymplectic form.

(2) On $\mathcal{A}m|m$ with coordinates (x^i,ξ^j),

$$\omega = \sum dx^i \wedge d\xi^i \tag{6.14}$$

defines an odd supersymplectic form

Let $\omega \in \Omega_L^2(M)$ be a 2-form on the supermanifold M and p∈M. Then ω(p)∈ $\mathrm{Alt}_L^2(T_pM)$ is nondegenerate if and only if the real 2-form $B\omega(Bp)=\omega(Bp)|_{B(TBpM)}\in \mathrm{Alt}^2(B(TB_{pM}))$ is nondegenerate.

A smooth vector field $X\in\mathfrak{X}(M)$ is called (globally) Hamiltonian if there is some function $H\in C^\infty(M,\mathcal{A})$ such that iXω=dH. For $f,g\in C^\infty(M,\mathcal{A})$ define the Super-Poisson bracket by

$$\{f,g\} = \langle X_f, X_g \mid \omega\rangle \in C^\infty(M,\mathcal{A}). \tag{6.15}$$

Fact. If ω is even, then $C^\infty(M,\mathcal{A})$ is a Lie superalgebra with respect to {·,·}. This is false if ω is odd.

Momentum Maps

Let φ:M×G→M be an action of the Lie supergroup G on the supersymplectic manifold (M,ω) which preserves ω. Recall that, in the ungraded case, a momentum map is an ℝ-linear map $\mathfrak{J}:g\to C^\infty(M)$ such that $X_{\mathfrak{J}(\mathfrak{X})=}\mathfrak{X}_M$.

Superversion. $\mathfrak{J}\in C^\infty(g\times M,\mathcal{A})$ is a momentum map for the action of G on M if $\mathfrak{J}$ is leftlinear in the first argument and

$$\left\langle (0,v) \mid d\hat{J}(\mathfrak{X},x) \right\rangle = \langle v, \mathfrak{X}_M(x) \mid \omega(x)\rangle \tag{6.16}$$

for all v∈TxM. Instead of $\mathfrak{J}\in C^\infty(g\times M,\mathcal{A})$, one can consider J:M→g* defined through $\langle\mathfrak{X}|J(x)\rangle=\mathfrak{J}(\mathfrak{X},x)$.

Theorem 6.4. Let $H\in C^\infty(M,\mathcal{A})$ be a Hamiltonian with vector field XH such that $[X_H,X_H]=0$. If H is G-invariant, then J is preserved by the flow φ of X_H; that is,

$$\mathbf{J} \circ \phi_{t,\tau} = \mathbf{J} \quad \forall (t,\tau) \in \mathcal{A}^{1|1} \text{ (where defined)}. \tag{6.17}$$

Suppose that the momentum map is Ad*-equivariant. Let $\mu \in Bg^{p(\omega)*}$ be a regular value of J. Let G_μ be the isotropy group of μ. Suppose that the quotient space $P_\mu = J^{-1}(\mu)/G_\mu$ can be given a supermanifold structure such that the projection $\pi\mu: J^{-1}(\mu) \to P_\mu$ is a surjective submersion.

Superreduction Theorem

The supermanifold Pμ has a unique supersymplectic form $\omega_\mu \in \Omega_L^2(P_\mu)$ with the property

$$\pi_\mu^* \omega_\mu = \imath_\mu^* \omega, \tag{6.18}$$

where $\pi\mu: J^{-1}(\mu) \to P\mu$ is the projection and $\imath\mu: J^{-1}(\mu) \to M$ is the inclusion. The form ωμ has the same parity as ω.

Example 6.5 (The Bose-Fermi Oscillator). We consider the phase space $P = \mathcal{A}^{2|2}$ with supersymplectic form $\omega = dp \wedge dq \mid (1/2)\ (d\xi_1 \wedge d\xi_1 + d\xi_2 \wedge \delta\xi_2)$. The Hamiltonian $H(q,p,\xi_1,\xi_2) = (1/2)(p^2+q^2) - \xi_1\xi_2$ defines the Hamiltonian vector field $XH = p\partial_q - q\partial_p - \xi_2\partial_{\xi 1} + \xi_1\partial_{\xi 2}$.

The SUSY algebra of the Bose-Fermi oscillator is given as follows. The Lie supergroup of invertible 2|2-supermatrices acts on $\mathcal{A}^{2|2}$ via

$$\mathcal{A}^{2|2} \times GL(2|2) \longrightarrow \mathcal{A}^{2|2}, \qquad (\mathbf{q}, G) \longmapsto \mathbf{q}^{ST} G. \tag{6.19}$$

The algebra of Bose-Fermi supersymmetry is the intersection of the Lie superalgebras of the stabilizer of ω and the stabilizer of H:

$$\mathfrak{bf}(2|2) = \mathfrak{stab}(H) \cap \mathfrak{stab}(\omega). \tag{6.20}$$

The SUSY algebra of the Bose-Fermi oscillator bf(2|2) is generated by A_1, A_2, C_1, C_2, where

$$A_i = \begin{pmatrix} 0 & 0 & \mathbf{e}_i^T \gamma^0 \\ 0 & 0 & -\mathbf{e}_i^T \\ \mathbf{e}_i & -\gamma^0 \mathbf{e}_i & 0 \end{pmatrix} \quad \text{for } i = 1,2,$$

$$C_1 = \frac{1}{2}\begin{pmatrix} -\gamma^0 & \mathbf{0} \\ \mathbf{0} & \gamma^0 \end{pmatrix}, \quad C_2 = \frac{1}{2}\begin{pmatrix} \gamma^0 & \mathbf{0} \\ \mathbf{0} & \gamma^0 \end{pmatrix}, \quad \gamma^0 = \begin{pmatrix} 0 & -1 \\ 1 & 0 \end{pmatrix}. \tag{6.21}$$

Momentum Map and Quotient

The action of BF(2|2) on $A^{2|2}$ admits a momentum map

$$\hat{J} : \mathfrak{bf}(2|2) \times \mathcal{A}^{2|2} \longrightarrow \mathcal{A}, \qquad (X, \mathbf{q}) \longmapsto \frac{1}{2}\mathbf{q}^{ST} X \Omega \mathbf{q}. \tag{6.22}$$

In components, we write $X = \sum_{i=1,2} a_i A_i + c_i C_i$. Then

$$\hat{J}(X,q,p,\xi) = (c_2 - c_1)\frac{1}{4}\left(p^2+q^2\right) - (c_2+c_1)\frac{1}{4}\xi^T\gamma^0\xi - \mathbf{a}^T\left(p\mathbb{I}_2 - q\gamma^0\right)\xi, \tag{6.23}$$

where $a=(a_1,a_2)^T$.

Now let $\mu \in B(\mathfrak{bf}(2|2)_0^*) \simeq \mathbb{R}^2$. The isotropy group of μ is the whole group BF(2|2). The action of BF(2|2) on $J^{-1}(\mu)$ is transitive; that is, the quotient space $J^{-1}(\mu)/BF(2|2)$ is a single point.

Example 6.6 (Wess-Zumino Model in 2+1-Dimensional Spacetime). Let $(\mathcal{A})$ be the Schwartz space of functions $\mathbb{R}^2 \rightarrow \mathcal{A}$. Consider the phase space $\mathcal{S}(\mathcal{A})^{2|2}=[\mathcal{S}(\mathcal{A})_0]\times[\mathcal{S}(\mathcal{A})_0]\times[\mathcal{S}(\mathcal{A})_1]^2 \ni(\phi,\pi,\psi)$ with supersymplectic form $\omega=d\pi\wedge d\phi+(1/2)d\psi\wedge d\psi$ and Hamiltonian

$$\mathcal{H}(\phi,\pi,\psi) = \frac{1}{2}\int_{\mathbb{R}^2}\left(\|\nabla\phi\|^2 + \pi^2 + \overline{\psi}\mathfrak{g}^i d_i\psi\right)d^2x, \tag{6.24}$$

where $\mathfrak{g}^1 = \begin{pmatrix}0&1\\1&0\end{pmatrix}$, $\mathfrak{g}^2 = \begin{pmatrix}1&0\\0&-1\end{pmatrix}$, and $\overline{\psi} = \psi^T\gamma^0$.

$$\dot{\phi} = \frac{\delta\mathcal{H}}{\delta\pi} = \pi, \qquad \dot{\pi} = -\frac{\delta\mathcal{H}}{\delta\phi} = \Delta\phi, \qquad \dot{\psi} = \frac{\delta\mathcal{H}}{\delta\psi} = \mathfrak{g}^0\mathfrak{g}^i\partial_i\psi. \tag{6.25}$$

This is equivalent to $(\partial_0^2 - \partial_1^2 - \partial_2^2)\phi = 0,\ \mathfrak{g}^\mu\partial_\mu\psi = 0$ which are the massless Klein-Gordon and Dirac equations in 2+1-dimensional spacetime.

Remark 6.7. The well-known SUSY algebra from the Lagrangian description can be "exported" to the Hamiltonian setup.

We reduce by an Abelian subgroup of the SUSY group:

$$\mathcal{S}(\mathcal{A})^{2|2} \times \mathcal{A}^{2|0} \longrightarrow \mathcal{S}(\mathcal{A})^{2|2},$$

$$(\Phi, r) \longmapsto [S_r(\Phi)](x) = \sum_k \frac{1}{k!}D^k\Phi(x+Br)[n,\dots,n], \tag{6.26}$$

where r = Br+n, is a "superspatial shift". The momentum map becomes

$$\mathrm{J}(\Phi) = \left(\int\left(\pi\partial_i\Phi + \frac{1}{2}\psi^T\partial_i\psi\right)\right)_{i=1,2}. \tag{6.27}$$

We determine the reduced phase space. The center of mass of $\Phi = (\phi,\pi,\psi) \in \mathcal{S}(\mathcal{A})^{2|2}$ is $C(\Phi) = (1/M(\Phi))\int_{\mathbb{R}^2} x\cdot|\Phi|^2 d^2x \in \mathcal{A}^{2|0}$, where $M(\Phi) = \int|\Phi|^2 d^2x \in \mathcal{A}_0$. We may identify the quotient space $J^{-1}(\mu)/\mathcal{A}^{2|0}$ with the subset

$$P_\mu = \left\{\tilde{\Phi} \in \mathbf{J}^{-1}(\mu) \mid C\left(\tilde{\Phi}\right) = 0\right\} \subseteq \mathcal{S}(\mathcal{A})^{2|2}. \tag{6.28}$$

Finally the reduced equations become

$$\begin{aligned} \dot{\phi} &= \pi + F^i(\phi, \pi, \psi)\partial_i\phi, \\ \dot{\pi} &= \Delta\phi + F^i(\phi, \pi, \psi)\partial_i\pi, \\ \dot{\psi} &= \mathfrak{g}^0\mathfrak{g}^i\partial_i\psi + F^i(\phi, \pi, \psi)\partial_i\psi, \end{aligned} \tag{6.29}$$

where $F^i(\Phi) = (1/M(\Phi)) \int x^i \cdot (\phi\pi + \pi\Delta\phi - \psi^T \mathfrak{g}^j \partial_j \psi) d^2x$.

See [41, 42, 49, 60–63].

REFERENCES

1. N. Bourbaki, Lie Groups and Lie Algebras, Herman, Paris, France, 1975.
2. H. H. Keller, Differential Calculus in Locally Convex Spaces, Lecture Notes in Mathematics, Vol. 417, Springer, Berlin, Germany, 1974.
3. Pressley and G. Segal, Loop Groups, Oxford Mathematical Monographs, Oxford University Press, New York, NY, USA, 1986.
4. J. Eichhorn and R. Schmid, "Form preserving diffeomorphisms on open manifolds," Annals of Global Analysis and Geometry, vol. 14, no. 2, pp. 147–176, 1996.
5. H. Omori, Infinite-Dimensional Lie Groups, vol. 158 of Translations of Mathematical Monographs, American Mathematical Society, Providence, RI, USA, 1997.
6. R. Schmid, "The geometry of BRS transformations," Illinois Journal of Mathematics, vol. 34, no. 1, pp. 87–97, 1990.
7. Banyaga, The Structure of Classical Diffeomorphism Groups, vol. 400 of Mathematics and Its Applications, Kluwer Academic Publishers, Dordrecht, The Netherlands, 1997.
8. D. Bleecker, Gauge Theory and Variational Principles, vol. 1 of Global Analysis Pure and Applied Series A, Addison-Wesley, Reading, Mass, USA, 1981.
9. N. Bourbaki, Variétés Différentielles et Analytiques. Fascicule de Résultats, Actualités Scientifiques et Industrielles, No. 1347, Hermann, Paris, France, 1971.
10. D. G. Ebin and J. Marsden, "Groups of diffeomorphisms and the motion of an incompressible fluid,"Annals of Mathematics, vol. 92, pp. 102–163, 1970.
11. G. A. Goldin, "Lectures on diffeomorphism groups in quantum physics," in Contemporary Problems in Mathematical Physics, pp. 3–93, World Scientific, Hackensack, NJ, USA, 2004.

12. R. S. Hamilton, "The inverse function theorem of Nash and Moser," Bulletin of the American Mathematical Society, vol. 7, no. 1, pp. 65–222, 1982.

13. V. G. Kac, Infinite-Dimensional Lie Algebras, Cambridge University Press, Cambridge, Mass, USA, 2nd edition, 1985.

14. S. Lang, Differential Manifolds, Addison-Wesley, Reading, Mass, USA, 1972.

15. J. E. Marsden, D. G. Ebin, and A. E. Fischer, "Diffeomorphism groups, hydrodynamics and relativity," in Proceedings of the 13th Biennial Seminar of the Canadian Mathematical Congress Differential Geometry and Applications, J. R. Vanstone, Ed., vol. 1, pp. 135–279, 1972.

16. P. W. Michor, Manifolds of Differentiable Mappings, vol. 3 of Shiva Mathematics Series, Shiva, Nantwich, UK, 1980.

17. P. J. Olver, Applications of Lie Groups to Differential Equations, vol. 107 of Graduate Texts in Mathematics, Springer, New York, NY, USA, 2nd edition, 1993.

18. R. Palais, Foundations of Global Nonlinear Analysis, Addison-Wesley, Reading, Mass, USA, 1968.

19. R. Schmid, Infinite-Dimensional Hamiltonian Systems, vol. 3 of Monographs and Textbooks in Physical Science. Lecture Notes, Bibliopolis, Naples, Italy, 1987.

20. R. Abraham and J. E. Marsden, Foundations of Mechanics, Benjamin/Cummings Publishing, Reading, Mass, USA, 2nd edition, 1978.

21. J. E. Marsden, Lectures on Mechanics, vol. 174 of London Mathematical Society Lecture Note Series, Cambridge University Press, Cambridge, UK, 1992

22. J. E. Marsden and T. S. Ratiu, Introduction to Mechanics and Symmetry: A Basic Exposition of Classical Mechanical System, vol. 17 of Texts in Applied Mathematics, Springer, New York, NY, USA, 1994.

23. J. Marsden and A. Weinstein, "Reduction of symplectic manifolds with symmetry," Reports on Mathematical Physics, vol. 5, no. 1, pp. 121–130, 1974.

24. R. Abraham, J. E. Marsden, and T. Ratiu, Manifolds, Tensor Analysis, and Applications, vol. 2 of Global Analysis Pure and Applied: Series B, Addison-Wesley, Reading, Mass, USA, 1983.

25. P. R. Chernoff and J. E. Marsden, Properties of Infinite Dimensional Hamiltonian Systems, Lecture Notes in Mathematics, Vol. 425, Springer, Berlin, Germany, 1974.

26. Y. Choquet-Bruhat, C. DeWitt-Morette, and M. Dillard-Bleick, Analysis, Manifolds and Physics, North-Holland, Amsterdam, The Netherlands, 2nd edition, 1982.

27. F. Gay-Balmaz and T. S. Ratiu, "Reduced Lagrangian and Hamiltonian formulations of Euler-Yang-Mills fluids," The Journal of Symplectic Geometry, vol. 6, no. 2, pp. 189–237, 2008.

28. F. Gay-Balmaz and T. S. Ratiu, "A new Lagrangian dynamic reduction in field theory," to appear inAnnales de l'Institut Fourier.

29. V. Guillemin and S. Sternberg, Symplectic Techniques in Physics, Cambridge University Press, Cambridge, Mass, USA, 1984.

30. R. Temam, Infinite-Dimensional Dynamical Systems in Mechanics and Physics, vol. 68 of Applied Mathematical Sciences, Springer, New York, NY, USA, 1988.

31. S. Vasylkevych and J. E. Marsden, "The Lie-Poisson structure of the Euler equations of an ideal fluid,"Dynamics of Partial Differential Equations, vol. 2, no. 4, pp. 281–300, 2005.

32. J. E. Marsden, A. Weinstein, T. Ratiu, R. Schmid, and R. G. Spencer, "Hamiltonian systems with symmetry, coadjoint orbits and plasma physics," in Proceedings of the IUTAM-ISIMM Symposium on Modern Developments in Analytical Mechanics, vol. 1, pp. 289–340, Atti della Accademia delle Scienze di Torino. Classe di Scienze Fisiche, Matematiche e Naturali, Torino, Italy, 1982.

33. M. Adams, T. Ratiu, and R. Schmid, "The Lie group structure of diffeomorphism groups and invertible Fourier integral operators, with applications," in Infinite-Dimensional Groups with Applications, vol. 4 of Math. Sci. Res. Inst. Publ., pp. 1–69, Springer, New York, NY, USA, 1985.

34. M. Adams, T. Ratiu, and R. Schmid, "A Lie group structure for pseudodifferential operators,"Mathematische Annalen, vol. 273, no. 4, pp. 529–551, 1986.

35. M. Adams, T. Ratiu, and R. Schmid, "A Lie group structure for Fourier integral operators,"Mathematische Annalen, vol. 276, no. 1, pp. 19–41, 1986.

36. E. Fischer and J. E. Marsden, "The Einstein equations of evolution—a geometric approach," Journal of Mathematical Physics, vol. 13, pp. 546–568, 1972.

37. F. Gay-Balmaz and T. S. Ratiu, "The Lie-Poisson structure of the LAE-α equation," Dynamics of Partial Differential Equations, vol. 2, no. 1, pp. 25–57, 2005.

38. J. E. Marsden and A. Weinstein, "The Hamiltonian structure of the Maxwell-Vlasov equations," Physica D, vol. 4, no. 3, pp. 394–406, 1982.

39. F. Trèves, Introduction to Pseudodifferential and Fourier Integral Operators: Pseudodifferential Operators, vol. 1 of The University Series in Mathematic, Plenum Press, New York, NY, USA, 1980.

40. F. Trèves, Introduction to Pseudodifferential and Fourier Integral Operators: Fourier Integral Operators, vol. 2 of The University Series in Mathematic, Plenum Press, New York, NY, USA, 1980.

41. P. Deligne, et al., Ed., Quantum Fields and Strings: A Course for Mathematicians, vol. 1, American Mathematical Society, Providence, RI, USA, 1999.

42. P. Deligne, et al., Ed., Quantum Fields and Strings: A Course for Mathematicians, vol. 2, American Mathematical Society, Providence, RI, USA, 1999.

43. C. Becchi, A. Rouet, and R. Stora, "Renormalization of gauge theories," Annals of Physics, vol. 98, no. 2, pp. 287–321, 1976.

44. R. Schmid, "Local cohomology in gauge theories, BRST transformations and anomalies," Differential Geometry and Its Applications, vol. 4, no. 2, pp. 107–116, 1994.

45. C. Chevalley and S. Eilenberg, "Cohomology theory of Lie groups and Lie algebras," Transactions of the American Mathematical Society, vol. 63, pp. 85–124, 1948.

46. R. Schmid, "BRST cohomology and anomalies," in Geometrical and Algebraic Aspects of Nonlinear Field Theory, North-Holland Delta Ser., pp. 159–172, North-Holland, Amsterdam, The Netherlands, 1989.

47. Einstein, Mein Weltbild, Ullstein, Frankfurt, Germany, 1934.

48. E. Verlinde, "On the origin of gravity and the laws of Newton,"

49. B. Kostant, "Graded manifolds, graded Lie theory, and prequantization," in Differential Geometrical Methods in Mathematical Physics, vol. 570 of Lecture Notes in Mathematics, pp. 177–306, Springer, Berlin, Germany, 1977.

50. E. Binz, J. Śniatycki, and H. Fischer, Geometry of Classical Fields, vol. 154 of North-Holland Mathematics Studies, North-Holland, Amsterdam, The Netherlands, 1988.

51. L. Bonora and P. Cotta-Ramusino, "Some remarks on BRS transformations, anomalies and the cohomology of the Lie algebra of the group of gauge transformations," Communications in Mathematical Physics, vol. 87, no. 4, pp. 589–603, 1982/83.

52. M. Dubois-Violette, "The Weil-BRS algebra of a Lie algebra and the anomalous terms in gauge theory,"Journal of Geometry and Physics, vol. 3, no. 4, pp. 525–565, 1986.

53. T. Eguchi, P. B. Gilkey, and A. J. Hanson, "Gravitation, gauge theories and differential geometry,"Physics Reports, vol. 66, no. 6, pp. 213–393, 1980.

54. L. D. Faddeev and A. A. Slavnov, Gauge Fields: Introduction to Quantum Theory, vol. 50 of Frontiers in Physics, Benjamin/Cummings Publishing, Reading, Mass, USA, 1980.

55. G. A. Goldin, R. Menikoff, and D. H. Sharp, "Diffeomorphism groups, gauge groups, and quantum theory," Physical Review Letters, vol. 51, no. 25, pp. 2246–2249, 1983.

56. G. A. Goldin, "The diffeomorphism group approach to nonlinear quantum systems," International Journal of Modern Physics B, vol. 6, no. 11-12, pp. 1905–1916, 1992.

57. C. Itzykson and J. B. Zuber, Quantum Field Theory, International Series in Pure and Applied Physic, McGraw-Hill, New York, NY, USA, 1980.

58. D. Kastler and R. Stora, "A differential geometric setting for BRS transformations and anomalies. I, II,"Journal of Geometry and Physics, vol. 3, no. 4, pp. 437–505, 1986.

59. B. Kostant and S. Sternberg, "Symplectic reduction, BRS cohomology, and infinite-dimensional Clifford algebras," Annals of Physics, vol. 176, no. 1, pp. 49–113, 1987.

60. T. Glimm, Supersymplectic reduction, Ph.D. thesis, Emory University, 2003, http://www.mathcs.emory.edu/~rudolf/Glimm.pdf.

61. G. M. Tuynman, Supermanifolds and Supergroups: Basic Theory, vol. 570 of Mathematics and Its Applications, Kluwer Academic Publishers, Dordrecht, The Netherlands, 2004.

62. B. DeWitt, Supermanifolds, Cambridge Monographs on Mathematical Physics, Cambridge University Press, Cambridge, Mass, USA, 2nd edition, 1992.

63. T. Glimm and R. Schmid, "On the supersymmetry group of the classical Bose-Fermi oscillator," Journal of Geometry and Symmetry in Physics, vol. 4, pp. 45–58, 2005.

2
CHAPTER

APPROXIMATE HAMILTONIAN SYMMETRY GROUPS AND RECURSION OPERATORS FOR PERTURBED EVOLUTION EQUATIONS

M. Nadjafikhah and A. Mokhtary

INTRODUCTION

The investigation of the exact solutions of nonlinear evolution equations has a fundamental role in the nonlinear physical phenomena. One of the significant and systematic methods for obtaining special solutions of systems of nonlinear differential equations is the classical symmetries method, also called Lie group analysis. This well-known approach originated at the end of nineteenth century from the pioneering work of Lie [1]. The fact that symmetry reductions for many PDEs cannot be determined, via the classical symmetry method, motivated the creation of several generalizations of the classical Lie group approach for symmetry reductions. Consequently, several alternative reduction methods have been proposed, going beyond Lie's classical procedure and providing further solutions. One of these techniques which is extremely applied particularly for nonlinear problems is perturbation analysis. It is worth mentioning that sometimes differential equations which appear in mathematical modelings are presented with terms involving a parameter called the perturbed term. Because of the instability of the Lie point symmetries with respect to perturbation of coefficients of differential equations, a new class of symmetries has been created for such equations, which are known as approximate (perturbed) symmetries. In the last century, in order to have the utmost result from the methods, combination of Lie symmetry method and perturbations are investigated and two different so-called approximate symmetry methods (ASMs) have been developed. The first method is due to Baikov et al. [2, 3]. The second procedure was proposed by Fushchich and Shtelen [4] and later was followed by Euler et al. [5, 6]. This method is generally based on the perturbation of dependent variables. In [7, 8], a comprehensive comparison of these two methods is presented.

Citation: M. Nadjafikhah and A. Mokhtary, "Approximate Hamiltonian Symmetry Groups and Recursion Operators for Perturbed Evolution Equations," Advances in Mathematical Physics, vol. 2013, Article ID 568632, 9 pages, 2013. doi:10.1155/2013/568632

As it is well known, Hamiltonian systems of differential equations are one of the famous and significant concepts in physics. These important systems appear in the various fields of physics such as motion of rigid bodies, celestial mechanics, quantization theory, fluid mechanics, plasma physics, and so forth. Due to the significance of Hamiltonian structures, in this paper, by applying the linear behavior of the Euler operator, characteristics, prolongation, and Fréchet derivative of vector fields, we have extended ASM on the Hamiltonian and bi-Hamiltonian systems of evolution equations, in order to investigate the interplay between approximate symmetry groups, approximate conservation laws, and approximate recursion operators.

The structure of the present paper is as follows. In Section 2, some necessary preliminaries regarding to the Hamiltonian structures are presented. In Section 3, a comprehensive investigation of the approximate Hamiltonian symmetries and approximate conservation laws associated to the perturbed evolution equations is proposed. Also, as an application of this procedure, approximate Hamiltonian symmetry groups, approximate bi-Hamiltonian structures, and approximate conservation laws of the Gardner equation are computed. In Section 4, the approximate recursion operators are studied and the proposed technique is implemented for the Gardner equation as an application. Finally, some concluding remarks are mentioned at the end of the paper.

PRELIMINARIES

In this section, we will mention some necessary preliminaries regarding Hamiltonian structures. In order to be familiar with the general concepts of the ASM, refer to [9]. It is also worth mentioning that most of this paper's definitions, theorems and techniques regarding Hamiltonian and bi-Hamiltonian structures are inspired from [10].

Let $M \subset X \times U$ denote a fixed connected open subset of the space of independent and dependent variables $x = (x^1, \dots, x^p)$ and $u = (u^1, \dots, u^q)$. The algebra of differential functions $(x, {}^{(n)}) = [u]$ over M is denoted by $\mathscr{A}$. We further define $\mathscr{A}^l$ to be the vector space of l-tuples of differential functions, $[u] = (P_1[u] \dots P_l[u])$, where each $P_j \in \mathscr{A}$.

A generalized vector field will be a (formal) expression of the form

$$\mathbf{v} = \sum_{i=1}^{p} \xi^i[u] \frac{\partial}{\partial x^i} + \sum_{\alpha=1}^{q} \phi_\alpha[u] \frac{\partial}{\partial u^\alpha} \tag{1}$$

in which ξ^i and ϕ_α are smooth differential functions. The Prolonged generalized vector field can be defined as follows:

$$\mathbf{pr}^{(n)}\mathbf{v} = \mathbf{v} + \sum_{\alpha=1}^{q} \sum_{\sharp J \le n} \phi_\alpha^J[u] \frac{\partial}{\partial u_J^\alpha}, \tag{2}$$

whose coefficients are determined by the formula

$$\phi_\alpha^J = D_J\left(\phi_\alpha - \sum_{i=1}^{p} \xi^i u_i^\alpha\right) + \sum_{i=1}^{p} \xi^i u_{J,i}^\alpha, \tag{3}$$

with the same notation as before. Given a generalized vector field v, its infinite prolongation (or briefly prolongation) is the formally infinite sum as follows:

$$\mathbf{pr}\,\mathbf{v} = \sum_{i=1}^{p} \xi^i \frac{\partial}{\partial x^i} + \sum_{\alpha=1}^{q} \sum_{J} \phi_\alpha^J \frac{\partial}{\partial u_J^\alpha}, \tag{4}$$

where each ϕ_α^J is given by (3), and the sum in (4) now extends over all multi-indices $J = (j_1,...,j_k)$ for $k \geq 0$, $1 \leq j_k \leq p$.

$$\Delta_\nu [u] = \Delta_\nu \left(x, u^{(n)}\right) = 0, \quad \nu = 1,\ldots,l, \tag{5}$$

if and only if

$$\mathbf{pr}\,\mathbf{v}\left[\Delta_\nu\right] = 0, \quad \nu = 1,\ldots,l, \tag{6}$$

for every smooth solution $u = f(x)$

Among all the generalized vector fields, those in which the coefficients $\xi^i[u]$ of the $\partial/\partial x^i$ are zero play a distinguished role. Let $[u] = (Q_1[u], \ldots, Q_q[u]) \in \mathscr{A}^q$ be a q-tuple of differential functions. The generalized vector field

$$\mathbf{v}_Q = \sum_{\alpha=1}^{q} Q_\alpha [u] \frac{\partial}{\partial u^\alpha} \tag{7}$$

is called an evolutionary vector field, and Q is called its characteristic.

A manifold M with a Poisson bracket is called a Poisson manifold, the bracket defining a Poisson structure on M. Let M be a Poisson manifold and $H{:}M \to \mathbb{R}$ be a smooth function. The Hamiltonian vector field associated with H is the unique smooth vector field $\hat{\mathrm{v}}_{\mathrm{H}}$ on M satisfying the following identity:

$$\hat{\mathbf{v}}_{\mathbf{H}} = \{F, H\} = -\{H, F\} \tag{8}$$

for every smooth function $F{:}M \to \mathbb{R}$. The equations governing the flow of $\hat{\mathrm{v}}_{\mathrm{H}}$ are referred to as Hamilton's equations for the "Hamiltonian" function H.

Let $x = (x^1,...,x^m)$ be local coordinates on M and $H(x)$ be a real-valued function. The following basic formula can be obtained for the Poisson bracket:

$$\{F, H\} = \sum_{i=1}^{m} \sum_{j=1}^{m} \left\{x^i, x^j\right\} \frac{\partial F}{\partial x^i} \frac{\partial H}{\partial x^j}. \tag{9}$$

In other words, in order to compute the Poisson bracket of any pair of functions in some given set of local coordinates, it suffices to know the Poisson brackets between the coordinate functions themselves. These basic brackets,

$$J^{ij}(x) = \left\{x^i, x^j\right\}, \quad i, j = 1,\ldots,m, \tag{10}$$

are called the structure functions of the Poisson manifold M relative to the given local coordinates and serve to uniquely determine the Poisson structure itself. For convenience, we assemble the structure functions into a skew-symmetric $m \times m$ matrix (x), called the structure matrix of M. Using ∇H to denote the (column) gradient vector for H, the local coordinate form (9) for the Poisson bracket can be written as

$$\{F, H\} = \nabla F \cdot J \nabla H \tag{11}$$

Therefore, in the given coordinate chart, Hamilton's equations take the form of

$$\frac{dx}{dt} = J(x)\,\nabla H(x)\,. \tag{12}$$

Alternatively, using (9), we could write this in the "bracket form" as follows:

$$\frac{dx}{dt} = \{x, H\}, \tag{13}$$

the ith component of the right-hand side being $\{^{xi}, H\}$. A system of first-order ordinary differential equations is said to be a Hamiltonian system if there is a Hamiltonian function $H(x)$ and a matrix of functions $J(x)$ determining a Poisson bracket (11) whereby the system takes the form (12).

If

$$\mathscr{D} = \sum_J P_J\,[u]\,D_J, \quad P_J \in \mathscr{A} \tag{14}$$

is a differential operator, its (formal) adjoint is the differential operator $\mathscr{D}^*$ which satisfies

$$\int_\Omega P \cdot \mathscr{D} Q\,dx = \int_\Omega Q \cdot \mathscr{D}^* P\,dx \tag{15}$$

for every pair of differential functions P, $Q \in \mathscr{A}$ which vanish when $u \equiv 0$. Also, for every domain $\Omega \subset Rp$ and every function $u = (x)$ of compact support in Ω. An operator $\mathscr{D}$ is self-adjoint if $D^* = \mathscr{D}$; it is skew-adjoint if $\mathscr{D}^* = -\mathscr{D}$.

The principal innovations needed to convert a Hamiltonian system of ordinary differential equations (12) to a Hamiltonian system of evolution equations are as follows (refer to [10] for more details):

(i) replacing the Hamiltonian function $\mathscr{H}$ (x) by a Hamiltonian functional H[u],

(ii) replacing the vector gradient operation $\nabla\mathscr{H}$ by the variational derivative $\delta\mathscr{H}$ of the Hamiltonian functional, and

(iii) replacing the skew-symmetric matrix $J(x)$ by a skewadjoint differential operator $\mathscr{D}$ which may depend on u.

The resulting Hamiltonian system will take the form of

$$\frac{\partial u}{\partial t} = \mathscr{D} \cdot \delta \mathscr{H}\,[u]\,. \tag{16}$$

Clearly, for a candidate Hamiltonian operator $\mathscr{D}$ the correct expression for the corresponding Poison bracket has the form of

$$\{\mathscr{P}, \mathscr{L}\} = \int \delta \mathscr{P} \cdot \mathscr{D} \delta \mathscr{L}\, dx, \tag{17}$$

whenever $\mathscr{P}, \mathscr{L} \in \mathscr{F}$ are functionals. Off course, the Hamiltonian operator D must satisfy certain further restrictions in order that (17) be a true Poisson bracket. A linear operator $\mathscr{D} : \mathscr{A}^q \to \mathscr{A}^q$ is called Hamiltonian if its Poisson bracket (17) satisfies the conditions of skew-symmetry and the Jacobi identity.

Proposition 1. Let $\mathscr{D}$ be a Hamiltonian operator with Poisson bracket (17). To each functional $\mathscr{H} = \int H\, dx \in \mathscr{F}$, there is an evolutionary vector field $\mathbf{pr}\,\hat{\mathbf{v}}_{\mathscr{H}}$, called the Hamiltonian vector field associated with H, which for all functionals $\mathscr{P} \in \mathscr{F}$ satisfies the following identity:

$$\mathbf{pr}\,\hat{\mathbf{v}}_{\mathscr{H}}\,(\mathscr{P}) = \{\mathscr{P}, \mathscr{H}\}\,. \tag{18}$$

APPROXIMATE HAMILTONIAN SYMMETRIES AND APPROXIMATE CONSERVATION LAWS

Consider a system of perturbed evolution equations:

$$\frac{\partial u}{\partial t} = P\,[u, \varepsilon] \tag{19}$$

in which $P[u, \varepsilon] = P(x, u^{(n)}, \varepsilon) \in \mathscr{A}^q,\ x \in \mathbb{R}^p,\ u \in \mathbb{R}^q$ and ε is a parameter

Substituting according to (19) and its derivatives, we see that any evolutionary symmetry must be equivalent to one whose characteristic $[u, \varepsilon] = (x, t, {}^{(m)}, \varepsilon)$ depends only on x, t, u, ε and the x-derivatives of u. On the other hand, (19) itself can be considered as the equations corresponding to the flow $\exp(t\mathrm{v}_p)$ of the evolutionary vector field with characteristic P. The symmetry criterion (6), which in this case is

$$D_t Q_\nu = \mathbf{pr}\,\mathbf{v}_Q\,(P_\nu) + o\,(\varepsilon^p)\,, \quad \nu = 1, \ldots, q, \tag{20}$$

can be readily seen to be equivalent to the following Lie bracket condition on the two approximate generalized vector fields. Indeed, this point generalizes the correspondence between symmetries of systems of first-order perturbed ordinary differential equations and the Lie bracket of the corresponding vector fields.

Considering the above assumptions, some useful relevant theorems and definitions could be rewritten as follows.

Proposition 2. An approximate evolutionary vector field v_Q is a symmetry of the system of perturbed evolution equations $u_t = [u, \varepsilon]$ if and only if

$$\frac{\partial \mathbf{v}_Q}{\partial t} + [\mathbf{v}_P, \mathbf{v}_Q] = o(\varepsilon^P) \tag{21}$$

holds identically in $(x, t, u^{(m)}, \varepsilon)$. (Here $\partial v_Q/\partial t$ denotes the evolutionary vector field with characteristic $\partial Q/\partial t$.).

Any approximate conservation law of a system of perturbed evolution equations takes the form of

$$D_t T + \mathbf{Div}\, X = o(\varepsilon^P), \tag{22}$$

in which Div denotes spatial divergence. Without loss of generality, the conserved density $(x, t, {}^{(n)},)$ can be assumed to depend only on x-derivatives of u. Equivalently, for $\Omega \subset X$, the functional

$$\mathcal{T}[t; u, \varepsilon] = \int_\Omega T\left(x, t, u^{(n)}, \varepsilon\right) dx \tag{23}$$

is a constant, independent of t, for all solutions u such that $T(x, t, u^{(n)}, \varepsilon) \to 0$ as $x \to \partial\Omega$. Note that if $T(x, t, u^{(n)}, \varepsilon)$ is any such differential function, and u is a solution of the perturbed evolutionary system $u_t = P[u, \varepsilon]$, then

$$D_t T \approx \partial_t T + \mathbf{pr}\, \mathbf{v}_p(T), \tag{24}$$

where $\partial t = \partial/\partial t$ denotes the partial t-derivative. Thus T is the density for a conservation law of the system if and only if its associated functional $\mathcal{T}$ satisfies the following identity:

$$\frac{\partial \mathcal{T}}{\partial t} + \mathbf{pr}\, \mathbf{v}_p(\mathcal{T}) = o(\varepsilon^P). \tag{25}$$

In the case that our system is of Hamiltonian form, the bracket relation (18) immediately leads to the Noether relation between approximate Hamiltonian symmetries and approximate conservation laws.

Definition 3. Let $\mathcal{D}$ be a $q \times q$ approximate Hamiltonian differential operator. An approximate distinguished functional for $\mathcal{D}$ is a functional $\mathcal{G} \in \mathcal{F}$ satisfying $\mathcal{D}\delta\mathcal{G} = o(\varepsilon^P)$ for all x, u.

In other words, the Hamiltonian system corresponding to a distinguished functional is completely trivial: $u_t = 0$.

Now, according to [10], the perturbed Hamiltonian version of Noether's theorem can be presented as follows.

Theorem 4. Let $u_t = \mathcal{D}\delta\mathcal{H}$ be a Hamiltonian system of perturbed evolution equations. An approximate Hamiltonian vector field $\hat{\mathbf{v}}_{\mathcal{P}}$ with characteristic $\mathcal{D}\delta\mathcal{P}, \mathcal{P} \in \mathcal{F}$ determines an approximate generalized symmetry group of the system if and only if there is an equivalent functional $\tilde{\mathcal{P}} \approx \mathcal{P} - \mathcal{G}$ differing only from $\mathcal{P}$ by a time-dependent approximate distinguished functional $\mathscr{G}[t;u,\varepsilon]$, such that $\tilde{\mathrm{P}}$ determines an approximate conservation law.

Example 5. The Gardner equation

$$u_t = 6\left(u + \varepsilon u^2\right)u_x - u_{xxx}, \tag{26}$$

can in fact be written in Hamiltonian form in two distinct ways. Firstly, we see

$$u_t = D_x\left(3u^2 + 2\varepsilon u^3 - u_{xx}\right) = \mathscr{D}\delta\mathscr{H}_1, \tag{27}$$

where $\mathscr{D} = D_x$ and

$$\mathscr{H}_1[u,\varepsilon] = \int\left(u^3 + \frac{\varepsilon}{2}u^4 + \frac{u_x^2}{2}\right)dx \tag{28}$$

The second Hamiltonian form is

$$u_t = \left(4uD_x + 2u_x + 3\varepsilon\left(uu_x + u^2D_x\right) - D_x^3\right)u = \mathscr{E}\delta\mathscr{H}_0, \tag{30}$$

in which

$$\mathscr{H}_0[u,\varepsilon] = \int\frac{1}{2}u^2dx \tag{31}$$

$\mathscr{E}$ is skew-adjoint and satisfies the Jacobi identity. Therefore it is Hamiltonian.

In [11], we have comprehensively analyzed the problem of approximate symmetries for the Gardner equation. We have shown that the approximate symmetries of the Gardner equation are given by the following generators:

$$\begin{aligned}
\mathbf{v}_1 &= \partial_x,\\
\mathbf{v}_2 &= \partial_t,\\
\mathbf{v}_3 &= 6t\partial_x + (2\varepsilon u - 1)\partial_u,\\
\mathbf{v}_4 &= \varepsilon\mathbf{v}_1,\\
\mathbf{v}_5 &= \varepsilon\mathbf{v}_2,\\
\mathbf{v}_6 &= \varepsilon\left(6t\partial_x - \partial_u\right) = \varepsilon\mathbf{v}_3,\\
\mathbf{v}_7 &= \varepsilon\left(x\partial_x + 3t\partial_t - 2u\partial_u\right),
\end{aligned} \tag{32}$$

with corresponding characteristics

$$Q_1 = u_x,$$

$$Q_2 = 6\left(u + \varepsilon u^2\right) u_x - u_{xxx},$$

$$Q_3 = 6tu_x + 1 - 2\varepsilon u,$$

$$Q_4 = \varepsilon Q_1 = \varepsilon u_x,$$

$$Q_5 = \varepsilon Q_2 = \varepsilon\left(6uu_x - u_{xxx}\right),$$

$$Q_6 = \varepsilon Q_3 = \varepsilon\left(6tu_x + 1\right),$$

$$Q_7 = \varepsilon\left(2u + xu_x + 3t\left(6uu_x - u_{xxx}\right)\right), \tag{33}$$

(up to sign).

For the first Hamiltonian operator $\mathscr{D} = D_x$, there is one independent nontrivial approximate distinguished functional, the mass $\mathscr{P}_0 = \mathscr{M} = \int udx$ which is consequently approximately conserved

For the above seven characteristics, we have

$$Q_i \approx D_x \delta \mathscr{P}_i, \quad i = 1, 2, 4, 5, 6, \tag{34}$$

with the following approximately conserved functionals:

$$\mathscr{P}_1 = \mathscr{H}_0[u, \varepsilon] = \int \frac{1}{2} u^2 dx,$$

$$\mathscr{P}_2 = \mathscr{H}_1[u, \varepsilon] = \int \left(u^3 + \frac{\varepsilon}{2} u^4 + \frac{1}{2} u_x^2\right) dx,$$

$$\mathscr{P}_4 = \varepsilon \mathscr{P}_1 = \int \frac{\varepsilon}{2} u^2 dx,$$

$$\mathscr{P}_5 = \varepsilon \mathscr{P}_2 = \varepsilon \int \left(u^3 + \frac{1}{2} u_x^2\right) dx,$$

$$\mathscr{P}_6 = \varepsilon \int \left(3tu^2 + xu\right) dx. \tag{35}$$

For the second Hamiltonian operator $\mathscr{E} = 4uD_x + 2u_x + 3\varepsilon(uu_x + u^2 D_x) - D_x^3,$

$$Q_i \approx \mathscr{E} \delta \overline{\mathscr{P}}_i, \quad i = 2, 4, 5, 7, \tag{36}$$

the following approximately conserved functionals are the corresponding approximate conservation laws:

$$\widetilde{\mathscr{P}}_2 = \mathscr{P}_1 = \int \frac{1}{2}u^2 dx,$$

$$\widetilde{\mathscr{P}}_4 = \frac{\varepsilon}{2}\mathscr{P}_0 = \frac{\varepsilon}{2}\int u\,dx,$$

$$\widetilde{\mathscr{P}}_5 = \varepsilon\widetilde{\mathscr{P}}_2 = \varepsilon\int \frac{1}{2}u^2 dx,$$

$$\widetilde{\mathscr{P}}_7 = \frac{1}{2}\mathscr{P}_6 = \frac{\varepsilon}{2}\int \left(3tu^2 + xu\right)dx. \tag{37}$$

In this case, nothing new is obtained. Note that the other approximate conservation law $\mathscr{P}_5$ did not arise from one of the geometrical symmetries. According to Theorem 4, however, there is an approximate Hamiltonian symmetry which gives rises to it, namely $\hat{\mathbf{v}}_{\mathscr{P}_5}$. The characteristic of this approximate generalized symmetry is

$$\overline{Q}_5 \approx \mathscr{E}\delta\mathscr{P}_5 = \mathscr{E}\varepsilon\left(3u^2 - u_{xx}\right)$$

$$\approx \varepsilon\left(u_{xxxxx} - 10uu_{xxx} - 20u_x u_{xx} + 30u^2 u_x\right). \tag{38}$$

Note that $\overline{Q}_5$ happens to satisfy the Hamiltonian condition (34) for $\mathscr{D}$ with the following functional:

Consequently, another approximate conservation law is provided for the Gardner equation

Keeping on this procedure recursively, further approximate conservation laws could be generated. But, this procedure will be done in the next section by applying approximate recursion operators

APPROXIMATE RECURSION OPERATORS

Definition 6. Let Δ be a system of perturbed differential equations. An approximate recursion operator for Δ is a linear operator $\mathscr{R} : \mathscr{A}^q \to \mathscr{A}^q$ in the space of q-tuples of differential functions with the property that whenever vQ is an approximate evolutionary symmetry of Δ, so $\mathbf{v}_{\widetilde{Q}}$ is with $\widetilde{Q} \approx \mathscr{R}Q$.

For nonlinear perturbed systems, there is an analogous criterion for a differential operator to be an approximate recursion operator, but to state it we need to introduce the notion of the (formal) Frechet derivative of a differential ′ function

Definition 7. Let $P[u,\varepsilon] = P(x,u^{(n)},\varepsilon) \in \mathscr{A}^r$ be an r-tuple of differential functions. The Frechet derivative ′ of P is the perturbed differential operator $\mathbf{D}_P : \mathscr{A}^q \to \mathscr{A}^r$ defined so that

$$\mathbf{D}_P(Q) = \left.\frac{d}{d\epsilon}\right|_{\epsilon=0} P\left[u + \epsilon Q\left[u,\varepsilon\right]\right] \tag{40}$$

for any $Q \in \mathscr{A}^q$.

Proposition 8. If $P \in \mathscr{A}^r$ *and* $Q \in \mathscr{A}^q$ then

$$\mathbf{D}_P(Q) \approx \mathbf{pr}\,\mathbf{v}_Q(P). \tag{41}$$

Theorem 9. Suppose that $\Delta[u, \varepsilon] = 0$ be a system of perturbed differential equations. If $\mathscr{R} : \mathscr{A}^q \to \mathscr{A}^q$ is a linear operator such that for all solutions u of Δ,

$$\mathbf{D}_\Delta \cdot \mathscr{R} \approx \widetilde{\mathscr{R}} \cdot \mathbf{D}_\Delta, \tag{42}$$

where $\mathscr{R} : \mathscr{A}^q \to \mathscr{A}^q$ is a linear differential operator, then $\mathscr{R}$ is an approximate recursion operator for the system.

Suppose that $\Delta[u, \varepsilon] = u_t - K[u, \varepsilon]$ is a perturbed evolution equation. Then $D\Delta = Dt - DK$. If $\mathscr{R}$ is an approximate recursion operator, then it is not hard to observe that the operator $\widetilde{\mathscr{R}}$ in (42) must be the same as $\mathscr{R}$. Therefore, the condition (42) in this case reduces to the commutator condition

$$\mathscr{R}_t \approx [\mathbf{D}_K, \mathscr{R}]$$

for an approximate recursion operator of a perturbed evolution equation

From (4), we can conclude that if $\mathscr{R}$ is an approximate recursion operator, then for all $l \geq 1$ in which $\varepsilon^l \mathscr{R} \neq 0$, $\varepsilon^l \mathscr{R}$ is an approximate recursion operator as follows:

$$\left(\varepsilon^l \mathscr{R}\right)_t = \varepsilon^l \mathscr{R}_t \approx \varepsilon^l [\mathbf{D}_K, \mathscr{R}] \approx \left[\mathbf{D}_K, \varepsilon^l \mathscr{R}\right].$$

In order to illustrate the significance of the above theorem, we discuss a couple of examples, including the potential Burgers' equation and the Gardner equation. In the first example, we apply some technical methods, used in Examples 5.8 and 5.30 of [10].

Example 10. Consider the potential Burgers' equation

$$u_t = u_{xx} + \varepsilon u_x^2. \tag{43}$$

As mentioned in [7], approximate symmetries of the potential Burgers' equation are given by the following twelve vector fields

$$\begin{aligned}
\mathbf{v}_1 &= \partial_x,\\
\mathbf{v}_2 &= \partial_t,\\
\mathbf{v}_3 &= x\partial_x + 2t\partial_t,\\
\mathbf{v}_4 &= 2t\partial_x - \left(xu - \varepsilon t\frac{u^2}{2}\right)\partial_u,\\
\mathbf{v}_5 &= \left(u - \varepsilon t\frac{u^2}{2}\right)\partial_u,\\
\mathbf{v}_6 &= 4xt\partial_x + 4t^2\partial_t - \left(x^2 + 2t\right)\left(u - \varepsilon t\frac{u^2}{2}\right)\partial_u,\\
\mathbf{v}_7 &= \varepsilon\mathbf{v}_1,\\
\mathbf{v}_8 &= \varepsilon\mathbf{v}_2,\\
\mathbf{v}_9 &= \varepsilon\left(x\partial_x + 2t\partial_t\right) = \varepsilon\mathbf{v}_3,\\
\mathbf{v}_{10} &= \varepsilon\left(2t\partial_x - xu\partial_u\right) = \varepsilon\mathbf{v}_4,\\
\mathbf{v}_{11} &= \varepsilon u\partial_u = \varepsilon\mathbf{v}_5,\\
\mathbf{v}_{12} &= \varepsilon\left(4xt\partial_x + 4t^2\partial_t - \left(x^2 + 2t\right)u\partial_u\right) = \varepsilon\mathbf{v}_6,
\end{aligned} \tag{44}$$

plus the infinite family of vector fields

$$\mathbf{v}_{f,g} = (f(x,t)(1-\varepsilon u) + \varepsilon g(x,t))\,\partial_u, \tag{45}$$

where f, g are arbitrary solutions of the heat equation $u_t = u_{xx}$

The corresponding characteristics of the first twelve approximate symmetries are

$$\begin{aligned}
Q_1 &= u_x,\\
Q_2 &= u_{xx} + \varepsilon u_x^2,\\
Q_3 &= xu_x + 2t\left(u_{xx} + \varepsilon u_x^2\right),\\
Q_4 &= xu + 2tu_x - \varepsilon t\frac{u^2}{2},\\
Q_5 &= u - \varepsilon t\frac{u^2}{2},\\
Q_6 &= \left(x^2 + 2t\right)\left(u - \varepsilon t\frac{u^2}{2}\right) + 4xtu_x + 4t^2\left(u_{xx} + \varepsilon u_x^2\right),
\end{aligned} \tag{46}$$

$$\begin{aligned}
Q_7 &= \varepsilon Q_1 = \varepsilon u_x,\\
Q_8 &= \varepsilon Q_2 = \varepsilon u_{xx},\\
Q_9 &= \varepsilon Q_3 = \varepsilon\left(xu_x + 2tu_{xx}\right),\\
Q_{10} &= \varepsilon Q_4 = \varepsilon\left(xu + 2tu_x\right),\\
Q_{11} &= \varepsilon Q_5 = \varepsilon u,\\
Q_{12} &= \varepsilon Q_6 = \varepsilon\left(\left(x^2 + 2t\right)u + 4xtu_x + 4t^2u_{xx}\right),
\end{aligned} \tag{47}$$

(up to sign).

Inspection of Q_1, Q_2, Q_7, Q_8 leads us to the conjecture that $\mathscr{R}_1 = D_x + \varepsilon u_x$ is an approximate recursion operator, since $Q_3 = \mathscr{R}_1 Q_1$, $Q_8 = \mathscr{R}_1 Q_7$, and so forth. To prove this, we note that the Frechet derivative for the right-hand side of ′ potential Burgers' equation is

$$D_K = D_x^2 + 2\varepsilon u_x D_x. \tag{48}$$

We must verify (4). The time derivative of the first approximate recursion operator $\mathscr{R}_1$ on the solutions of the potential Burgers' equation is the multiplication operator as follows:

$$(\mathscr{R}_1)_t = (D_x + \varepsilon u_x)_t = \varepsilon u_{xt} = \varepsilon\left(u_{xxx} + 2\varepsilon u_x u_{xx}\right) = \varepsilon u_{xxx}. \tag{49}$$

On the other hand, the commutator is computed using Leibniz' rule for differential operators:

$$[D_K, \mathscr{R}_1] = \varepsilon u_{xxx}. \tag{50}$$

Comparing these two verifies (4) and proves that $\mathscr{R}_1$ is an approximate recursion operator for the potential Burgers' equation.

There is thus an infinite hierarchy of approximate symmetries, with characteristics $\mathcal{R}_1^k Q_1, k = 0,1,2,\ldots$. For example, the next characteristic after Q_{12} in the sequence is

$$\mathcal{R}_1 Q_{12} = \varepsilon\left(\left(x^2 + 6t\right)u_x + 2x\left(u + 2tu_{xx}\right) + 4t^2 u_{xxx}\right). \tag{51}$$

To obtain the characteristics depending on x and t, we require a second approximate recursion operator, which by inspection, we guess to be

$$\mathcal{R}_2 = t\mathcal{R}_1 + \frac{x}{2}. \tag{52}$$

Using the fact that $\mathcal{R}_1$ satisfies (4), we readily find

$$(\mathcal{R}_2)_t = t(\mathcal{R}_1)_t + \mathcal{R}_1 = t\left[D_K, \mathcal{R}_1\right] + \mathcal{R}_1, \tag{53}$$

whereas

$$\begin{aligned}\left[D_K, \mathcal{R}_2\right] &= t\left[D_K, \mathcal{R}_1\right] + \left[D_x^2 + 2\varepsilon u_x D_x, \frac{1}{2}x\right] \\ &= t\left[D_K, \mathcal{R}_1\right] + \left(D_x + \varepsilon u_x\right) = t\left[D_K, \mathcal{R}_1\right] + \mathcal{R}_1,\end{aligned} \tag{54}$$

proving that $\mathcal{R}_2$ is also an approximate recursion operator. There is thus a doubly infinite hierarchy of approximate generalized symmetries of potential Burgers' equation, with characteristics $\mathcal{R}_2^l \mathcal{R}_1^k Q_1, k, l \geq 0$. For instance, $Q_2 = \mathcal{R}_1 Q_1, Q_3 = 2\mathcal{R}_2\mathcal{R}_1 Q_1$ and so on.

Example 11. Consider the Gardner equation, which was shown to have two Hamiltonian structures with

$$\mathcal{D} = D_x, \qquad \mathcal{E} = 4uD_x + 2u_x + 3\varepsilon\left(uu_x + u^2 D_x\right) - D_x^3. \tag{55}$$

Hence, the operator connecting our hierarchy of approximate Hamiltonian symmetries is

$$\mathcal{R} = \mathcal{E} \cdot \mathcal{D}^{-1} = 4u + 3\varepsilon u^2 + (2 + 3\varepsilon u)\, u_x D_x^{-1} - D_x^2. \tag{56}$$

Therefore, our results on approximate bi-Hamiltonian systems will provide ready-made proofs of the existence of infinitely many approximate conservation laws and approximate symmetries for the Gardner equation

Note that the Frechet derivative for the right-hand side of ′ Gardner's equation is

$$\mathbf{D}_K = 6(1 + 2\varepsilon u)\, u_x + 6\left(u + \varepsilon u^2\right) D_x - D_x^3,$$

$$\begin{aligned}\mathcal{R}_t &= (4 + 6\varepsilon u)\, u_t + \left(2u_{xt} + 3\varepsilon u_t u_x + 3\varepsilon u u_{xt}\right) D_x^{-1} \\ &= 12uu_x(2 + 5\varepsilon u) - (4 + 6\varepsilon u)\, u_{xxx} \\ &\quad + \Big(6uu_{xx}(2 + 5\varepsilon u) + 12u_x^2(1 + 5\varepsilon u) \\ &\qquad - u_{xxxx}(2 + 3\varepsilon u) - 3\varepsilon u_x u_{xxx}\Big) D_x^{-1}.\end{aligned} \tag{57}$$

Theorem 12. Let $\overline{Q}_0 = \varepsilon u_x$. For each $k \geq 0$, the differential polynomial $\overline{Q}_k = \mathscr{R}^k \overline{Q}_0$ is a total x-derivative, $\overline{Q}_k = D_x R_k$, and hence we can recursively define $\overline{Q}_{k+1} = \mathscr{R}\overline{Q}_k$. *Each* $\overline{Q}_k$ is the characteristic of an approximate symmetry of the Gardner equation.

Proof. To prove this theorem, we apply the similar method applied in Theorem 5.31 of [10].

We proceed by induction on k, so suppose that $\overline{Q}_k = \mathscr{R}^k \overline{Q}_0$ for some $R_k \in \mathscr{A}$. From the form of the approximate recursion operator,

$$\overline{Q}_{k+1} = \varepsilon\left(4u\overline{Q}_k + 2u_x D_x^{-1}\overline{Q}_k - D_x^2\overline{Q}_k\right)$$
$$= \varepsilon D_x\left(2uD_x^{-1}\overline{Q}_k + 2D_x^{-1}\left(u\overline{Q}_k\right) - D_x\overline{Q}_k\right). \tag{58}$$

If we can prove that for some differential polynomial $S_k \in \mathscr{A}$, $u\overline{Q}_k = D_x S_k$, we will indeed have proved that $\overline{Q}_{k+1} = D_x R_{k+1}$, where R_{k+1} is the above expression in brackets. Consequently, the induction step will be completed.

To prove this fact, note that the formal adjoint of the approximate recursion operator $\varepsilon\mathscr{R}$ is

$$\varepsilon\mathscr{R}^* = \varepsilon\left(4u - 2D_x^{-1} \cdot u_x - D_x^2\right) = D_x^{-1}\varepsilon\mathscr{R}D_x. \tag{59}$$

We apply this in order to integrate the expression $u\overline{Q}_k$, by parts, so

$$\overline{Q}_k = u\mathscr{R}^k\left[\varepsilon u_x\right] = u_x \cdot \left(\varepsilon\mathscr{R}^*\right)^k [u] + D_x A_k \tag{60}$$

for some differential function $A_k \in \mathscr{A}$. On the other hand, using a further integration by parts, for some $B_k \in \mathscr{A}$ the following identity holds:

$$u_x \cdot \left(\varepsilon\mathscr{R}^*\right)^k [u] = u_x \cdot D_x^{-1}\ \varepsilon\mathscr{R}\left[u_x\right]$$
$$= u_x \cdot D_x^{-1}\overline{Q}_k$$
$$= -u\overline{Q}_k + D_x B_k. \tag{61}$$

Substituting into the previous identity, we conclude

$$u\overline{Q}_k = D_x S_k, \quad \text{where } S_k = \frac{1}{2}\left(A_k + B_k\right), \tag{62}$$

which proves our claim.

Definition 13. A pair of skew-adjoint $q \times q$ matrix of differential operators $\mathscr{D}$ and $\mathscr{E}$ is said to form an approximately Hamiltonian pair if every linear combination $a\mathscr{D} + b\mathscr{E}$, $a, b \in \mathbb{R}$, is an approximate Hamiltonian operator. A system of perturbed evolution equations is an approximate biHamiltonian system if it can be written in the form of

$$\frac{\partial u}{\partial t} = K_1[u, \varepsilon] \approx \mathscr{D}\delta\mathscr{H}_1 \approx \mathscr{E}\delta\mathscr{H}_0, \tag{63}$$

where $\mathscr{D}, \mathscr{E}$ form an approximately Hamiltonian pair, and $\mathscr{H}_0$ and $\mathscr{H}_1$ are appropriate Hamiltonian functionals.

Lemma 14. If $\mathscr{D}, \mathscr{E}$ are skew-adjoint operators, then they form an approximately Hamiltonian pair if and only if $\mathscr{D}, \mathscr{E}$ and $\mathscr{D} + \mathscr{E}$ are all approximate Hamiltonian operators.

Corollary 15. Let $\mathscr{D}$ *and* $\mathscr{E}$ be Hamiltonian differential operators. Then $\mathscr{D}, \mathscr{E}$ form an approximately Hamiltonian pair if and only if

$$\mathbf{pr}\,\mathbf{v}_{\mathscr{D}\theta}(\Theta_{\mathscr{E}}) + \mathbf{pr}\,\mathbf{v}_{\mathscr{E}\theta}(\Theta_{\mathscr{D}}) = o(\varepsilon^P), \tag{64}$$

where

$$\Theta_{\mathscr{D}} = \frac{1}{2}\int \{\theta \wedge \mathscr{D}\theta\}\,dx, \qquad \Theta_{\mathscr{E}} = \frac{1}{2}\int \{\theta \wedge \mathscr{E}\theta\}\,dx \tag{65}$$

are the functional bi-vectors representing the respective Poisson brackets.

Example 16. Consider the approximate Hamiltonian operators $\mathscr{D}, \mathscr{E}$ associated with the Gardner equation. We have

$$\begin{aligned}\mathbf{pr}\,\mathbf{v}_{\mathscr{D}\theta} &= \sum_{\alpha,J} D_J(\mathscr{D}\theta)_\alpha \frac{\partial}{\partial u_J^\alpha} \\ &= \sum_{\alpha,J} D_J\left(\sum_{\beta=1}^{q} \mathscr{D}_{\alpha\beta}\theta^\beta\right)\frac{\partial}{\partial u_J^\alpha}\end{aligned} \tag{66}$$

in the case of the second approximate Hamiltonian operator for the Gardner equation, we have

$$\mathbf{pr}\,\mathbf{v}_{\mathscr{E}\theta}(u) = \mathscr{E}\theta, \qquad \mathbf{pr}\,\mathbf{v}_{\mathscr{E}\theta}(u^2) = 2u\mathscr{E}\theta,$$

$$\mathbf{pr}\,\mathbf{v}_{\mathscr{E}\theta}(\Theta_{\mathscr{D}}) = \mathbf{pr}\,\mathbf{v}_{\mathscr{E}\theta}\int \frac{1}{2}\{\theta \wedge \theta_x\}\,dx = o(\varepsilon^P) \tag{67}$$

trivially, by the properties of the wedge product, it is deduced that

$$\begin{aligned}&\mathbf{pr}\,\mathbf{v}_{\mathscr{D}\theta}(\Theta_{\mathscr{E}}) \\ &= \mathbf{pr}\,\mathbf{v}_{\mathscr{D}\theta}\int\left\{\left(2u + \frac{3\varepsilon}{2}u^2\right)\theta \wedge \theta_x + \frac{1}{2}\theta_x \wedge \theta\right\}dx \\ &\approx \int \{(2 + 3\varepsilon u)\theta_x \wedge \theta \wedge \theta_x\} = o(\varepsilon^P).\end{aligned} \tag{68}$$

Thus $\mathscr{D}$ and $\mathscr{E}$ form an approximately Hamiltonian pair.

Definition 17. A differential operator $\mathscr{D} : \mathscr{A}^r \to \mathscr{A}^s$ is approximately degenerate if there is a nonzero differential operator $\widetilde{\mathscr{D}} : \mathscr{A}^s \to \mathscr{A}$ such that $\widetilde{\mathscr{D}} \cdot \mathscr{D} \equiv o(\varepsilon^P)$.

Now, according to [10], we are in a situation to state the main theorem on approximate bi-Hamiltonian systems

Theorem 18. Let

$$u_t = K_1[u, \varepsilon] \approx \mathscr{D}\delta\mathscr{H}_1 \approx \mathscr{E}\delta\mathscr{H}_0 \tag{69}$$

be an approximate bi-Hamiltonian system of perturbed evolution equations. Assume that the operator D of the approximately Hamiltonian pair is approximate nondegenerate. Let

$\mathcal{R} = \mathcal{E} \cdot \mathcal{D}^{-1}$ be the corresponding approximate recursion operator, and let $K_0 \approx \mathcal{D}\delta\mathcal{H}_0$. Assume that for each $n = 1, 2, \ldots$ one can recursively define

$$K_n \approx \mathcal{R}K_{n-1}, \quad n \geq 1, \tag{70}$$

meaning that for each n, K_{n-1} lies in the image of D. Then there exists a sequence of functionals $\mathcal{H}_0, \mathcal{H}_1, \mathcal{H}_2, \ldots$,... such that

(i) for each $n \geq 1$ the perturbed evolution equation

$$u_t = K_n[u, \varepsilon] \approx \mathcal{D}\delta\mathcal{H}_n \approx \mathcal{E}\delta\mathcal{H}_{n-1} \tag{71}$$

is an approximate bi-Hamiltonian system;

(ii) the corresponding approximate evolutionary vector fields $\mathbf{v}_n = \mathbf{v}_{Kn}$ all mutually commute

$$[\mathbf{v}_n, \mathbf{v}_m] = o(\varepsilon^p), \quad n, m \geq 0; \tag{72}$$

(iii) the approximate Hamiltonian functionals $\mathcal{H}_n$ are all in involution with respect to either Poisson bracket:

$$\{\mathcal{H}_n, \mathcal{H}_n\}_{\mathcal{D}} = o(\varepsilon^p) = \{\mathcal{H}_n, \mathcal{H}_n\}_{\mathcal{E}}, \quad n, m \geq 0, \tag{73}$$

and hence provide an infinite collection of approximate conservation laws for each of the approximate biHamiltonian systems (63).

We have seen that given an approximate bi-Hamiltonian system, the operator $\mathcal{R} = \mathcal{E} \cdot \mathcal{D}^{-1}$, when applied successively to the initial equation $K_0 = \mathcal{D}\delta\mathcal{H}_0$, produces an infinite sequence of approximate generalized symmetries of the original system (subject to the technical assumptions contained in Theorem 18). It is still not clear that $\mathcal{R}$ is a true approximate recursion operator for the system, in the sense that whenever $\mathbf{v}_Q$ is an approximate generalized symmetry, so is $\mathbf{v}_{\mathcal{R}Q}$. So far, we only know it for approximate symmetries with $Q=K_n$ for some n. In order to establish this more general result, we need a formula for the infinitesimal change of the approximate Hamiltonian operator itself under a Hamiltonian flow

Lemma 19. Let $u_t = K \approx \mathcal{D}\delta\mathcal{H}$ be an approximate Hamiltonian system of perturbed evolution equations with corresponding vector field $\mathbf{v}_K = \hat{\mathbf{v}}_{\mathcal{H}}$. Then

$$\mathbf{pr}\,\hat{\mathbf{v}}_{\mathcal{H}}(\mathcal{D}) \approx \mathbf{D}_K \cdot \mathcal{D} + \mathcal{D} \cdot \mathbf{D}_K^*. \tag{74}$$

Theorem 20. Let $u_t = K \approx \mathcal{D}\delta\mathcal{H}_1 \approx \mathcal{E}\delta\mathcal{H}_0$ be an approximate bi-Hamiltonian system of perturbed evolution equations. Then the operators $\mathcal{R}_l = \varepsilon^l\mathcal{E} \cdot \mathcal{D}^{-1}, 0 \leq l \leq p,$, are approximate recursion operators for the system

Judging from $\mathcal{R}_l^p = o(\varepsilon^p)$ when $l \neq 0$, this type of approximate recursion operators have less significance than $\mathcal{R}_0$.

Example 21. The approximate recursion operators of the Gardner equation are

$$\mathcal{R}_0 = \mathcal{E} \cdot \mathcal{D}^{-1} = 4u + 2u_x D_x^{-1} + 3\varepsilon\left(uu_x D_x^{-1} + u^2\right) - D_x^2,$$

$$\mathcal{R}_1 = \varepsilon\left(4u + 2u_x D_x^{-1} - D_x^2\right) \tag{75}$$

and we can apply $\mathcal{R}_0$ to the right-hand side of the Gardner equation to obtain the approximate symmetries. The first step in this recursion is the flow

$$\begin{aligned} u_t &\approx \mathcal{E}\delta\mathcal{H}_1 \approx \mathcal{D}\delta\mathcal{H}_2 \\ &\approx u_{xxxxx} - 10uu_{xxx} - 20u_x u_{xx} + 30u^2 u_x \\ &\quad + \varepsilon\left(55u^3 u_x - 39uu_x u_{xx} - 9u^2 u_{xxx} - 12u_x^3\right), \end{aligned} \tag{76}$$

which is not approximately total derivative, so we cannot reapply the approximate recursion operator to get a meaningful approximate generalized symmetry.

But if we set

$$\overline{K}_1[u,\varepsilon] = Q_5 = \varepsilon K_1[u,\varepsilon] = \varepsilon\left(6uu_x - u_{xxx}\right),$$

$$\overline{\mathcal{H}}_0 = \overline{\mathcal{P}}_5 = \varepsilon\mathcal{H}_0, \qquad \overline{\mathcal{H}}_1 = \mathcal{P}_5 = \varepsilon\mathcal{H}_1, \tag{77}$$

then we can apply $\mathcal{R}_0$ successively to $\overline{K}_1$ in order to obtain the approximate symmetries. The first phase become

$$\begin{aligned} u_t &\approx \mathcal{E}\delta\overline{\mathcal{H}}_1 \approx \mathcal{D}\delta\overline{\mathcal{H}}_2 \approx \mathcal{R}_0\overline{K}_1 \\ &\approx \varepsilon\left(u_{xxxxx} - 10uu_{xxx} - 20u_x u_{xx} + 30u^2 u_x\right) \end{aligned} \tag{78}$$

in which

$$\overline{\mathcal{H}}_2 = \overline{\mathcal{P}}_5 = \frac{\varepsilon}{2}\int\left(u_{xx}^2 - 5u^2 u_{xx} + 5u^4\right)dx \tag{79}$$

is another approximate conservation law.

Now, for $\overline{K}_2 = \mathcal{R}_0\overline{K}_1$ we have

$$\begin{aligned} u_t &\approx \mathcal{E}\delta\overline{\mathcal{H}}_2 \approx \mathcal{D}\delta\overline{\mathcal{H}}_3 \approx \mathcal{R}_0\overline{K}_2 \\ &\approx \varepsilon\left(-u_{xxxxxxx} + 14uu_{xxxxx} + 42u_x u_{xxxx}\right) \\ &\qquad + 70\varepsilon\left(u_{xx}u_{xxx} - u^2 u_{xxx} + 2u^3 u_x - 4uu_x u_{xx} - u_x^3\right), \end{aligned} \tag{80}$$

where

$$\overline{\mathcal{H}}_3 = 7\varepsilon\int\left(\frac{u_{xxx}^2}{14} + uu_{xx}^2 + 5u^2\, u_x^2 + u^5\right)dx \tag{81}$$

is a further approximate conservation law

CONCLUDING REMARKS

Sometimes, differential equations appearing in mathematical modelings are written with terms involving a small parameter which is known as the perturbed term. Taking into account the instability of the Lie point symmetries with respect to perturbation of coefficients of differential equations, the approximate (perturbed) symmetries for such equations are obtained. Different methods for computing the approximate symmetries of a system of differential equations are available in the literature [2–4].

The approximate symmetry method proposed by Fushchich and Shtelen [4] is based on a perturbation of dependent variables. This method has so many advantages such as producing more approximate group-invariant solutions, consistence with the perturbation theory, solving singular perturbation problems [7, 8], and close relationship with approximate homotopy symmetry method [12]. But despite the above-mentioned benefits, this procedure converts a perturbed evolution equation to an equivalent perturbed evolutionary system. In his case, obtaining the corresponding Hamiltonian formulation will be hard. Due to the increase of the dimensions of Hamiltonian operators $\mathscr{D}, \mathscr{E}$, computation of the approximate recursion operator $\mathscr{R} = \mathscr{E} \cdot \mathscr{D}^{-1}$ is difficult.

Since prolongation and Fréchet derivative of vector fields are linear, both of the approximate symmetry methods can be extended on the Hamiltonian structures. But due to the significance of vector fields in Hamiltonian and bi-Hamiltonian systems, the approximate symmetry method proposed by Baikov et al. [2, 3] seems to be more consistent.

ACKNOWLEDGMENTS

It is a pleasure to thank the anonymous referees for their constructive suggestions and helpful comments which have improved the presentation of the paper. The authors wish to express their sincere gratitude to Fatemeh Ahangari for her useful advice and suggestions.

REFERENCES

1. S. Lie, "On integration of a class of linear partial differential equations by means of definite integrals,"Archiv der Mathematik, vol. 6, pp. 328–368, 1881.
2. V. A. Baikov, R. K. Gazizov, and N. Kh. Ibragimov, "Approximate symmetries of equations with a small parameter," Matematicheskiĭ Sbornik, vol. 136, no. 4, pp. 435–450, 1988.
3. V. A. Baikov, R. K. Gazizov, and N. H. Ibragimov, "Approximate transformation groups and deformations of symmetry Lie algebras," in CRC Handbook of Lie Group Analysis of Differential Equations, N. H. Ibragimov, Ed., vol. 3, chapter 2, CRC Press, Boca Raton, Fla, USA, 1996.
4. W. I. Fushchich and W. M. Shtelen, "On approximate symmetry and approximate solutions of the nonlinear wave equation with a small parameter," Journal of Physics A, vol. 22, no. 18, pp. L887–L890, 1989.

5. N. Euler, M. W. Shul'ga, and W.-H. Steeb, "Approximate symmetries and approximate solutions for a multidimensional Landau-Ginzburg equation," Journal of Physics A, vol. 25, no. 18, pp. L1095–L1103, 1992.
6. M. Euler, N. Euler, and A. Köhler, "On the construction of approximate solutions for a multidimensional nonlinear heat equation," Journal of Physics A, vol. 27, no. 6, pp. 2083–2092, 1994.
7. M. Pakdemirli, M. Yürüsoy, and İ. T. Dolapçi, "Comparison of approximate symmetry methods for differential equations," Acta Applicandae Mathematicae, vol. 80, no. 3, pp. 243–271, 2004.
8. R. Wiltshire, "Two approaches to the calculation of approximate symmetry exemplified using a system of advection-diffusion equations," Journal of Computational and Applied Mathematics, vol. 197, no. 2, pp. 287–301, 2006.
9. N. H. Ibragimov and V. F. Kovalev, Approximate and Renormgroup Symmetries, Nonlinear Physical Science, Springer, Berlin, Germany, 2009.
10. P. J. Olver, Applications of Lie Groups to Differential Equations, vol. 107 of Graduate Texts in Mathematics, Springer, New York, NY, USA, 2nd edition, 1993
11. M. Nadjafikhah and A. Mokhtary, "Approximate symmetry analysis of Gardner equation," http://arxiv.org/abs/1212.3604.
12. Z. Zhang, "Approximate homotopy series solutions of perturbed PDEs via approximate symmetry method," http://128.84.158.119/abs/1112.4225v2.

3
CHAPTER

DUAL SOLUTIONS FOR NONLINEAR FLOW USING LIE GROUP ANALYSIS

Muhammad Awais, Tasawar Hayat Sania Irum and Salman Saleem

INTRODUCTION

The non-Newtonian fluids at present are acknowledged more appropriate for scientific and technological applications than the materials obeying the Newtonian law of viscosity. However, there exists no single constitutive equation by which all the non-Newtonian materials can be described. Researchers and applied mathematicians have proposed a variety of constitutive equations in view of their rheological properties describing the non-Newtonian behavior. Such constitutive equations give rise to several complexities in the momentum equations which arises various challenges for researchers to model the problems in mathematical form and to evaluate/analyze the computed solutions mathematically and physically. The classifications of non-Newtonian fluids in general are classified into three types known as the differential, rate and integral. Available information on the topic indicates that much attention of the researchers has been accorded to the flows of second, third and fourth order fluids i.e. the subclasses of differential type materials. Undoubtedly these subclasses predicts the normal stress, shear thickening and thinning features but cannot describe relaxation time effect which many polymers show at low molecular weight. The relaxation time effect can be examined using Maxwell model (the simplest subclass of rate type fluid). The literature for flows of Maxwell fluid is quite sizeable. Few recent studies and several references on rheology of Maxwell fluid can be seen in the investigations [1–15].

As per our knowledge no attempt has been made for the Lie group analysis and the development of dual solutions for magnetohydrodynamic (MHD) [16–21] flow of Maxwell fluid over a permeable shrinking surface. Numerical solutions are presented for different cases of wall mass transfer. Graphical results are presented for the variations of different sundry parameters.

Citation: Awais M, Hayat T, Irum S, Saleem S (2015) Dual Solutions for Nonlinear Flow Using Lie Group Analysis. PLoS ONE 10(11): e0142732. doi:10.1371/journal.pone.0142732

Mathematical statement

Let us consider an upper-convected Maxwell (UCM) fluid flowing over a permeable shrinking sheet in a two-dimensional steady and incompressible field. The uniform magnetic field of strength B_0 is applied along the $\bar{y}$—axis whereas the induced magnetic field is neglected due to the assumption of very small magnetic Reynold number whereas an upper-convected Maxwell (UCM) fluid occupies the region $\bar{y} \succ 0$. The laws of conservations of mass and momentum for MHD boundary layer flow of an upper-convected Maxwell (UCM) fluid over a permeable shrinking surface are given by

$$\frac{\partial \bar{u}}{\partial \bar{x}} + \frac{\partial \bar{v}}{\partial \bar{y}} = 0, \tag{1}$$

$$\bar{u}\frac{\partial \bar{u}}{\partial \bar{x}} + \bar{v}\frac{\partial \bar{u}}{\partial \bar{y}} + \lambda\left(\bar{u}^2\frac{\partial^2 \bar{u}}{\partial \bar{x}^2} + \bar{v}^2\frac{\partial^2 \bar{u}}{\partial \bar{y}^2} + 2\bar{u}\bar{v}\frac{\partial^2 \bar{u}}{\partial \bar{x}\partial \bar{y}}\right) = \nu\frac{\partial^2 \bar{u}}{\partial \bar{y}^2} - \frac{\sigma B_0^2}{\rho}\left(\bar{u} + \lambda\bar{v}\frac{\partial \bar{u}}{\partial \bar{y}}\right), \tag{2}$$

along with the boundary conditions

$$\bar{u} = \bar{U}_w(\bar{x}) = -C_1\bar{x}, \quad \bar{v} = \bar{v}_w(\bar{x}), \quad \text{at } \bar{y} = 0,$$

$$\bar{u} \to 0, \quad \text{as } \bar{y} \to \infty, \tag{3}$$

in which $\bar{u}$ and $\bar{v}$ are the velocity components along $\bar{x}$ and $\bar{y}$ directions respectively, ρrepresented the density, ν the kinematic viscosity, σ the electrical conductivity and $\bar{U}_w(\bar{x})$ the shrinking velocity with $C_1 > 0$ is shrinking constant.

By using the following dimensionless quantities

$$x = \frac{C_1\bar{x}}{U_1}, \quad y = \sqrt{\frac{C_1}{\nu}}\bar{y}, \quad u = \frac{\bar{u}}{U_1}, \quad v = \frac{\bar{v}}{\sqrt{C_1\nu}}, \tag{4}$$

Eqs (1–3) take the following forms

$$\frac{\partial u}{\partial x} + \frac{\partial v}{\partial y} = 0, \tag{5}$$

$$u\frac{\partial u}{\partial x} + v\frac{\partial u}{\partial y} + \beta\left(u^2\frac{\partial^2 u}{\partial x^2} + v^2\frac{\partial^2 u}{\partial y^2} + 2uv\frac{\partial^2 u}{\partial x\partial y}\right) = \frac{\partial^2 u}{\partial y^2} - M^2\left(u + \beta v\frac{\partial u}{\partial y}\right), \tag{6}$$

$$u = -x, \quad v = \frac{v_w}{\sqrt{C_1\nu}}, \quad \text{at } y = 0,$$

$$u \to 0, \quad \text{as } y \to \infty, \tag{7}$$

where the Hartman number M and the Deborah number β are defined as follows:

$$M^2 = \frac{\sigma B_0^2}{\rho C_1}, \quad \beta = \lambda C_1. \tag{8}$$

The stream function ψ(x, y) is introduced such as

$$u = \frac{\partial \psi}{\partial y}, \quad v = -\frac{\partial \psi}{\partial x}, \tag{9}$$

Making use of Eq (9), law of conservation of mass (Eq (5)) is identically satisfied whereas Eqs (6) and (7) become

$$\psi_y\psi_{xy} - \psi_x\psi_{yy} + \beta(\psi_y^2\psi_{xxy} + \psi_x^2\psi_{yyy} - 2\psi_x\psi_y\psi_{xyy}) = \psi_{yyy} - M^2(\psi_y - \beta\psi_x\psi_{yy}), \quad (10)$$

$$\psi_y = -x, \quad \psi_x = -\frac{v_w}{\sqrt{C_1 \nu}}, \qquad \text{at } y = 0,$$

$$\psi_y \to 0, \qquad \text{as } y \to \infty. \quad (11)$$

SCALING GROUP OF TRANSFORMATION

In order to proceed for the Lie group analysis we consider

$$\Gamma = \begin{cases} x^* = xe^{\varepsilon\alpha_1}, \quad y^* = ye^{\varepsilon\alpha_2}, \quad \psi^* = \psi e^{\varepsilon\alpha_3}, \\ u^* = ue^{\varepsilon\alpha_4}, \quad v^* = ve^{\varepsilon\alpha_5} \end{cases} \quad (12)$$

Making use of Eq (12), the coordinates (x,y,ψ,u,v,θ,φ) transform into the coordinates (x^*, y^*, ψ^*,u^*, v^*, θ^*, φ^*) whereas Eqs (10 and 11) take the forms

$$e^{\varepsilon(-2\alpha_3+2\alpha_2+\alpha_1)}(\psi^*_{y^*}\psi^*_{x^*y^*} - \psi^*_{x^*}\psi^*_{y^*y^*}) - e^{\varepsilon(-\alpha_3+3\alpha_2)}\psi^*_{y^*y^*y^*}$$

$$+\beta e^{\varepsilon(-3\alpha_3+3\alpha_2+2\alpha_1)}\psi^{*2}_{y^*}\psi^*_{x^*y^*x^*} + \psi^{*2}_{x^*}\psi^*_{y^*y^*y^*} - 2\psi^*_{x^*}\psi^*_{y^*}\psi^*_{x^*y^*y^*}$$

$$+M^2(e^{\varepsilon(-\alpha_3+\alpha_2)}\psi^*_{y^*} + e^{\varepsilon(-2\alpha_3+2\alpha_2+\alpha_1)}\beta\psi^*_{x^*}\psi^*_{y^*y^*}) = 0, \quad (13)$$

$$e^{\varepsilon(-\alpha_3+\alpha_2)}\psi^*_{y^*} = -e^{-\varepsilon\alpha_1}x^*, \quad e^{\varepsilon(-\alpha_3+\alpha_1)}\psi^*_{x^*} = -\frac{v_w}{\sqrt{C_1\nu}}, \qquad \text{at } y^* = 0,$$

$$e^{\varepsilon(-\alpha_3+\alpha_2)}\psi^*_{y^*} \to 0, \qquad \text{when } y^* \to \infty. \quad (14)$$

It is noted that the system remains invariant under the group of transformations Γ and we get the following relation among parameters

$$-2\alpha_3 + 2\alpha_2 + \alpha_1 = -3\alpha_3 + 3\alpha_2 + 2\alpha_1 = -\alpha_3 + 3\alpha_2 = -\alpha_3 + \alpha_2,$$

$$-\alpha_3 + \alpha_1 = 0,$$

$$-\alpha_3 + \alpha_2 = -\alpha_1. \quad (15)$$

By solving the system given in Eq (15) we obtain

$$\alpha_1 = \alpha_3 = \alpha_4 \quad \text{and} \quad \alpha_2 = \alpha_5 = 0. \quad (16)$$

Making use of Eq (16), the set of transformations Γ reduces to

$$\Gamma = \begin{cases} x^* = xe^{\varepsilon\alpha_1}, \quad y^* = y, \quad \psi^* = \psi e^{\varepsilon\alpha_1}, \\ u^* = ue^{\varepsilon\alpha_1}, \quad v^* = v. \end{cases} \quad (17)$$

Expanding by Taylor's method in powers of ε and retaining terms up to the order ε, we get

$$\Gamma = \begin{cases} x^* - x = x\varepsilon\alpha_1, & \psi^* - \psi = \psi\varepsilon\alpha_1, \\ u^* - u = u\varepsilon\alpha_1, & v^* - v = y^* - y = 0. \end{cases} \quad (18)$$

In terms of differentials we get

$$\frac{dx}{\alpha_1 x} = \frac{dy}{0} = \frac{d\psi}{\alpha_1 \psi} = \frac{du}{\alpha_1 u} = \frac{dv}{0}. \quad (19)$$

Solving the above equations we acquire

$$y^* = \eta, \qquad \psi^* = x^* f(\eta). \quad (20)$$

It is noted that with the help of the above relations, Eq (13) becomes

$$f''' + ff'' - f'^2 - \beta(f^2 f''' - 2ff'f'') - M^2(f' + \beta ff'') = 0, \quad (21)$$

along with the boundary conditions

$$f = S, \quad f' = -1, \quad \text{at} \quad \eta = 0,$$
$$f' \to 0 \quad \text{as} \quad \eta \to \infty. \quad (22)$$

SOLUTION METHODOLOGY

Shooting Method:

The differential Eq (21) along with conditions (22) is solved numerically using an efficient approach namely shooting method. Runge-Kutta fourth-order algorithm combined with Newtons' method is utilized to approximate the shoot values in order to match at a finite value of $\eta \to \infty$ say η_∞. For this we first write

$$f' = f_1, \quad f_1' = f_2, \quad f_2' = \frac{f_1^2 - (M^2\beta + 1)ff_2 - 2\beta ff_1 f_2 + M^2 f_1}{1 - \beta f^2},$$

with conditions

$$f(0) = S, \quad f_1(0) = -1,$$
$$f_2(0) = ?$$

It is noted that in order to compute the solution of the above differential equation as an initial value problem, we require the value of $f_2(0)$, whereas no such value is given initially. In order to compute the desired result we select an initial guess and then applied the fourth-order Runge-Kutta method to approximate the value up to the desired accuracy of 10^{-5}.

RESULTS AND DISCUSSION

This part of the paper corresponds to some graphical observations so the one can analyze the physical insight of the problem. Figs 1–3 are prepared to show the streamlines for different cases of suction and injection. It is observed from Fig 1 that when the suction/injection phenomenon is absent then the viscous boundary layer is very much thinner and the flow near the shrinking wall is dominant whereas injection at the wall adds up an extra force which causes the boundary layer to become thicker as shown in Fig 2. Fig 3 presents the streamlines for the suction case which are found to be qualitatively opposite when compared to the injection case. Larger suction in combination with the shrinking acts as a supporting agent to suction the fluid out of the system. Fig 4 portrays the dual nature of solutions for the magnetohydrodynamic (MHD) vs hydrodynamic cases. It is observed that for hydrodynamic case, the two-dimensional flow of upper-convected Maxwell (UCM) fluid over a permeable wall is possible only when the wall mass transfer satisfies the inequality $S \geq 1.8018$ (see Fig 4). Moreover it is elucidated from this plot that for the magnetohydrodynamic case, the wall mass suction inequality reduces to the value $S \geq 1.648$ due to the decrease in the intrinsic fluidity of the material. Fig 5 presents the difference between the flow of Newtonian and Maxwell fluids for some values of suction parameter S. It is observed that the dual nature of solution for Newtonian fluid model ($\beta = 0.0$) exists only when the wall mass transfer satisfies the inequality $S \geq 1.7264$ whereas for Maxwell fluid model ($\beta > 0.0$) the inequality reduces to $S \geq 1.6$. Moreover dual nature of the solutions for both hydrodynamic and hydromagnetic flow situations are presented and found quite different. It is observed that the value of skin friction increase rapidly for the first solution whereas it decreases for the second solution but the variations in first solution is quite significant when compared with the other. Physically we can say that the increase in Deborah number β results into a decrease in a vorticity generation of a Maxwell fluid. Since the vorticity generation due to shrinking for Maxwell fluid is not as much as of Newtonian fluid therefore the boundary layer separation for Maxwell fluid is delayed and the similar solution exists for lesser amount of wall mass transfer. Moreover it is also noted that for the Newtonian fluid flow, both solutions are closed to each other and nearly parallel to x-axis which shows the linear behavior of the skin friction whereas for the non-Newtonian fluid, the variations in both solutions are quite significant and found quite opposite. Figs 6 and 7 present the effects of Deborah number β on the velocity fields f' and f respectively. It is observed that the dual solutions for velocity against different values of β exist. It is observed that the non-Newtonian fluid behaves much like liquids (such as paints, polymer solutions etc.) for smaller value of Deborah number however for larger values of Deborah number, fluid transforms from liquids to the position of becoming a viscoelastic solid (such as toothpaste, jelly or rubber etc.). Thus the flow retards by increasing Deborah number which is quite in agreement with the Fig 6. It is also elucidated from Figs 6 and 7 that velocity increases with an increase in Deborah number β for the first solution whereas from second solution, the oscillatory effects and cross over is found. The effects of wall mass suction (due to the permeability of the surface) on the flow are demonstrated in the Figs 8 and 9. The dual solutions exist for different values of S and both solutions show quite opposite behavior. The magnitude of first solution enhances with an increasing in the suction rate whereas the magnitude of second solution decreases. Figs 10 and 11 present the influence of magnetic fieldM on the velocity. It is observed that dual nature of solutions exists for different values of M. It is also illustrated that the first solution is dominant in comparison to the second solution (see Figs10 and 11). It is perceived that when any fluid is subjected to a magnetic field then its apparent viscosity increases to the point of becoming a viscoelastic solid. Importantly, the velocity and the rheology of the fluid can be controlled very accurately by varying the magnetic field intensity. The outcome of which is that the fluid›s ability to transmit force can be controlled with the help of an electromagnet which gives rise to its many possible control-based applications including MHD power generation, electromagnetic casting of metals, MHD ion propulsion etc. Moreover nomenclature have been identified in Table 1.

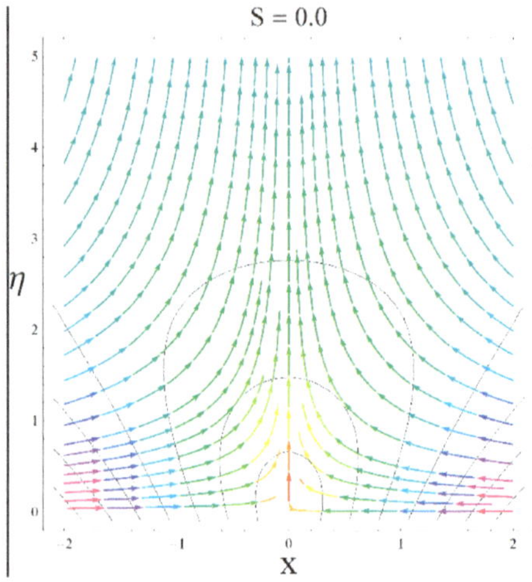

Figure 1: Streamlines without suction/injection.

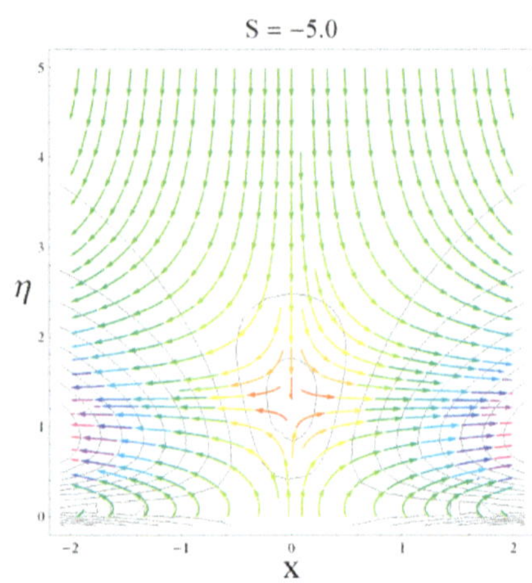

Figure 2: Stream lines for injection case.

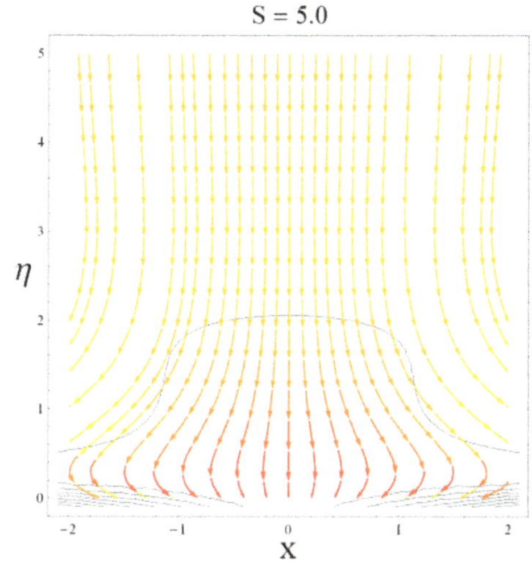

Figure 3: Streamlines for suction case.

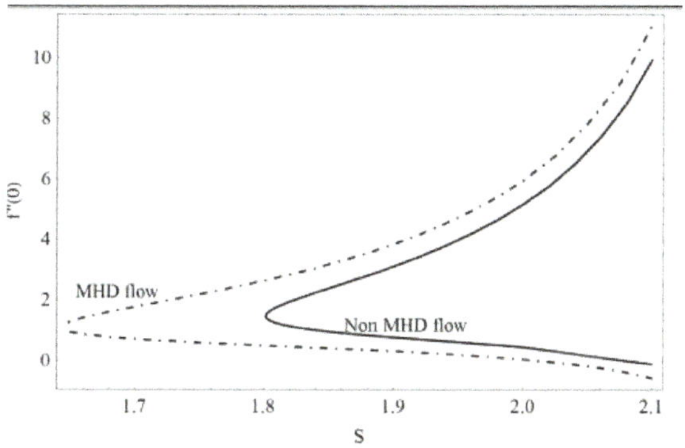

Figure 4: Dual solutions: MHD Vs hydrodynamic case.

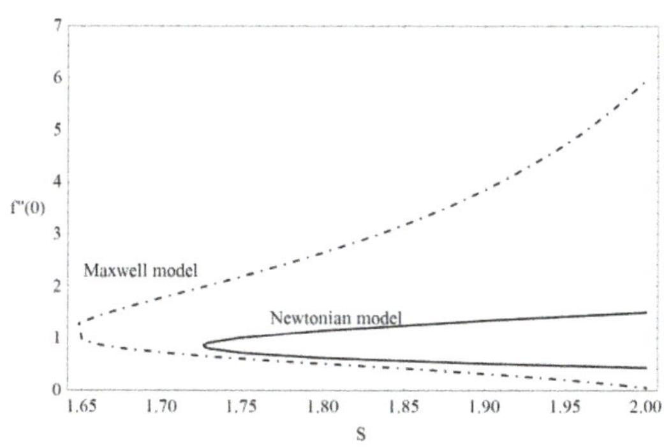

Figure 5: Dual solutions: Newtonian vs Maxwell model.

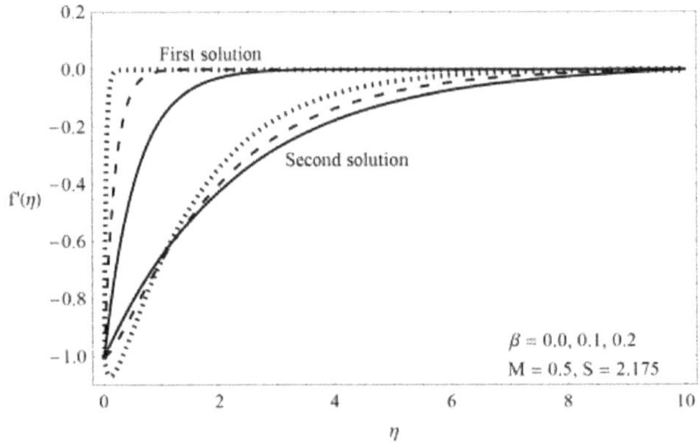

Figure 6: Dual solutions (f'(η)) for Deborah number.

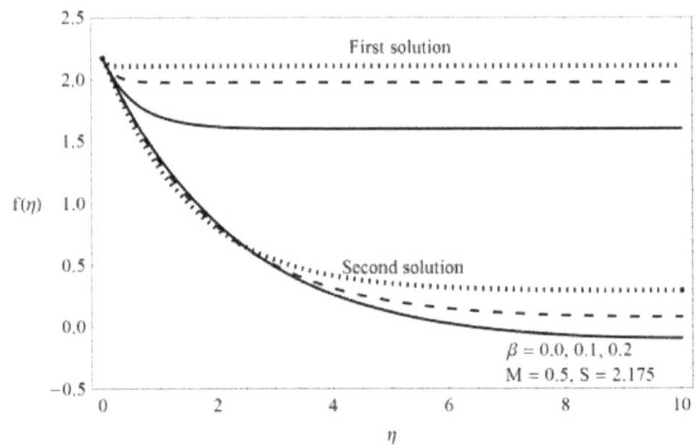

Figure 7: Dual solutions (f(η)) for Deborah number.

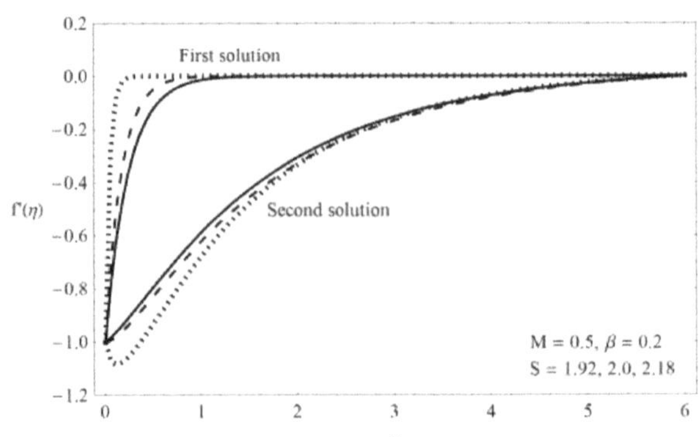

Figure 8: Dual solutions (f'(η)) for suction.

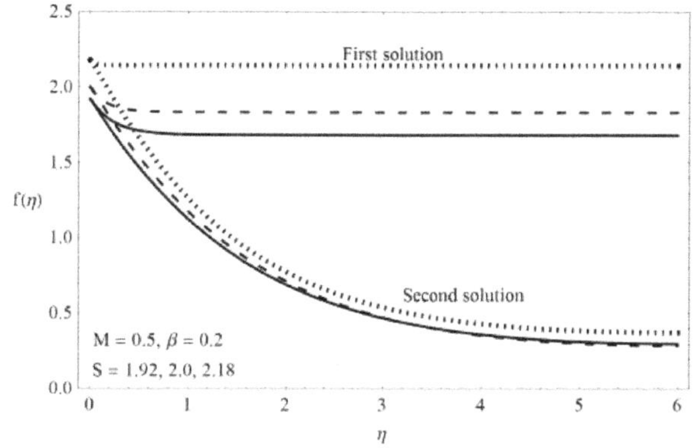

Figure 9: Dual solutions (f(**η**)) for suction.

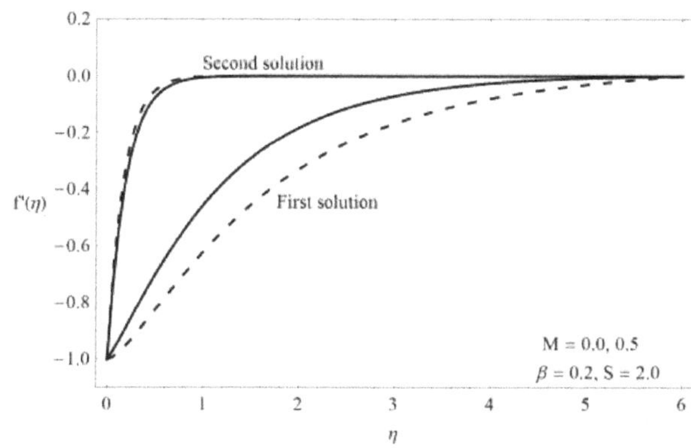

Figure 10: Dual solutions (f'(**η**)) for MHD case.

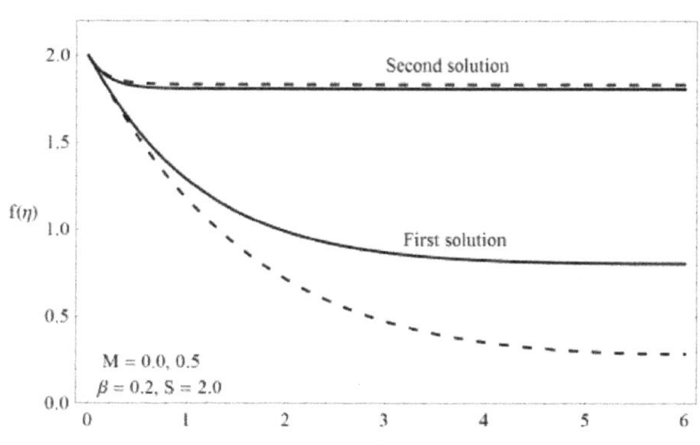

Figure 11: Dual solutions (f(**η**)) for MHD case.

Table 1: Nomenclature

B_0	applied magnetic field	*Greeks symbol*	
$\hat{u}, \hat{v}$	velocity components	λ	convection parameter
$\hat{x}, \hat{y}$	Cartesian coordinates	ψ	stream function
U_w	shrinking velocity	B	Deborah number
V_w	porosity of wall	ρ	density
M	Hartman number	Γ	scaling set
C_1	dimensional constant	η	similarity variable
f	dimensionless stream function	σ	electrical conductivity
S	suction/blowing	v	kinematical viscosity

doi:10.1371/journal.pone.0142732.t001

FINAL OUTCOMES

The main results of the present analysis are mentioned below:

Larger values of Deborah number β results in to steady flow of non-Newtonian fluid with lessor amount of wall mass transfer.

Momentum boundary layer decreases with Deborah number in both first and second solutions.

Wall mass transfer has opposite effects on the viscous boundary layer in first and second solutions.

Magnetic field retards the rheology and the dynamics of the flow field.

Suction and blowing phenomena show the opposite behavior on the streamlines.

ACKNOWLEDGMENTS

We sincerely thank the reviewer for their constructive comments. More the first author would like to acknowledge the support of Dr. Atiq Ur Rehman (Assistant Professor, Department of Mathematics, CIIT Attock) for his useful suggestions regarding improvement of the graphical illustrations.

REFERENCES

1. Ramesh GK, Gireesha BJ (2014) Influence of heat source/sink of a Maxwell fluid over a stretching surface with convective boundary conditions in the presence of nanoparticles, Ain Shams Eng., 5: 991–998.
2. Abbasbandy S, Naz R, Hayat T, Alsaedi A (2014), Numerical and analytical solutions Falkner-Skan flow of MHD Maxwell fluid, Appl. Math. Comp., 242: 569–575.
3. Zhao M, Wang S, Zhang Q (2014) Onset of triply diffusive convection in a Maxwell fluid saturated porous layer, Appl. Math. Mod., 38: 2345–2352.

4. Mukhopadhyay S, Bhattacharyya K (2012) Unsteady flow of a Maxwell fluid over a stretching surface in presence of chemical reaction, J. Egyptian Math. Soc., 20: 229–234.

5. Yang D, Zhu KQ (2010) Start-up flow of a viscoelastic fluid in a pipe with a fractional Maxwell's model, Comp. Math. Appl., 60: 2231–2238.

6. Wang S, Tan W (2011) Stability analysis of soret-driven double-diffusive convection of Maxwell fluid in a porous medium, Int. J. Heat Fluid Flow, 32: 88–94.

7. Awada FG, Sibandaa P, Motsab SS (2010) On the linear stability analysis of a Maxwell fluid with double-diffusive convection, Appl. Math. Mod., 34: 3509–3517.

8. Hayat T, Sajjad R, Abbas Z, Sajid M, Hendi AA (2011) Radiation effects on MHD flow of Maxwell fluid in a channel with porous medium, Int. J. Heat Mass transfer, 54: 854–862.

9. Zheng L, Li C, Zhang X, Gao Y (2011) Exact solutions for the unsteady rotating flows of a generalized Maxwell fluid with oscillating pressure gradient between coaxial cylinders, Comp. Math. Appl., 62: 1106–1115.

10. Jamil M, Fetecau C (2010) Helical flows of Maxwell fluid between coaxial cylinders with given shear stresses on the boundary, Nonlinear Analysis: Real World Applications, 11: 4302–4311.

11. Fetecau C, Athar M, Fetecau C (2009) Unsteady flow of a generalized Maxwell fluid with fractional derivative due to a constantly accelerating plate, Comp. Math. Appl., 57: 596–603.

12. Zierep J, Fetecau C (2007) Energetic balance for the Rayleigh-Stokes problem of a Maxwell fluid, Int. J. Engng. Sci., 45: 617–627.

13. Hayat T, Awais M, Qasim M, Hendi AA (2011) Effects of mass transfer on the stagnation point flow of an upper-convected Maxwell (UCM) fluid, Int. J. Heat Mass Transfer, 54: 3777–3782.

14. Hayat T, Awais M (2011) Three-dimensional flow of upper-convected Maxwell (UCM) fluid, Int. J. Numer. Meth. Fluids, 66: 875–884.

15. Awais M, Hayat T, Alsaedi A, Asghar S (2014) Time-dependent three-dimensional boundary layer flow of a Maxwell fluid, Comp. Fluids, 91: 21–27.

16. Sadeghy K, Khabazi N, Taghavi SM (2007) Magnetohydrodynamic (MHD) flows of viscoelastic fluid in converging/diverging channels. Int. J. Engng. Sci., 45: 923–938.

17. Rashidi MM, Erfani E (2012) Analytical method for solving steady MHD convective and slip flow due to a rotating disk with viscous dissipation and Ohmic heating, Eng. Comm., 29: 562–579.

18. Rashidi MM, Kavyani N, Abelman S (2014) Investigation of entropy gegeration in MHD ansd slip flow over a rotating porous disk with variable properties, Int. J. Heat Mass Trans., 73: 892–917.

19. Hayat T, Ali S, Awais M, Alhuthali MS (2015), Newtonian heating in stagnation point flow of Burgers fluid, App. Math. Mech., 36: 61–68.

20. Ellahi R, Hussain F (2015) Simultaneous effects of MHD and partial slip on peristalsis flow of Jeffery fluid in a rectangular duct, J. Mag. Mag. Mat., 393: 284–292.

21. Turkyilmazoglu M (2014) MHD fluid flow and heat transfer due to a shrinking rotating disk, Comp. Fluids, 92: 51–56.

4
CHAPTER

SYMPLECTIC INTEGRATORS TO STOCHASTIC HAMILTONIAN DYNAMICAL SYSTEMS DERIVED FROM COMPOSITION METHODS

Tetsuya Misawa

INTRODUCTION

The theory of Hamiltonian dynamical systems is understood as one of the fundamental tools for the description of dynamical phenomena treated in physics, engineering, and economics. To investigate the behavior of such a system, numerical simulation is often carried out. Then, as one of the important subjects in the numerical approach, attention has been paid to how to construct numerical schemes preserving Hamiltonian dynamical structures, that is, "symplectic integration schemes" (Hairer et al. [1] and references therein). The symplectic schemes are superior to numerical realization of the phase flows of Hamiltonian dynamical systems on long time intervals, and hence one can apply them to various dynamical systems described in the framework of classical Hamiltonian mechanics.

As an extension of this topic, Milstein et al. [2] have proposed symplectic integrators for "stochastic Hamiltonian dynamical systems (e.g., Misawa [3])" which are governed by stochastic differential equations (SDEs) (Ikeda and Watanabe [4]). The stochastic Hamiltonian dynamical systems are viewed as "open" Hamiltonian systems perturbed by random fluctuations from an external world; from the viewpoint of finance, such systems may also be regarded as dynamical ones describing some risky assets. For the stochastic systems, they have formulated "symplectic schemes of Runge-Kutta type" which are analogous to those in deterministic cases. On the other hand, symplectic integrators for deterministic Hamiltonian systems are often produced through the so-called "composition methods (operator splitting methods)", which are formulated on the basis of Lie algebraic theory and Lie group theory (McLachlan and Quispel [5]). The method is outlined as follows:

Citation: Tetsuya Misawa, "Symplectic Integrators to Stochastic Hamiltonian Dynamical Systems Derived from Composition Methods," Mathematical Problems in Engineering, vol. 2010, Article ID 384937, 12 pages, 2010. doi:10.1155/2010/384937

Let X denote vector fields on some space with coordinates x, with flows exp(Tx), that is, the solutions of differential equations of the form ˙x(t) = X(x) are given in the form x(t=) exp(tX)(x(0)). Assume that one can write X = A + B in such a way that exp(tA) and exp(tB) can both be calculated explicitly. (More generally, this can be relaxed by approximations of the exponential maps). In the most elementary case, the method gives the approximation for x(t) through

$$\tilde{x}(t) = \exp(tA)\exp(tB)(x) = x(t) + O\left(t^2\right); \tag{1.1}$$

the last equality is obtained by using the Baker-Campbell-Hausdorff (BCH) formula, which is well known in Lie algebraic theory. The advantage of this method is that geometric properties of the true flow exp(tX) are often preserved. If X, A, and B are Hamiltonian vector fields, then both true flow and the approximating flow are symplectic. Therefore, it seems to be quite natural to derive symplectic integrators for stochastic Hamiltonian systems by using these methods.

The purpose of the present paper is to propose some symplectic numerical integration schemes for stochastic Hamiltonian dynamical systems. Then, on the basis of the author's results on composition methods for SDEs (Misawa [6]), the numerical approximation errors of the new stochastic schemes are estimated.

This paper is organized as follows. In Section 2, we first review the composition methods for SDEs and the theorem on error estimation to the schemes. In Section 3, some symplectic schemes by composition methods are proposed for stochastic Hamiltonian dynamical systems described by SDE with a one-dimensional Wiener process. The approximation error of numerical solutions through the schemes is also estimated. In Section 4, in order to examine the superiority of the new schemes, we compare the numerical solutions from the proposed schemes with those from the stochastic Euler-Maruyama scheme through an illustrative example. Some concluding remarks are given in the end of the section.

Finally, we would like to quote here another approach to composition methods for SDEs by K. Burrage and P. M. Burrage [7]. Furthermore, the other contributions related to our topic can also be found in the recent articles by Malham and Wiese [8] and Bou-Rabee and Owhadi [9].

COMPOSITION METHODS FOR NUMERICAL INTEGRATION OF SDES

Now, we start with a review of composition methods for numerical integration of SDEs proposed by Misawa [6]. Let (Ω, F, P) be a probability space. We consider an n-dimensional stochastic differential equation of the Stratonovich type associated with a one-dimensional Wiener process W as follows:

$$dS(t) = b(S(t))dt + g(S(t)) \circ dW(t), \tag{2.1}$$

where b and g are n-dimensional C_∞ vector functions, respectively. By using these functions, we define the differential operators X_0 and X_1 as

$$X_0 = \sum_{i=1}^{n} b^i \partial_i, \qquad X_1 = \sum_{i=1}^{n} g^i \partial_i, \tag{2.2}$$

where $\partial_i = \partial/\partial S_i$. According to Kunita [10], a solution of the SDE under an initial value S(0) = s can be formally represented in terms of an exponential map derived from a "stochastic" vector field Y_t as follows:

$$S(t) = (\exp Y_t)(s), \tag{2.3}$$

where $Y_t(\omega)$ is the vector field for each t and almost surely (a.s). $\omega \in \Omega$ given by Yt

$$Y_t = tX_0 + W(t)X_1 + \frac{1}{2}\left(\int_0^t u dW(u) - \int_0^t W(u)du\right)[X_0, X_1] + \sum_{J;3\leq|J|}\left\{\sum_{\Delta J}^{*} c_{\Delta J} W^{\Delta J}(t)\right\}X^J, \tag{2.4}$$

and $X^J = [\cdots[X_{j_1}, X_{j_2}]\cdots]X_{j_m}]$ $(J = (j_1, \ldots, j_m))$, where $[X_0, X_1]$ is a Lie bracket defined as $X_0X_1 - X_1X_0$. Here, $\Sigma_{\Delta J}^{*}$ is the sum for all single and double divided indices of a multi-index J, and $W^{\Delta J}(t)$ and $c_{\Delta J}$ are multiple Wiener-Stratonovich integrals on time interval [0, t] and the coefficients, respectively, which are determined through a rule for the divided indices of J. Moreover, |J| denotes the length of J. We note that 2.2 with 2.4 means that the solution $S(t, \omega)$ equals $\phi(1, s, \omega)$ a.s., where $\phi(\tau, s, \omega)$ the solution of the ordinary differential equation

$$\frac{d\phi(\tau)}{d\tau} = Y_t(\omega)(\phi(\tau)), \quad \phi(0) = s, \tag{2.5}$$

regarding t and ω as parameters

Now, we proceed to a numerical integration of the stochastic equation (2.1) on the discretized time series by using the representation of solutions to SDEs mentioned above. It adopts a uniform discretization of the time interval [0, T] with step size

$$\Delta t = \frac{T}{N} \tag{2.6}$$

for a fixed natural number N. Let $t_n = n\Delta t$ $(n = 0, 1, 2, \ldots, N)$ be the nth step point. Then, for all n ∈ {0,...,N}, we abbreviate $S_n = S(t_n)$ and set $S_0 = S(0) = s_0$. Moreover, we use ΔW_n for $n = 0, 1, \ldots, N$ to denote the increments $W(t_{n+1}) - W(t_n)$; ; they are independent Gaussian random variables with mean 0 and variance Δt, that is, N(0, Δt)-distributed random variables.

On account of (2.3), we find a series of numerical solutions S_n $(n = 0, 1, \ldots, N)$ to SDE (2.1) by using

$$S_{n+1} = \exp(Y_{n\Delta t})(S_n) \quad (n = 0, 1, 2, \ldots, N-1), \tag{2.7}$$

formally, where $Y_{n\Delta t}$ is a vector field derived by replacing all the multiple Wiener integrals on time [0, t] by those on time interval Δt in (2.4). In the theory of ordinary differential equations, $(Y_{n\Delta t})\cdot(\,)$ is often called the time- Δtmap or exponential map. However, it is usually difficult to find the explicit form of the exponential map, and hence, we need to build an approximation for (2.7). Hence, we formulate a new stochastic numerical scheme as the following two procedures, which are composed of the truncation of the vector field (2.4) and a composition method (or operator-splitting method) applied to the exponential map derived from the truncated vector field.

Procedure. We replace the vector field $Y_{n\Delta t}$ in (2.7) by a "truncated" vector field $\widehat{Y}_{n\Delta t}$ with a truncation order which is defined as

$$\hat{Y}_{n\Delta t} = \sum_{J;1\leq|J|\leq\gamma} H^J(\Delta t)X^J, \tag{2.8}$$

where H^J (Δt) is a polynomial function of multiple Wiener integrals for the vector field XJ with a multi-index J. Then, we define a numerical sequence $(\hat{S}_n)_{n=0}^{N}$ through

$$\hat{S}_{n+1} = \exp\left(\hat{Y}_{n\Delta t}\right)\left(\hat{S}_n\right) \quad (n = 0, 1, \ldots, N-1), \tag{2.9}$$

Where $\hat{S}_0 = S(0) = s_0$.

Procedure. For $\hat{S}_{n+1} = \exp(\hat{Y}_{n\Delta t})(\hat{S}_n)$, we apply a "composition method" in a way analogous to that in the theory of ordinary differential equations. Suppose that the vector field $\widehat{Y}_{n\Delta t}$ is of the form

$$\widehat{Y}_{n\Delta t} = A_{n\Delta t} + B_{n\Delta t}, \tag{2.10}$$

where $\exp(A_{n\Delta t})$ and $\exp(B_{n\Delta t})$ can both be explicitly calculated through (2.15). Then an approximation to the exponential map $\widehat{Y}_{n\Delta t}$ is given by $\exp(A_{n\Delta t})\exp(B_{n\Delta t})$. Hence, the sequence of $(\hat{S}_n)_{n=0}^{N}$ in Procedure 1 is approximated by

$$\tilde{S}_{n+1} = \exp(A_{n\Delta t})\exp(B_{n\Delta t})\left(\tilde{S}_n\right) \quad (n = 0, 1, \ldots, N-1), \tag{2.11}$$

Where $\hat{S}_0 = S(0) = s_0$.

We regard $(\tilde{S}_n)_{n=0}^{N}$ as a numerical approximation to the exact discretized solutions $(S_n)_{n=C}^{N}$.

We turn to estimate the approximation error of the numerical scheme described previously. In stochastic numerics, it is usually measured through "global errors in the mean-square sense" for the scheme, and hence we first review the following definition (Gard [11], Kloeden and Platen [12]).

Definition 2.1. Suppose that $S(t)$ and $(\tilde{S}_n)_{n=0}^{N}$ are an exact solution and the numerical approximation solutions to SDE (2.1), respectively. Moreover, let $E_{0,\xi}$ be the expectation conditioned on starting at ξ at "initial time" τ. Then the global error order λ is defined by

$$E_{0,s}\left[\left|\bar{S}_N - S_N\right|^2\right] = O\left((\Delta t)^{2\lambda}\right) \quad (\Delta t \downarrow 0), \tag{2.12}$$

where $S_N = S(N\Delta t)$, $N\Delta t = T$, and $|\cdot|$ denotes the Euclidean norm on the space R^d. Here, the condition in the expectation (2.12) means $S_0 = \bar{S}_0 = s$.

It is obvious that the accuracy of the numerical scheme improves with increasing global order

We remark that in the framework of "global error order" defined by Saito and Mitsui [13], the global order for the numerical solutions $\bar{S}_n$ satisfying (2.12) is given by 2λ.

It is evident that such an error estimation order depends on a truncated order of $Y_{\Delta t}$ and the way of decomposition of the truncated vector field $\widehat{Y}_{\Delta t}$. In [6], Misawa gives a way to calculate the global error order of stochastic numerical schemes determined through the two procedures mentioned above.

Theorem 2.2 (see [6, Theorem 3.2]). Let γ, α, and β be a truncation order of $\widehat{Y}_{n\Delta t}$ described in Procedure 1, the least values of "mean square orders" of multiple Wiener integrals which are included in coefficients of the operators $A_{n\Delta t}$ and $B_{n\Delta t}$, respectively. The mean square order s of a multiple Wiener integral Δt on time interval mentioned above is defined as $E[|I(\Delta t)|^2] = O((\Delta t)^s)$ $(\Delta t \downarrow 0)$. Then

$$E\left[\left|\bar{S}_N - S_N\right|^2 \mid X_0 = \bar{X}_0 = x\right] = O\left((\Delta t)^{\delta-1}\right) \quad (\Delta t \downarrow 0) \tag{2.13}$$

holds, where δ = min(γ + 2, α + β). Therefore, the global error order of stochastic numerical schemes provided by Procedures 1 and 2 is equal to (δ – 1)/2. As

As a preparation for later discussions, we review some examples of our schemes and the global error orders given in [6].

Example 2.3. Suppose that a truncated vector field $\widehat{Y}_{n\Delta t}$ in Procedure 1 is given by

$$\widehat{Y}_{n\Delta t} = \Delta t X_0 + \Delta W_n X_1. \tag{2.14}$$

We further set $A_{n\Delta t} = \Delta t X_0$ and $B_{n\Delta t} = \Delta W_n X_1$ in the decomposition in Procedure2 and assume that the explicit forms of both exponential maps for them are obtained. Then, the numerical scheme (2.11) can be put into the following form.

Scheme 1. One has

$$\tilde{S}_{n+1} = \exp(\Delta t X_0)\exp(\Delta W_n X_1)\left(\tilde{S}_n\right). \tag{2.15}$$

In this case, γ, α and β are equal to 1, 2 and 1, respectively, and therefore the above theorem provides the global order of Scheme 1 as 1.

Example 2.4: For $\widehat{Y}_n(\Delta t)$ in Example 2.3, we set $A_{n\Delta t} = \Delta t X_0^A + \Delta W_n X_1^A$ and $B_{n\Delta t} = \Delta t X_0^B + \Delta W_n X_1^B$, where $X_0 = X_0^A + X_0^B$ and $X_1 = X_1^A + X_1^B$. We assume that $[X_1^A, X_1^B] \neq 0$ and that the explicit forms of both exponential maps for them are obtained. Then, the numerical scheme (2.11) can be put into the following form.

Scheme 2. One has

$$\tilde{S}_{n+1} = \exp\left(\Delta t X_0^A + \Delta W_n X_1^A\right) \exp\left(\Delta t X_0^B + \Delta W_n X_1^B\right)\left(\tilde{S}_n\right). \tag{2.16}$$

In this case, α and β are both equal to 1; hence the global order of the scheme (2.10) for the above $A_{n\Delta t}$ and $B_{n\Delta t}$ becomes 0.5. Note that this order is equal to that of the Euler-Maruyama scheme for SDEs which is the most famous scheme in stochastic numerics.

Remark 2.5. By further manipulating the BCH formula to eliminate higher-order terms, we can obtain various schemes which give higher-order approximations to the exponential map. For example, the scheme corresponding to "leapfrog", which is well known in deterministic numerical analysis, is given by

$$\exp((\Delta t)(X + Y)) = \exp\left(\frac{\Delta t Y}{2}\right) \exp(\Delta t X) \exp\left(\frac{\Delta t Y}{2}\right) + O\left((\Delta t)^3\right). \tag{2.17}$$

In a way analogous to that in (2.11), we define a stochastic leapfrog scheme as follows:

$$\tilde{S}_{n+1} = \exp\left(\frac{B_{n\Delta t}}{2}\right) \exp(A_{n\Delta t}) \exp\left(\frac{B_{n\Delta t}}{2}\right)\left(\tilde{S}_n\right) \quad (n = 0, 1, \ldots, N-1). \tag{2.18}$$

Then, using the BCH formula, we can produce numerical schemes having a better error order than that of scheme (2.11). In detail, see Remark 3.3 and Example 3.2 in [6]

SYMPLECTIC INTEGRATORS OF STOCHASTIC HAMILTONIAN DYNAMICAL SYSTEMS

Now, we proceed to derive symplectic integrators for stochastic Hamiltonian dynamical systems in terms of composition methods mentioned in the previous section. We start with the definition of stochastic Hamiltonian dynamical systems defined in [3]. Consider the following 2l-dimensional stochastic dynamical system with one Wiener process:

$$d\begin{pmatrix} x^i(t) \\ x^{\ell+i}(t) \end{pmatrix} = \begin{pmatrix} \partial_{\ell+i} H_0(x(t)) \\ -\partial_i H_0(x(t)) \end{pmatrix} dt + \begin{pmatrix} \partial_{\ell+i} H_1(x(t)) \\ -\partial_i H_1(x(t)) \end{pmatrix} \circ dW(t) \quad (i = 1, \ldots, \ell), \tag{3.1}$$

Where $x = (x^k)_{k=1}^{2\ell}$ and $\partial_j = \partial/\partial x^j$ $(j = 1, 2, \ldots, 2\ell)$, respectively. In (3.1), $H_\alpha(x)$ $(\alpha = 0, 1)$ are smooth scalar functions on $R^{2\ell}$. Formally, one may regard this as a Hamiltonian dynamical system

$$\frac{d}{dt}\begin{pmatrix} x^i \\ x^{\ell+i} \end{pmatrix} = \begin{pmatrix} \partial_{\ell+i}\widehat{H}(x) \\ -\partial_i\widehat{H}(x) \end{pmatrix} \quad (i = 1, \ldots, \ell) \tag{3.2}$$

with a "randomized" Hamiltonian $\widehat{H}$ given by

$$\widehat{H} = H_0 + H_1\gamma_t, \tag{3.3}$$

where γt is a one-dimensional Gaussian white noise. Hence, we call Hence, we call (3.1) and Hα (α = 0, 1) an l -dimensional stochastic Hamiltonian dynamical system and the Hamiltonian, respectively. In the following, for simplicity, we set $\ell = 1$ and denote $x^1(t)$ and $x^2(t)$ by $q(t)$ and $p(t)$, respectively, that is, we treat the following stochastic Hamiltonian dynamical system:

$$d\begin{pmatrix} q(t) \\ p(t) \end{pmatrix} = \begin{pmatrix} \frac{\partial}{\partial p}H_0(p(t), q(t)) \\ -\frac{\partial}{\partial q}H_0(p(t), q(t)) \end{pmatrix} dt + \begin{pmatrix} \frac{\partial}{\partial p}H_1(p(t), q(t)) \\ -\frac{\partial}{\partial q}H_1(p(t), q(t)) \end{pmatrix} \circ dW(t). \tag{3.4}$$

Let $t_n = n\Delta t \ (n = 0, 1, 2, \ldots, N)$ be the nth step-point in a uniform discretization of the time interval $[0, T]$ with step size $\Delta t = T/N$ for fixed natural number N. Suppose that $\{(p_n, q_n)\}_{n=0}^{N}$ is a series of the numerical solutions for the above Hamiltonian system derived by some numerical scheme on the above time discretization. In the same manner as that in the case of the numerical scheme on deterministic Hamiltonian systems, we call the numerical scheme for stochastic Hamiltonian dynamical systems "symplectic", which produces numerical solutions satisfying

$$\det\left(\frac{\partial(p_{n+1}, q_{n+1})}{\partial(p_n, q_n)}\right) = 1 \tag{3.5}$$

on any step point n. In consideration of the advantages of using symplectic schemes for usual deterministic systems, we also may expect to obtain more stable numerical orbits by using these schemes than by using ordinary numerical schemes, for example, the Euler scheme.

Now, we produce the symplectic schemes by using the composition methods mentioned in Section 2. First, as an example of symplectic scheme on the basis of Scheme 1, we can provide the following scheme.

Scheme 3. One has

$$\begin{pmatrix} \widetilde{q}_{n+1} \\ \widetilde{p}_{n+1} \end{pmatrix} = \exp(A_{n\Delta t})\exp(B_{n\Delta t})\begin{pmatrix} q_n \\ p_n \end{pmatrix}, \tag{3.6}$$

where a truncated vector field $\widehat{Y}_{n\Delta t}$ shown by (2.14) and the splitting vecter fields $A_{n\Delta t}$ and $B_{n\Delta t}$ are given by

$$\widehat{Y}_{n\Delta t} = \Delta t X_0 + \Delta W_n X_1 = (\partial_p H_0 \partial_q - \partial_q H_0 \partial_p)\Delta t + (\partial_p H_1 \partial_q - \partial_q H_1 \partial_p)\Delta W_n, \tag{3.7}$$

$$A_{n\Delta t} = \Delta t X_0 = (\partial_p H_0 \partial_q - \partial_q H_0 \partial_p)\Delta t, \tag{3.8}$$

$$B_{n\Delta t} = \Delta W_n X_1 = (\partial_p H_1 \partial_q - \partial_q H_1 \partial_p)\Delta W_n, \tag{3.9}$$

respectively.

It is easy to see that the above composition scheme is symplectic, since for each fixed stochastic parameter ω, $A_{n\Delta t}$, and $B_{n\Delta t}$ are Hamiltonian vector fields with the Hamiltonians $H_0\Delta t$ and $H_1\Delta W_n$ on time interval $[t_n, t_{n+1}]$, respectively, and hence the composition of their exponential maps is symplectic (McLachlan and Quispel [5]). Moreover, as mentioned in Section 2, the global error order of this scheme is 1.

$$\begin{pmatrix} \tilde{q}_{n+1} \\ \tilde{p}_{n+1} \end{pmatrix} = \exp(A_{n\Delta t})\exp(B_{n\Delta t})\begin{pmatrix} q_n \\ p_n \end{pmatrix}, \tag{3.10}$$

where for the same truncated vector field $\widehat{Y}_{n\Delta t}$ as Scheme 3, we choose $A_{n\Delta t}$ and $B_{n\Delta t}$ as follows:

$$A_{n\Delta t} = \Delta t X_0^A + \Delta W_n X_1^A = \Delta t\left(\partial_p H_0^A \partial_q - \partial_q H_0^A \partial_p\right) + \Delta W_n\left(\partial_p H_1^A \partial_q - \partial_q H_1^a \partial_p\right), \tag{3.11}$$

$$B_{n\Delta t} = \Delta t X_0^B + \Delta W_n X_1^B = \Delta t\left(\partial_p H_0^B \partial_q - \partial_q H_0^B \partial_p\right) + \Delta W_n\left(\partial_p H_1^B \partial_q - \partial_q H_1^B \partial_p\right), \tag{3.12}$$

Under the conditions $H_0^A + H_0^B = H_0$ and $H_1^A + H_1^B = H_1$.

We can also prove that this scheme is simplistic, since for each fixed stochastic parameter ω, $A_{n\Delta t}$, and $B_{n\Delta t}$ are also Hamiltonian vector fields with the Hamiltonians $\Delta t H_0^A + \Delta W_n H_1^A$ and $\Delta t H_0^B + \Delta W_n H_1^B$ on time interval $[t_n, t_{n+1}]$. We note that the global error order of this scheme is 0.5 as mentioned in Section 2.

Remark 3.1. If concrete Hamiltonians H_0 and H_1 are given, one can also check the condition (3.5) for each scheme mentioned above by direct calculation.

ILLUSTRATIVE EXAMPLES

In this section, as an illustrative example of Schemes 3 and 4, we will show the numerical solutions to a stochastic harmonic oscillator system by the symplectic schemes and compare them with the numerical results by the Euler scheme.

Let us consider the following stochastic Hamiltonians dynamical system (3.4) with the Hamiltonian $H_0 = (1/2)p^2 + (1/2)q^2$ and $H_1 = \sigma H_0$:

$$d\begin{pmatrix} q(t) \\ p(t) \end{pmatrix} = \begin{pmatrix} p(t) \\ -q(t) \end{pmatrix} dt + \sigma\begin{pmatrix} p(t) \\ -q(t) \end{pmatrix} \circ dW(t), \tag{4.1}$$

where σ is a constant It is evident that this is a stochastic system with the "conserved quantity" H0 [3, 6]; that is, the solution $(q(t), p(t))$ randomly moves on a circle determined by $H_0(q(t), p(t)) = H_0(q(0), p(0)) = (1/2)(q(0)^2 + p(0)^2)$ on p-q plane. We call such a system a stochastic harmonic oscillator system.

Now, in order to investigate the advantage of our symplectic schemes, we will produce numerical solutions to this system by the new schemes and the "Euler scheme" [11–13]. Then, by straightforward calculation of the definitions of the schemes together with the exponential maps (2.5), the Euler scheme, Schemes 3 and4 under setting H_0^A, H_1^A, H_0^B, and H_1^B as $H_0^A = (1/\sigma)H_1^A = (1/2)p^2$ and $H_0^B = (1/\sigma)H_1^B = (1/2)q^2$ for this system are given as follows.

$$\begin{pmatrix} q_{n+1} \\ p_{n+1} \end{pmatrix} = \begin{pmatrix} q_n \\ p_n \end{pmatrix} + \begin{pmatrix} p_n - \frac{1}{2}q_n \\ -q_n - \frac{1}{2}p_n \end{pmatrix} \Delta t + \begin{pmatrix} p_n \\ -q_n \end{pmatrix} \sigma \Delta W_n. \tag{4.2}$$

Scheme 5 (Scheme 3 for (4.1)). One has

$$\begin{pmatrix} q_{n+1} \\ p_{n+1} \end{pmatrix} = \begin{pmatrix} \cos(\Delta t) & -\sin(\Delta t) \\ \sin(\Delta t) & \cos(\Delta t) \end{pmatrix} \begin{pmatrix} \cos(\sigma\Delta W_n) & -\sin(\sigma\Delta W_n) \\ \sin(\sigma\Delta W_n) & \cos(\sigma\Delta W_n) \end{pmatrix} \begin{pmatrix} q_n \\ p_n \end{pmatrix} \tag{4.3}$$

Scheme 6 (Scheme 4 for (4.1)). One has

$$\begin{pmatrix} q_{n+1} \\ p_{n+1} \end{pmatrix} = \begin{pmatrix} \hat{p}_n(\Delta t + \sigma\Delta W_n) + \hat{q}_n \\ \hat{p}_n \end{pmatrix}, \tag{4.4}$$

where

$$\begin{pmatrix} \hat{q}_n \\ \hat{p}_n \end{pmatrix} = \begin{pmatrix} q_n \\ -q_n(\Delta t + \sigma\Delta W_n) + p_n \end{pmatrix}. \tag{4.5}$$

Here we note that we can directly examine the symplectic condition (3.5) which holds for Schemes 5 and 6mentioned above.

In these schemes, Δ_{W_n} are numerically realized in terms of the independent N(0, 1) random numbers γ_n as [12]

$$\Delta W_n = \gamma_n \sqrt{\Delta t}. \tag{4.6}$$

Moreover, we set σ, Δt, and N as σ = 0.2, Δt = 0.05, and N = 1000, respectively, and we choose the initial values of p_0 and q_0 as 1 and 0, respectively.

Figure 1 denotes the numerical time series of "p" by these schemes, and Figures 2(a)–2(c) display sample paths of numerical solutions from Euler scheme andSchemes 6 and 5, respectively. Here, as mentioned, it should be remembered that the exact solution to the original stochastic system should run on the circle $H_0(q(t), p(t)) = (1/2)(q(0)^2 + p(0)^2) = 1$, , that is, a unit circle on the p-q plane. Figures 1 and 2(a), however, indicate that a series of the numerical solutions by the Euler scheme gradually goes far from the circle. In contrast, Figures 1, 2(b), and 2(c) show that the numerical solutions by our symplecticSchemes 5 and 6move around the circle stably, and especially, a series of the solution byScheme 5runs completely on the circle. Thus, as in the deterministic case, the new stochastic symplectic schemes also numerically preserve the Hamiltonian dynamical structure; this is the superiority of the new stochastic schemes over the ordinary stochastic schemes

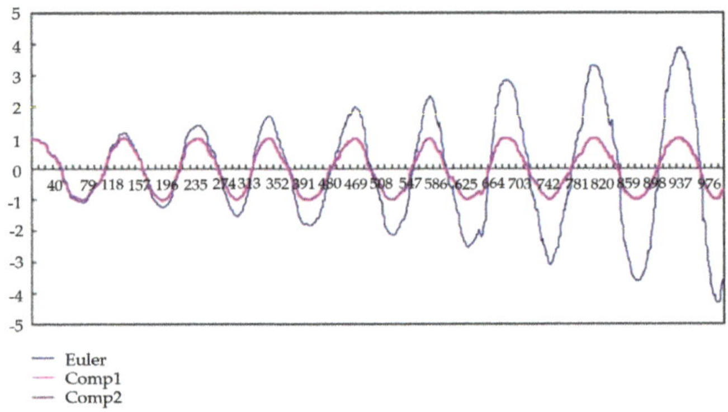

Figure 1: Numerical sample paths of "p" of a stochastic harmonic oscillator system (SHOS) via Euler scheme, Composition 1 (Scheme 6) and Composition 2 (Scheme 5) (dt = 0.05, sigma = 0.2).

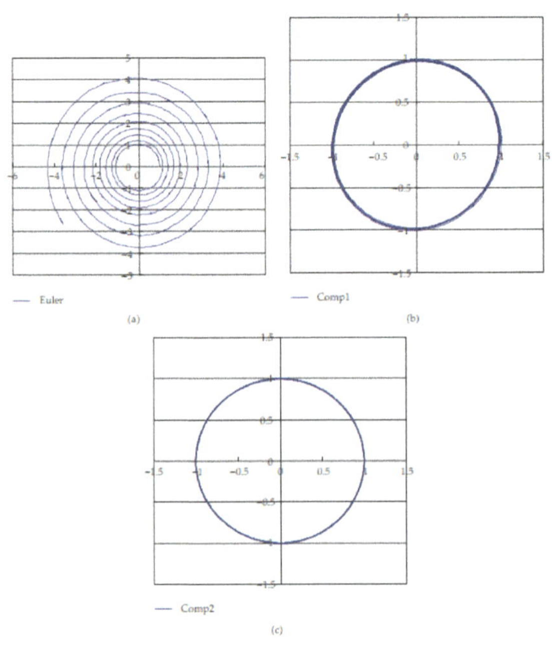

Figure 2: (a) A numerical sample path of SHOS in phase space via Euler scheme. (b) A numerical sample path of SHOS in phase space via Composition 1 (Scheme 6). (c) A numerical sample path of SHOS in phase space via Composition 2 (Scheme 5).

Finally, we give some concluding remarks on our method.

(1) In this paper, we have treated the stochastic Hamiltonian dynamical systems with a one-dimensional Wiener process. However, it often happens that the error order of a numerical method collapses, if there is more than one Wiener process. Hence, it would be important to investigate the error orders of our schemes in the case of stochastic Hamiltonian dynamical systems with a "multidimensional Wiener process". Therefore, we should be able to improve our composition methods and offer new schemes with high order for SDEs with a multi-dimensional Wiener process.

(2) Moreover, we have referred only to a single composition method to SDEs so far. It seems that one would readily be able to extend the method to a multi-composition one, since such compositions are obtained by iterations of single compositions. Using the multicompositions, we may easily produce a symplectic scheme for the more complicated stochastic Hamiltonian systems.

In future papers, the topics mentioned above will be reported.

ACKNOWLEDGMENT

This work was supported by Grants-in-Aid for Scientific Research nos. 18540134 and 21540140 from the Japan Society for the Promotion of Science.

REFERENCES

1. E. Hairer, C. Lubich, and G. Wanner, Geometric Numerical Integration, vol. 31 of Springer Series in Computational Mathematics, Springer, Berlin, Germany, 2nd edition, 2006, Structure-Preserving Algorithms for Ordinary Differential Equation.
2. G. N. Milstein, Yu. M. Repin, and M. V. Tretyakov, "Numerical methods for stochastic systems preserving symplectic structure," SIAM Journal on Numerical Analysis, vol. 40, no. 4, pp. 1583–1604, 2002.
3. T. Misawa, "Conserved quantities and symmetries related to stochastic dynamical systems," Annals of the Institute of Statistical Mathematics, vol. 51, no. 4, pp. 779–802, 1999.
4. N. Ikeda and S. Watanabe, Stochastic Differential Equations and Diffusion Processes, vol. 24 of North-Holland Mathematical Library, North-Holland, Amsterdam, The Netherlands, 2nd edition, 1989.
5. R. I. McLachlan and G. R. W. Quispel, "Splitting methods," Acta Numerica, vol. 11, pp. 341–434, 2002.
6. T. Misawa, "A Lie algebraic approach to numerical integration of stochastic differential equations," SIAM Journal on Scientific Computing, vol. 23, no. 3, pp. 866–890, 2001.
7. K. Burrage and P. M. Burrage, "High strong order methods for non-commutative stochastic ordinary differential equation systems and the Magnus formula," Physica D, vol. 133, no. 1–4, pp. 34–48, 1999.

8. S. J. A. Malham and A. Wiese, "Stochastic Lie group integrators," SIAM Journal on Scientific Computing, vol. 30, no. 2, pp. 597–617, 2008.
9. N. Bou-Rabee and H. Owhadi, "Stochastic variational integrators," IMA Journal of Numerical Analysis, vol. 29, no. 2, pp. 421–443, 2009.
10. H. Kunita, "On the representation of solutions of stochastic differential equations," in Séminaire de Probabilités XIV 1978/79, vol. 784 of Lecture Notes in Mathematics, pp. 282–304, Springer, Berlin, Germany, 1980.
11. T. C. Gard, Introduction to Stochastic Differential Equations, vol. 114 of Monographs and Textbooks in Pure and Applied Mathematics, Marcel Dekker, New York, NY, USA, 1988.
12. P. E. Kloeden and E. Platen, Numerical Solution of Stochastic Differential Equations, vol. 23 of Applications of Mathematics (New York), Springer, Berlin, Germany, 1992.
13. Y. Saito and T. Mitsui, "Simulation of stochastic differential equations," Annals of the Institute of Statistical Mathematics, vol. 45, no. 3, pp. 419–432, 1993.

5
CHAPTER

SIMILARITY SOLUTIONS OF MHD MIXED CONVECTION FLOW WITH VARIABLE REACTIVE INDEX, MAGNETIC FIELD, AND VELOCITY SLIP NEAR A MOVING HORIZONTAL PLATE: A GROUP THEORY APPROACH

W. A. Khan, Md. Jashim Uddin, and A. I. Md. Ismail

INTRODUCTION

A convection situation involving both free and forced convection is known as mixed convection and has been an important topic because of its application in electronic equipment cooled by a fan and flows in the ocean and in the atmosphere ([1, 2]). In mixed convection flows, the forced and the free convection effects are of comparable magnitude. Thus, mixed convection occurs if the effect of buoyancy forces on a forced flow or vice versa is significant. The laminar mixed convection takes place in various applications in thermal engineering and science and has drawn attention from researchers in the past decades. Examples of these applications include atmospheric boundary layer flows, solar energy systems, boilers, compact heat exchangers, and cooling of electronic devices [3, 4]. Many studies exist for the mixed convection boundary layer flow about vertical/inclined/horizontal plate and wedge surfaces immersed in a viscous fluid. The details of mixed convection can be found in books by Chen and Armaly [5], Jaluria [6], Gebhart et al. [7], Bejan [8], Pop and Ingham [9] and in the articles in the literatures by Lloyd and Sparrow [10], Schneider [11], Wilks [12], Yao [13], Ramachandran et al. [14], Moulic and Yao, [15], Daskalakis et al. [16], Fan et al. [17], Magyari et al. [18], Hassanien et al. [19], Aydin and Kaya [20], Ishak [21], Moulic and Yao [22], Datta et al. [23], Ishak et al. [24], Magyari [25].

In all these studies, conventional no slip boundary conditions were applied at the surface. However, fluid flows in micro/nanoscale such as micronozzles, micropumps,

Citation: W. A. Khan, Md. Jashim Uddin, and A. I. Md. Ismail, "Similarity Solutions of MHD Mixed Convection Flow with Variable Reactive Index, Magnetic Field, and Velocity Slip Near a Moving Horizontal Plate: A Group Theory Approach," Mathematical Problems in Engineering, vol. 2012, Article ID 183029, 15 pages, 2012. doi:10.1155/2012/183029 microturbines, microheat exchangers, microvalves, turbines, sensors, and actuators are important for micro and nanoscience and the conventional no slip boundary condition at the solid fluid interface must be replaced with the slip condition [26]. The slip condition states that the tangential component of the velocity at the surface is equal to the velocity gradient normal to the surface ([27, 28]). Slips may occur on a stationary/ moving boundary when the fluid is in particulate form such as in the form of emulsions, suspensions, foams, and polymer solutions [29, 30]. An excellent review for the fundamentals, theory, and applications of these microscale systems may be found in the text by Duncan and Peterson [31] Gad-el-Hak [32]. Slips effect on flow field has been studied by various authors; Fang and Lee [33], Martin and Boyd [34], Matthews and Hill [35], Wang [36]. Cao and Baker [37] studied laminar mixed convection over an isothermal vertical plate with first-order momentum and thermal discontinuities at the wall. The hydrodynamic and thermal slip flow boundary layer over a flat plate with constant heat flux boundary condition has been investigated by Hak [27] who concluded that as the slip parameter increases, the slip velocity increases and the wall shear stress decreases.

Due to the nonlinearities of the governing equations, no analytical solution is available and the nonlinear equations usually solved numerically subject to initial/ boundary conditions. The solution of differential equations with/without initial/ boundary conditions is presented by different authors such as Ames [38], Boyce [39], Na [40], Zauderer [41], Seshadri and Na [42], Kincaid and Cheney [43], Strauss [44], Aziz [45], and White and Subramanian [46].

We study combined convective Newtonian fluid flow adjacent to a moving horizontal permeable flat plate with higher order chemical reaction taking into account velocity slip and convective surface boundary conditions. The governing boundary layer equations are converted into a two point boundary value problem using similarity transformations developed by group method and the transformed equations are solved by means of an implicit finite difference numerical method. Graphs are plotted for velocity, temperature and concentration profiles as well as the skin friction factor, the rate of heat transfer and the rate of mass transfer to exhibits the influences of the parameters involving in our problem.

MATHEMATICAL FORMULATIONS OF THE PROBLEM

Consider a continuous moving permeable horizontal flat plate which moves with a nonuniform velocity U(x) as shown in Figure 1. A variable magnetic field of strength B(x) is applied to the normal direction of the plate. The bottom surface of the plate is heated by convection from a hot fluid of temperature $T_f(x)$ which provides a heat transfer coefficient h_f (x). The viscous dissipation in the energy equation is neglected. Assume that fluid properties are constant accepts in buoyancy term. We will further assume the magnetic Reynolds number; the electric field owing to the polarization of charges and Hall effects are negligible. Under the forgoing assumptions the governing equations in dimensional form can be written as [47, 48]

$$\frac{\partial u}{\partial x}+\frac{\partial v}{\partial y}=0,$$

$$u\frac{\partial u}{\partial x}+v\frac{\partial u}{\partial y}=-\frac{1}{\rho}\frac{\partial p}{\partial x}+\nu\frac{\partial^2 u}{\partial y^2}-\frac{\sigma B^2(x)}{\rho}u$$

$$-\frac{1}{\rho}\frac{\partial p}{\partial y}+g\,\beta_T(T-T_\infty)+g\,\beta_C(C-C_\infty)=0,$$

$$u\frac{\partial T}{\partial x}+v\frac{\partial T}{\partial y}=\alpha\frac{\partial^2 T}{\partial y^2},$$

$$u\frac{\partial C}{\partial x}+v\frac{\partial C}{\partial y}=D\frac{\partial^2 C}{\partial y^2}-k_0(C-C_\infty)^n. \tag{2.1}$$

The boundary conditions are [49]

$$u=U(x)+N_1(x)\nu\frac{\partial u}{\partial y},\qquad v=-v_w(x),\qquad -k\frac{\partial T}{\partial y}=h_f(x)\left[T_f(x)-T_w\right],$$

$$C=C_w(x)\quad \text{at } y=0,$$

$$u\longrightarrow 0,\qquad v\longrightarrow 0,\qquad T\longrightarrow T_\infty,\qquad C\longrightarrow C_\infty\quad \text{as } y\longrightarrow\infty, \tag{2.2}$$

where $v_w(x)$: velocity normal to the plate, ν: coefficient of kinematic viscosity, ρ: density of the fluid, σ: electric conductivity, p: pressure, k: thermal conductivity, α: thermal diffusivity, D: mass diffusivity of species in fluid, βT : volumetric thermal coefficient, β_C: volumetric concentration coefficient, g: acceleration due to gravity, k0: reaction rate, n: order of chemical reaction, and $N_1(x)$: velocity slip factor with dimension (velocity) $^{-1}$.

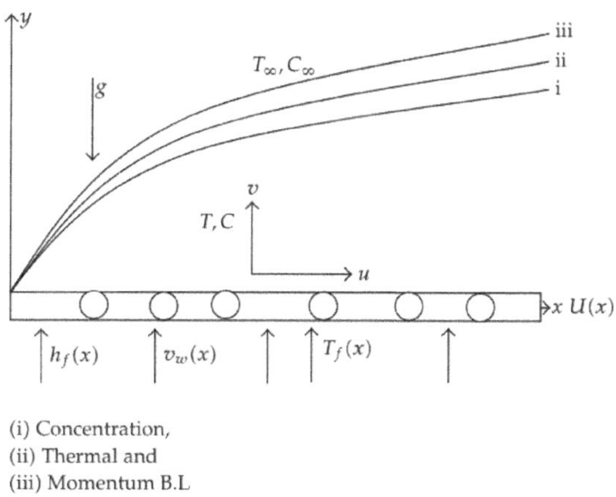

Figure 1: Physical configuration and coordinate system the problem.

Introducing stream function ψ, dimensionless temperature function θ and concentration function φ are defined by

$$u=\frac{\partial\psi}{\partial y},\qquad v=-\frac{\partial\psi}{\partial x},\qquad \theta=\frac{T-T_\infty}{T_f-T_\infty},\qquad \phi=\frac{C-C_\infty}{C_w-C_\infty}. \tag{2.3}$$

Using (2.3) we get from (2.1)

$$\frac{\partial\psi}{\partial y}\frac{\partial^2\psi}{\partial x\partial y}-\frac{\partial\psi}{\partial x}\frac{\partial^2\psi}{\partial y^2}=-\frac{1}{\rho}\frac{\partial p}{\partial x}+\nu\frac{\partial^3\psi}{\partial y^3}-\frac{\sigma B^2(x)}{\rho}\frac{\partial\psi}{\partial y}, \tag{2.4}$$

$$-\frac{1}{\rho}\frac{\partial p}{\partial y}+g\beta_T\theta\Delta T+\beta_C\phi\Delta C=0, \tag{2.5}$$

$$\frac{\partial\psi}{\partial y}\frac{\partial}{\partial x}(\theta\Delta T)-\frac{\partial\psi}{\partial x}\frac{\partial}{\partial y}(\theta\Delta T)=\alpha\frac{\partial^2}{\partial y^2}(\theta\Delta T), \tag{2.6}$$

$$\frac{\partial\psi}{\partial y}\frac{\partial}{\partial x}(\phi\Delta C)-\frac{\partial\psi}{\partial x}\frac{\partial}{\partial y}(\phi\Delta C)=D\frac{\partial^2}{\partial y^2}(\phi\Delta C)-k_0(\phi\Delta C)^n. \tag{2.7}$$

Eliminating the pressure gradient terms from (2.4) and (2.5) by cross-differentiation, we get

$$\begin{aligned}
&\nu\frac{\partial^4\psi}{\partial y^4}+\frac{\partial\psi}{\partial x}\frac{\partial^3\psi}{\partial y^3}-\frac{\partial\psi}{\partial y}\frac{\partial^3\psi}{\partial x\partial y^2}-g\beta_T(\Delta T\theta)_x-g\beta_T(\Delta C\phi)_x-\frac{\sigma B^2(x)}{\rho}\psi_{yy}=0,\\
&\alpha\frac{\partial^2}{\partial y^2}(\theta\Delta T)+\frac{\partial\psi}{\partial x}\frac{\partial}{\partial y}(\theta\Delta T)-\frac{\partial\psi}{\partial y}\frac{\partial}{\partial x}(\theta\Delta T)=0,\\
&D\frac{\partial^2}{\partial y^2}(\phi\Delta C)+\frac{\partial\psi}{\partial x}\frac{\partial}{\partial y}(\phi\Delta C)-\frac{\partial\psi}{\partial y}\frac{\partial}{\partial x}(\phi\Delta C)-k_0(\phi\Delta C)^n=0.
\end{aligned} \tag{2.8}$$

The boundary conditions (2.2) become

$$\begin{aligned}
&\frac{\partial\psi}{\partial y}=U(x)+\nu N_1(x)\frac{\partial^2\psi}{\partial y^2},\quad -\frac{\partial\psi}{\partial x}=v_w(x)\frac{\partial\theta}{\partial y}=-\frac{h_f(x)}{k}[1-\theta(0)],\quad \phi=1\ \text{ at } y=0,\\
&\frac{\partial\psi}{\partial y}\longrightarrow 0,\quad \frac{\partial\psi}{\partial x}\longrightarrow 0,\quad \theta\longrightarrow 0,\quad \phi\longrightarrow 0\ \text{ as } y\longrightarrow\infty.
\end{aligned} \tag{2.9}$$

APPLICATION OF GROUP THEORY

We will now search group invariant solutions (similarity solutions) of (2.8) to (2.9) under a particular continuous one-parameter group. We define the following one-parameter group transformations Γ:

$$\begin{aligned}
&x^*=e^{\varepsilon\alpha_1}x,\quad y^*=e^{\varepsilon\alpha_2}y,\quad \psi^*=e^{\varepsilon\alpha_3}\psi,\quad \theta^*=e^{\varepsilon\alpha_4}\theta,\quad \phi^*=e^{\varepsilon\alpha_5}\phi,\quad B^*=e^{\varepsilon\alpha_6}B\\
&\Delta T^*=e^{\varepsilon\alpha_7}\Delta T,\quad \Delta C^*=e^{\varepsilon\alpha_8}\Delta C,\quad U=e^{\varepsilon\alpha_9}U,\quad v_w^*=e^{\varepsilon\alpha_{10}}v_w,\quad h_f^*=e^{\varepsilon\alpha_{11}}h_f
\end{aligned} \tag{3.1}$$

Here ε is the parameter of the group and α_i (i 1, 2, 3,..., 11) are the arbitrary real numbers whose interrelationship will be determined by our investigation. Transformations in (3.1) may be considered as point transformations which transform the coordinates

$$(x,y,\psi,\theta,\phi,B,\Delta T,\Delta C,U,v_w,h_f)\ \text{to}\ \left(x^*,y^*,\psi^*,\theta^*,\phi^*,B^*,\Delta T^*,\Delta C^*,U^*,v_w^*,h_f^*\right). \tag{3.2}$$

We now find the relationship among the exponents α_i such that (2.8) with the boundary conditions in (2.9) are invariant (the structure of the equations remains same before and after the transformations) under the transformation group in (3.1). Substituting transformations in (3.1) into (2.8) and equating various exponents of e for constant conformally invariant, we get the following algebraic equations:

$$4\alpha_2 - \alpha_3 = \alpha_1 + 3\alpha_2 - 2\alpha_3 = \alpha_1 - \alpha_4 - \alpha_7 = \alpha_1 - \alpha_5 - \alpha_8 = 2\alpha_2 - \alpha_3 - 2\alpha_6,$$
$$2\alpha_2 - \alpha_4 - \alpha_7 = \alpha_1 + \alpha_2 - \alpha_3 - \alpha_4 - \alpha_7,$$
$$2\alpha_2 - \alpha_5 - \alpha_8 = \alpha_1 + \alpha_2 - \alpha_3 - \alpha_5 - \alpha_8 = n(-\alpha_5 - \alpha_8). \quad (3.3)$$

Using (3.1) in boundary conditions (2.9) and equating various exponents of e, we get

$$\alpha_2 - \alpha_3 = -\alpha_9, \qquad \alpha_1 - \alpha_3 = -\alpha_{10}, \qquad \alpha_2 = -\alpha_{11}. \quad (3.4)$$

Without loss of generality, we may put $\alpha_4 = \alpha_5 = 0$.

Solving (3.3) and (3.4) we have the following relationship among the exponents

$$\alpha_1 = \frac{(7-5n)}{4\alpha_8}, \quad \alpha_2 = \frac{(1-n)}{2\alpha_8}, \quad \alpha_3 = \frac{(5-3n)}{4\alpha_8}, \quad \alpha_4 = 0, \quad \alpha_5 = 0,$$
$$\alpha_7 = \alpha_8, \quad \alpha_9 = \frac{(3-n)}{4\alpha_8}, \quad \alpha_6 = \alpha_{10} = \alpha_{11} = \frac{(n-1)}{2\alpha_8}. \quad (3.5)$$

It can be easily verified that (2.8) and the boundary conditions (2.9) are invariant under the transformations in (3.1) subject to the conditions in (3.5).

Similarity Transformations

Using the relationships in (3.5), we have from (3.1)

$$x^* = e^{\varepsilon(7-5n)/4\alpha_8}x, \quad y^* = e^{\varepsilon(1-n)/2\alpha_8}y, \quad \psi^* = e^{\varepsilon(5-3n)/4\alpha_8}\psi, \quad \theta^* = \theta, \quad \phi = \phi,$$
$$B^* = e^{\varepsilon(n-1)/2\alpha_8}B, \quad \Delta T^* = e^{\varepsilon\alpha_8}\Delta T, \quad \Delta C^* = e^{\varepsilon\alpha_8}\Delta C, \quad U^* = e^{\varepsilon(3-n)/4\alpha_8}U,$$
$$v_w^* = e^{\varepsilon\alpha_8(n-1)/2}v_w, \quad h_f^* = e^{\varepsilon\alpha_8(n-1)/2}h_f. \quad (3.6)$$

Expanding each of the transformations in (3.6) in powers of ε, keeping the first order term and neglecting higher order terms, we get the following characteristic equations:

$$\frac{4dx}{(7-5n)x} = \frac{2dy}{(1-n)y} = \frac{4d\psi}{(5-3n)\psi} = \frac{d\theta}{0} = \frac{d\phi}{0} = \frac{2dB}{(n-1)B}$$
$$= \frac{d(\Delta T)}{\Delta T} = \frac{d(\Delta C)}{\Delta C} = \frac{4dU}{(n-3)U} = \frac{2dv_w}{(n-1)v_w} = \frac{2dh_f}{(n-1)h_f}. \quad (3.7)$$

Solving the system of first order linear differential equations in (3.7), we get

$$\eta^* = \beta_1 y x^{(2n-2)/(7-5n)}, \quad \psi^* = \beta_2 x^{(5-3n)/(7-5n)} f(\eta),$$
$$\theta^* = \theta, \quad \phi^* = \phi, \quad B^* = \beta_3 x^{(2n-2)/(7-5n)},$$
$$\Delta T^* = \beta_4 x^{4/(7-5n)}, \quad \Delta C^* = \beta_5 x^{4/(7-5n)}, \quad U^* = \beta_6 x^{(3-n)/(7-5n)},$$
$$v_w^* = \beta_7 x^{(2n-2)/(7-5n)}, \quad h_f^* = \beta_8 x^{(2n-2)/(7-5n)}, \quad (3.8)$$

where β_i (i (,2,3,... 8) are real constants. To get dimensionless form of the transformations, we define the following dimensionless transformations:

$$\eta = \sqrt{\text{Re}}\,\frac{y}{L}\left(\frac{x}{L}\right)^{(2n-2)/(7-5n)}, \quad \psi = \nu\sqrt{\text{Re}}\left(\frac{x}{L}\right)^{(5-3n)/(7-5n)} f(\eta),$$
$$\theta = \theta(\eta), \quad \phi = \phi(\eta),$$
$$B = B_0\left(\frac{x}{L}\right)^{(2n-2)/(7-5n)}, \quad \Delta T = (\Delta T)_0\left(\frac{x}{L}\right)^{4/(7-5n)}, \quad \Delta C = (\Delta C)_0\left(\frac{x}{L}\right)^{4/(7-5n)},$$
$$U = U_0\left(\frac{x}{L}\right)^{(3-n)/(7-5n)}, \quad v_w = v_0\left(\frac{x}{L}\right)^{(2n-2)/(7-5n)}, \quad h_f(x) = h_0\left(\frac{x}{L}\right)^{(2n-2)/(7-5n)}, \tag{3.9}$$

where $f(\eta), \theta(\eta)$ and $\phi(\eta)$ represent dimensionless velocity, temperature, and concentration functions, respectively. B_0, $(\Delta T)_0$, $(\Delta C)_0$, U_0, v_0 and $_{h0}$ are constants.

Using (3.9), (2.8) become

$$f'''' + \frac{5-3n}{7-5n} f f''' - \frac{n+1}{7-5n} f' f'' - M f' - \lambda_1\left[\frac{2n-2}{7-5n}\eta\theta' + \frac{4}{7-5n}\theta\right]$$
$$- \lambda_2\left[\frac{2n-2}{7-5n}\eta\phi' + \frac{4}{7-5n}\phi\right] = 0,$$
$$\text{Pr}^{-1}\theta'' + \frac{5-3n}{7-5n} f\theta' - \frac{4}{7-5n} f'\theta = 0,$$
$$\text{Sc}^{-1}\phi'' + \frac{5-3n}{7-5n} f\phi' - \frac{4}{7-5n} f'\phi - K\phi^n = 0, \tag{3.10}$$

subject to the boundary conditions

$$f(0) = \frac{7-5n}{5-3n} fw, \quad f'(0) = 1 + af''(0), \quad \theta'(0) = -\gamma[1-\theta(0)], \quad \phi(0) = 1,$$
$$f(\infty) = f'(\infty) = \theta(\infty) = \phi(\infty) = 0, \tag{3.11}$$

where $fw = v_0 L/\nu\sqrt{\text{Re}}$ is the suction/injection parameter, γ Lh_0/Re k is the convective heat transfer parameter, λ_1 Gr/Re$^{2.5}$ is the thermal buoyancy parameter, λ_2 Gc/Re$^{2.5}$ is the concentration buoyancy parameter, $M = \sigma B_0^2 L^2/\rho\nu$ Re is the magnetic parameter, $K = L^2 k_0 (\Delta C)_0^{n-1}/\nu$ Re is the reaction parameter, $\text{Gr} = g\,\beta_c(\Delta T)_0 L^3/\nu^2$ is the thermal Grashof number, $\text{Gc} = g\,\beta_c\,(\Delta T)_0 L^3/\nu^2$ is the solutal Grashof number, and $a = \nu\sqrt{\text{Re}}\,N_0/L$ is the velocity slip parameter. Note that similarity solutions will exist if $N1(x)=N_0 x^{(2-2n)/(7-5n)}$, N_0 is a constant. The

The physical quantities of interest are the local skin friction coefficient C_{fx}, local Nusselt number Nu_x (rate of heat transfer), and local Sherwood number Sh_x (rate of mass transfer), which are defined as

$$C_{fx} = \frac{-\tau_W}{\rho U^2}, \quad \text{Nu}_x = \frac{-xq_w}{k\,\Delta T}, \quad \text{Sh}_x = \frac{-m_w x}{D\,\Delta C}, \tag{3.12}$$

where τ_w is the wall shear stress, q_w is the wall heat flux, m_w is the quantity of mass transfer through the unit area of the surface, which are given by

$$\tau_w = \mu\left(\frac{\partial u}{\partial y}\right)_{y=0}, \quad q_w = k\left(\frac{\partial T}{\partial y}\right)_{y=0}, \quad m_w = D\left(\frac{\partial C}{\partial y}\right)_{y=0}. \tag{3.13}$$

It can be shown that physical quantities are proportional to $-f''(0)$, $-\theta'(0)$ and $-\phi'(0)$.

RESULTS AND DISCUSSION

Equations (3.10) with boundary conditions (3.11) were solved using the dsolve command in MAPLE 14 with numeric option. Depending upon the nature of the ordinary differential equations, MAPLE uses a suitable scheme based on trapezoid or midpoint rule with Richardson extrapolation or deferred correction enhancement. In order to justify the accuracy of our numerical method, we compared our results with available data in Table 1 and found a good agreement. The influence of different parameters on the dimensionless velocity, temperature, and concentration is shown in Figures 2, 3, and 4, respectively. Figure 2(a) exhibits the influence of thermal buoyancy parameter λ_1 and suction/injection parameter f_w, whilst Figure 2(b) shows the effects of solutal buoyancy parameter λ_2 and suction/injection parameter f_w, on the dimensionless fluid velocity $f'(\eta)$ profiles. We further observed that velocity increases with buoyancy parameters λ_1 and λ_2. Figure 2(c) describes the influence of magnetic field parameter M on velocity $f'(\eta)$ profiles. It is found that as M increases, the fluid velocity decreases. The effect of velocity slip parameter on the dimensionless velocity is depicted in Figure 2(d). It is observed that an increase in the velocity slip reduces the dimensionless velocity at the surface. In all cases suction increases the velocity whilst injection decreases the velocity, as expected.

Table 1: Comparison of results for $f''(0)$ and $-\phi'(0)$ for different values of parameters.

	Pr = 0.72, Pr = 0.72, $\lambda_1 = 1$, $\lambda_2 = 1$, $K = 1$			
Sc	Fan et al. [17]		Present results	
	$f''(0)$	$\phi'(0)$	$f''(0)$	$-\phi'(0)$
0.1	0.9098	0.13678	0.9099	0.1368
0.72	0.47511	0.28697	0.4751	0.287
5	0.23157	0.77288	0.2316	0.7729
	Pr = 0.72, Pr = 0.72, Sc = 0.72, $\lambda_2 = 1$, $K = 1$			
λ_1	Fan et al. [17]		Present results	
0	0.06194	0.29757	0.0619	0.2976
0.4	0.24155	0.29252	0.2415	0.2925
1	0.47511	0.28697	0.4751	0.287
	Pr = 0.72, Pr = 0.72, $\lambda_1 = 1$, Sc = 0.72, $K = 1$			
λ_2	Fan et al. [17]		Present results	
0	0.14425	0.29487	0.1442	0.2949
0.4	0.2835	0.29133	0.2835	0.2913
1	0.47511	0.28697	0.4751	0.287
	Pr = 0.72, Pr = 0.72, $\lambda_1 = 1$, $\lambda_2 = 1$, Sc = 0.72			
K	Fan et al. [17]		Present results	
0	0.53159	0.00015	0.5316	0.00015
1	0.47511	0.28697	0.4751	0.287
3	0.42378	0.61989	0.4238	0.6199

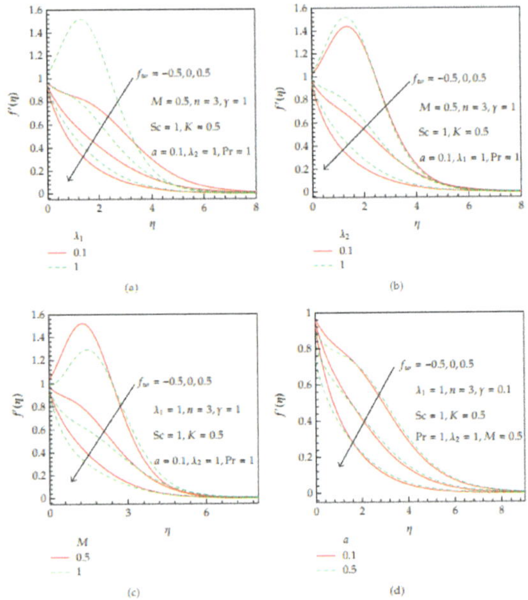

Figure 2: Effects of different parameters on the dimensionless velocity.

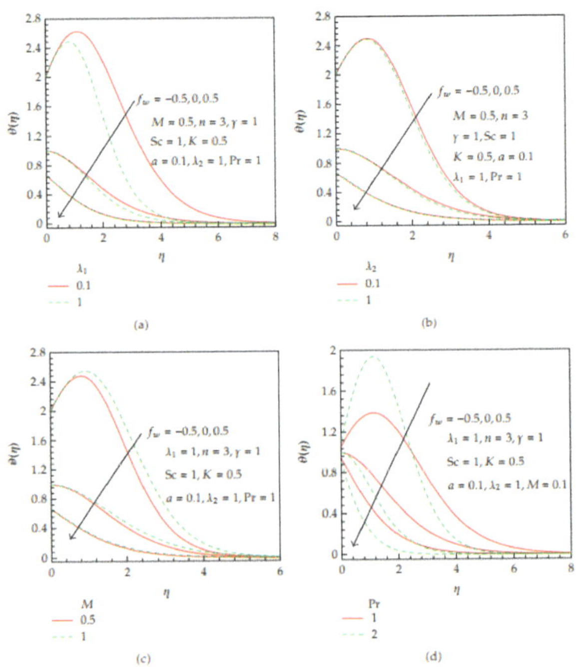

Figure 3: Effects of different parameters on the dimensionless temperature.

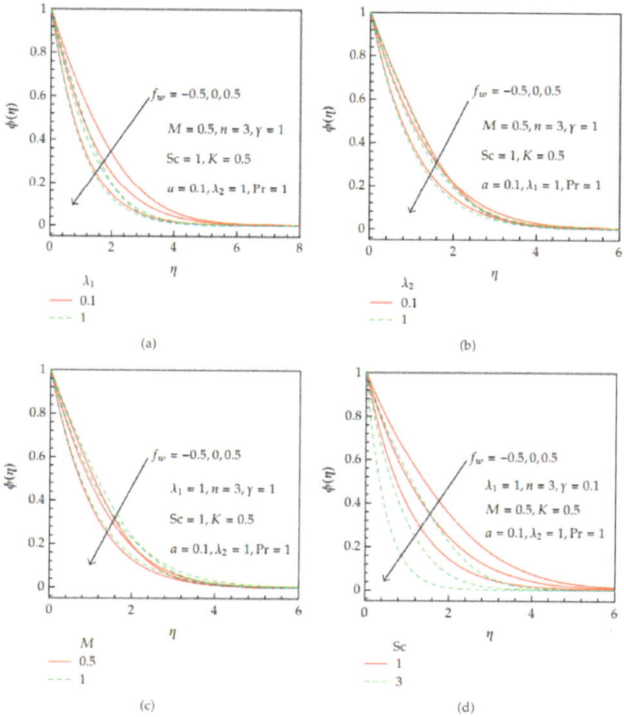

Figure 4: Effects of different parameters on the dimensionless concentration.

Figures 3(a), 3(b), 3(c), and 3(d) display effects of various parameters on the dimensionless temperature function. The dimensionless temperature decreases with suction and increases with injection, as expected. Note that as we increase the values of λ_1 and λ_2, the values of $\theta(\eta)$ decrease (Figures 3(a) and 3(b)). It is found that magnetic field increases temperature whilst Prandtl number decreases the temperature. Figures 4(a), 4(b), 4(c), and 4(d) exhibit influence of suction/injection f_w, buoyancies λ_1, λ_2, the magnetic field M and the Schmidt number Sc on the dimensionless concentration $\varphi(\eta)$ function.

The dimensionless concentration decreases with the increase of f_w. We also notice from Figures 4(a) and 4(b) as we increase the values of λ_1 and λ_2, values of $\varphi(\eta)$ decrease. Figure 4(c) shows that the increasing values of magnetic field parameter lead to increase the concentration whilst Figure 4(d) shows that increasing values of Schmidt number Sc leads to decrease the concentration, as expected.

Figures 5(a) and 5(b) depict the variation of the buoyancies λ_1, λ_2, the slip a and the suction/injection f_w parameters on the dimensionless shear stress. It is observed that suction enhanced the values of the dimensionless shear stress. We also see that the velocity slip a and the parameters λ_1, λ_2 increase the values of the dimensionless shear stress fall. The effect of Prandtl number Pr, convective heat transfer parameter γ, order of chemical reaction n, and the suction/injection parameter f_w on the dimensionless rate of heat transfer is shown in Figures 6(a) and 6(b). From Figure 6(a) we can conclude that as Pr, γ and fw increase, the values of the dimensionless rate of heat transfer increase. An increase in the order of reaction causes decrease in the dimensionless rate of heat transfer (Figure 6(b)). The variation of the dimensionless rate of mass transfer with different parameters is shown in Figures 7(a) and 7(b). Like the dimensionless rate of heat transfer, the dimensionless rate of mass transfer increases with an increase in Sc, K, and fw (Figure 7(a)) and decreases with an increase in the order of reaction n.

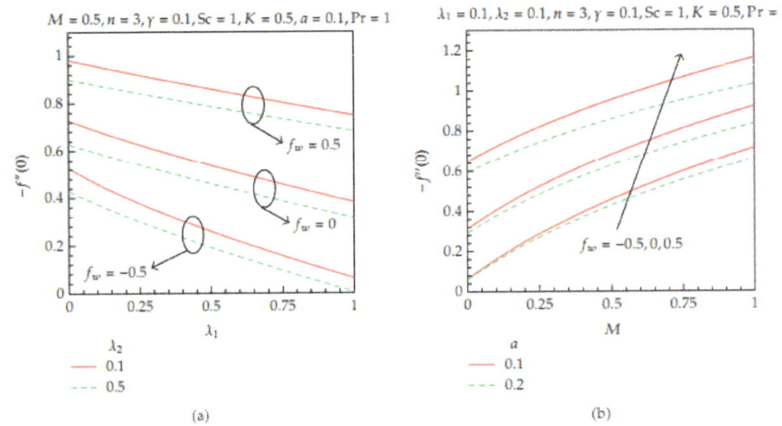

Figure 5: Effects of different parameters on the dimensionless shear stress.

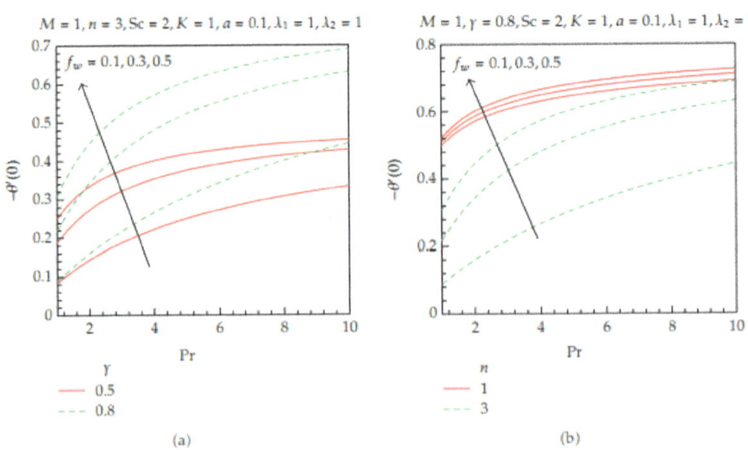

Figure 6: Effects of different parameters on the dimensionless rate of heat transfer.

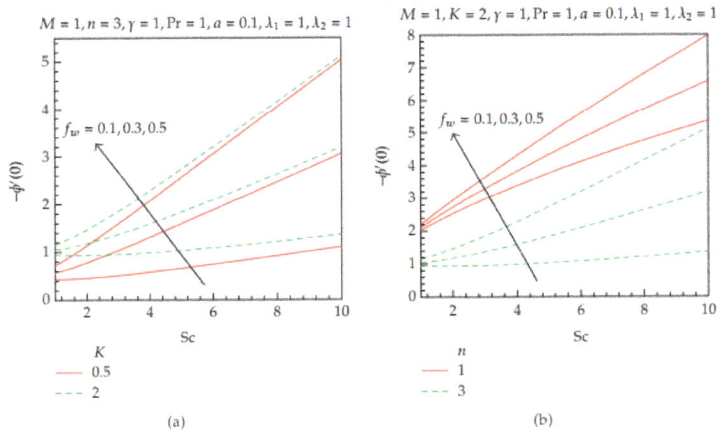

Figure 7: Effects of different parameters on the dimensionless rate of mass transfer.

CONCLUSIONS

MHD boundary layer equations for mixed convection of Newtonian fluids along a moving horizontal plate with velocity slip and thermal convective boundary condition are transformed into similarity equations using one-parameter continuous group method and then solved numerically using an implicit finite difference numerical method. Our analysis revealed that $-f''(0)$ increases with suction fw and decreases with a, λ_1, and λ2 whilst the values of $-\theta'(0)$ enhance with fw, Pr, and γ. The values of $-\theta'(0)$ enhance with fw, Sc and k.

REFERENCES

1. T. K. Sengupta, S. Unnikrishnan, S. Bhaumik, P. Singh, and S. Usman, "Linear spatial stability analysis of mixed convection boundary layer over a heated plate," Progress in Applied Mathematics, vol. 1, pp. 71–89, 2011.
2. F. M. Ali, R. Nazar, N. M. Arifin, and I. Pop, "MHD mixed convection boundary layer flow toward a stagnation point on a vertical surface with induced magnetic field," Journal of Heat Transfer, vol. 133, no. 2, Article ID 022502, pp. 1–6, 2011.
3. S. V. Subhashini, N. Samuel, and I. Pop, "Effects of buoyancy assisting and opposing flows on mixed convection boundary layer flow over a permeable vertical surface," International Communications in Heat and Mass Transfer, vol. 38, no. 4, pp. 499–503, 2011.
4. F. P. Incropera, "Buoyancy effects in double-diffusive mixed convection Flows," in Proceedings of the 8th International Heat Transfer Conference, C. L. Tein, V. P. Carey, and J. K. Ferrel, Eds., pp. 121–130.
5. T. S. Chen and B. S. Armaly, "Mixed convection in external flow," in Handbook of Single-Phase Convective Heat Transfer, S. Kakac, R. K. Shah, and W. Aung, Eds., Wiley, New York, NY, USA, 1987.

6. Y. Jaluria, "Basic of Natural ConvectionS," in Handbook of Single-Phase Convective Heat Transfer, Kakac, R. K. Shah, and W. Aung, Eds., Wiley, New York, NY, USA, 1987.
7. B. Gebhart, Y. Jaluria, R. L. Mahajan, and B. Sammarkia, Buoyancy-Induced Flows and Transport, Chapter 10, Hemisphere, Washington, DC, USA, 1988.
8. Bejan, Convection Heat Transfer, Wiley, New York, NY, USA, 1995.
9. Pop and D. B. Ingham, Convective Heat Transfer, Pergamon, Amsterdam, The Netherlands, 2001.
10. R. Lloyd and E. M. Sparrow, "Combined forced and free convection flow on vertical surfaces,"International Journal of Heat and Mass Transfer, vol. 13, no. 2, pp. 434–438, 1970.
11. W. Schneider, "A similarity solution for combined forced and free convection flow over a horizontal plate," International Journal of Heat and Mass Transfer, vol. 22, no. 10, pp. 1401–1406, 1979.
12. G. Wilks, "Combined forced and free convection flow on vertical surfaces," International Journal of Heat and Mass Transfer, vol. 16, no. 10, pp. 1958–1964, 1973.
13. S. Yao, "Two-dimensional mixed convection along a flat plate," Journal of Heat Transfer, vol. 109, no. 2, pp. 440–445, 1987.
14. N. Ramachandran, T. S. Chen, and B. F. Armaly, "Mixed convection in stagnation flows adjacent to vertical surfaces," Journal of Heat Transfer, vol. 110, no. 2, pp. 373–377, 1988.
15. G. S. Moulic and L. S. Yao, "Mixed convection along a wavy surface," ASME Journal of Heat Transfer, vol. 111, pp. 974–979, 1989.
16. J. Daskalakis, N. Kafoussias, A. Lewkowicz, and E. W. Williams, "Similarity solution for free and forced convection hydromagnetic flow over a horizontal semi-infinite plate through a non-homogeneous porous medium," Astrophysics and Space Science, vol. 151, no. 2, pp. 217–226, 1989.
17. J. R. Fan, J. M. Shi, and X. Z. Xu, "Similarity solution of mixed convection over a horizontal moving plate," Heat and Mass Transfer, vol. 32, no. 3, pp. 199–206, 1997.
18. E. Magyari, I. Pop, and B. Keller, "Mixed convection boundary-layer flow past a horizontal permeable flat plate," Fluid Dynamics Research, vol. 31, no. 3, pp. 215–225, 2002.
19. A. Hassanien, A. H. Essawy, and N. M. Moursy, "Variable viscosity and thermal conductivity effects on combined heat and mass transfer in mixed convection over a UHF/UMF wedge in porous media: the entire regime," Applied Mathematics and Computation, vol. 145, no. 2-3, pp. 667–682, 2003.
20. O. Aydin and A. Kaya, "Mixed convection of a viscous dissipating fluid about a vertical flat plate," Applied Mathematical Modelling, vol. 31, no. 5, pp. 843–853, 2007.

21. Ishak, "Mixed convection boundary layer flow over a horizontal plate with thermal radiation," Heat and Mass Transfer, vol. 46, no. 2, pp. 147–151, 2009.
22. S. G. Moulic and L. S. Yao, "Mixed convection along a semi-infinite vertical flat plate with uniform surface heat flux," Journal of Heat Transfer, vol. 131, no. 2, pp. 1–8, 2009.
23. P. Datta, S. V. Subhashini, and R. Ravindran, "Influence of surface mass transfer on mixed convection flows over non-isothermal horizontal flat plates," Applied Mathematical Modelling, vol. 33, no. 3, pp. 1285–1294, 2009.
24. Ishak, R. Nazar, N. Bachok, and I. Pop, "MHD mixed convection flow near the stagnation-point on a vertical permeable surface," Physica A, vol. 389, no. 1, pp. 40–46, 2010.
25. E. Magyari, "Comment on "Mixed convection boundary layer flow over a horizontal plate with thermal radiation" by A. Ishak, Heat Mass Transfer," Heat and Mass Transfer, vol. 46, no. 8-9, pp. 809–810, 2010.
26. M. Jiji, Heat Convection, cshapter 11, Springer, 2nd edition, 2009.
27. G. M. Hak, "Flow physics in the MEMS handbook," in MEMS Handbook, M. Gad-el-Hak, Ed., chapter 4, CRC Press, Boca Raton, Fla, USA, 2002.
28. Aziz, "Hydrodynamic and thermal slip flow boundary layers over a flat plate with constant heat flux boundary condition," Communications in Nonlinear Science and Numerical Simulation, vol. 15, no. 3, pp. 573–580, 2010.
29. Yoshimura and R. K. Prud'homme, "Wall slip corrections for Couette and parallel disk viscometers,"Journal of Rheology, vol. 32, no. 1, pp. 53–67, 1988.
30. M. Rahman, "Locally similar solutions for hydromagnetic and thermal slip flow boundary layers over a flat plate with variable fluid properties and convective surface boundary condition," Meccanica, vol. 46, no. 5, pp. 1127–1143, 2011.
31. G. P. Duncan and G. P. Peterson, "Review of Microscale Heat Transfer," Applied Mechanics Review, vol. 47, pp. 397–428, 1994.
32. Gad-el-Hak, "Fluid mechanics of microdevices—the freeman scholar lecture," ASME Journal of Fluids Engineering, vol. 121, no. 1, pp. 5–33, 1999.
33. T. Fang and Chia-f.F. Lee, "A moving-wall boundary layer flow of a slightly rarefied gas free stream over a moving flat plate," Applied Mathematics Letters, vol. 18, no. 5, pp. 487–495, 2005.
34. M. J. Martin and I. D. Boyd, "Falkner-Skan flow over a wedge with slip boundary conditions," Journal of Thermophysics and Heat Transfer, vol. 24, no. 2, pp. 263–270, 2010.
35. M. T. Matthews and J. M. Hill, "Micro/nano thermal boundary layer equations with slip-creep-jump boundary conditions," IMA Journal of Applied Mathematics, vol. 72, no. 6, pp. 894–911, 2007.
36. Y. Wang, "Analysis of viscous flow due to a stretching sheet with surface slip and suction," Nonlinear Analysis. Real World Applications, vol. 10, no. 1, pp. 375–380, 2009.

37. Cao and J. Baker, "Slip effects on mixed convective flow and heat transfer from a vertical plate,"International Journal of Heat and Mass Transfer, vol. 52, no. 15-16, pp. 3829–3841, 2009.

38. W. F. Ames, Nonlinear Partial Differential Equations in Engineering, Academic Press, New York, NY, USA, 1965.

39. W. E. Boyce, Elementary Differential Equations and Boundary Value Problems, John Wiley and Sons, New York, NY, USA, 1977.

40. T. Y. Na, Computational methods in engineering boundary value problems, vol. 145 of Mathematics in Science and Engineering, Academic Press, New York, NY, USA, 1979.

41. E. Zauderer, Partial Differential Equations of Applied Mathematics, Pure and Applied Mathematics, John Wiley & Sons, New York, NY, USA, 1983.

42. R. Seshadri and T. Y. Na, Group Invariance in Engineering Boundary Value Problems, Springer, New York, NY, USA, 1985.

43. Kincaid and W. Cheney, Numerical Analysis, Brooks/Cole Publishing, Pacific Grove, Calif, USA, 1991.

44. W. A. Strauss, Partial Differential Equations. An Introduction, John Wiley & Sons, New York, NY, USA, 1992.

45. Aziz, Heat Conduction with Maple, R.T Edwards, 2005.

46. R. E. White and V. R. Subramanian, Computational Methods InChemical Engineering With Maple, Springer, 2010.

47. C. Slattery, Advanced Transport Phenomena, Cambridge University Press, Cambridge, UK, 1999.

48. F. P. Incropera, D. P. Dewitt, T. L. Bergman, and A. S. Lavine, Fundamentals of Heat and Mass Transfer, John Wiley, New York, NY, USA, 6th edition, 2007.

49. S. Mukhopadhyay and H. I. Andersson, "Effects of slip and heat transfer analysis of flow over an unsteady stretching surface," Heat and Mass Transfer, vol. 45, no. 11, pp. 1447–1452, 2009.

6

CHAPTER

BOUNDARY LAYER EQUATIONS AND LIE GROUP ANALYSIS OF A SISKO FLUID

Gözde Sarı, Mehmet Pakdemirli, Tasawar Hayat, and Yiğit Aksoy

INTRODUCTION

Due to the inadequacy of Newtonian fluid model which predicts a linear relationship between the shear stress and velocity gradient, many non-Newtonian fluid models were proposed to explain the complex behavior. Usually, the stress constitutive relations of such models inherit complexities which lead to highly nonlinear equations of motion with many terms. To simplify the extremely complex equations, one alternative is to use boundary layer theory which is known to effectively reduce the complexity of Navier-Stokes equations and reduce drastically the computational time. Since there are many non-Newtonian models and new models are being proposed continuously, boundary layer theory for each proposed model also appears in the literature. It is beyond the scope of this work to review vast literature on the boundary layers of non-Newtonian fluids. A limited work on the topic can be found in [1–19].

In this work, boundary layer theory is developed for the Sisko fluid, a non-Newtonian fluid model which combines the features of viscous and power law models. A complete symmetry analysis of the boundary layer equations is presented. Using one of the symmetries, the partial differential system is transformed into an ordinary differential system. The resulting ordinary differential system is numerically solved by a finite difference algorithm. Effect of non-Newtonian parameters on the velocity profiles is shown in the graphs.

As a non-Newtonian model, Sisko [20] proposed a new model which is later called sisko fluid. Some of the recent work on Sisko fluids is as follows: an analytical solution was presented using homotopy analysis method for the flow of a magnetohydrodynamic

Citation: Gözde Sarı, Mehmet Pakdemirli, Tasawar Hayat, and Yiğit Aksoy, "Boundary Layer Equations and Lie Group Analysis of a Sisko Fluid," Journal of Applied Mathematics, vol. 2012, Article ID 259608, 9 pages, 2012. doi:10.1155/2012/259608

Sisko fluid through a porous medium [21]. Wire coating analysis was investigated by withdrawal from a bath of Sisko fluid [22]. The equation modeling thin film flow of a Sisko fluid on a moving belt qualitatively was analyzed and a series solution was found using homotopy analysis method [23]. Unsteady unidirectional flow of a Sisko fluid bounded by a suddenly moved plate was studied [24]. Unsteady flow of a Sisko fluid over a moving wall with suction or blowing was considered [25]. A numerical study for the unsteady flow of a magnetohydrodynamic Sisko fluid in annular pipe was presented [26]. Heat transfer was also incorporated into the annular pipe problem [27]. To the best of authors' knowledge, boundary layer analysis and symmetry reductions of Sisko fluids do not exist in the literature.

BOUNDARY LAYER EQUATIONS

The Cauchy stress tensor for Sisko fluid is

$$\mathbf{T} = -p\mathbf{I} + \mathbf{S}, \tag{2.1}$$

where

$$\mathbf{S} = \left[a + b \left| \sqrt{\frac{1}{2}\operatorname{tr}(\mathbf{A}_1^2)} \right|^{n-1} \right] \mathbf{A}_1,$$
$$\mathbf{A}_1 = \mathbf{L} + \mathbf{L}^T, \qquad \mathbf{L} = \operatorname{grad} \mathbf{V}. \tag{2.2}$$

V is the velocity vector and A1 is the first Rivlin Ericksen tensor. a and b are the material constants. The model is a combination of viscous and power-law models. For a 0, the model exhibits power-law behavior whereas, for b 0, the flow is Newtonian. Steady-state, two-dimensional, incompressible equations of motion including mass conservation can be written as

$$\frac{\partial u^*}{\partial x^*} + \frac{\partial v^*}{\partial y^*} = 0$$
$$\rho\left(u^*\frac{\partial u^*}{\partial x^*} + v^*\frac{\partial u^*}{\partial y^*}\right) = -\frac{\partial p^*}{\partial x^*} + \frac{\partial S_{xx}}{\partial x^*} + \frac{\partial S_{xy}}{\partial y^*}$$
$$\rho\left(u^*\frac{\partial v^*}{\partial x^*} + v^*\frac{\partial v^*}{\partial y^*}\right) = -\frac{\partial p^*}{\partial y^*} + \frac{\partial S_{xy}}{\partial x^*} + \frac{\partial S_{yy}}{\partial y^*}, \tag{2.3}$$

where x^* is the spatial coordinate along the surface, y^* is vertical to it, u^* and v^* are the velocity components in the x^* and y^* coordinates. The shear stress components are inserted into the equations of motion and the usual boundary layer assumptions are made, that is, $x^* \sim O(1),\ y^* \sim O(\delta),\ u^* \sim O(1),\ v^* \sim O(\delta)$. The highest order terms are retained and the momentum equations become

$$\rho\left(u^*\frac{\partial u^*}{\partial x^*} + v^*\frac{\partial u^*}{\partial y^*}\right) = -\frac{\partial p^*}{\partial x^*} + a\frac{\partial^2 u^*}{\partial y^{*2}} + bn\left(\frac{\partial u^*}{\partial y^*}\right)^{n-1}\frac{\partial^2 u^*}{\partial y^{*2}}$$
$$0 = \frac{\partial p^*}{\partial y^*} \tag{2.4}$$

from which dependence of pressure on y^* is eliminated. Dimensionless variables and parameters are defined as follows:

$$x = \frac{x^*}{L}, \quad y = \frac{y^*}{L}, \quad u = \frac{u^*}{V}, \quad v = \frac{v^*}{V}, \quad p = \frac{p^*}{\rho V^2}\varepsilon_1 = \frac{a}{\rho VL}, \quad \varepsilon_2 = \frac{bnV^{n-2}}{\rho L^n}, \tag{2.5}$$

where L is a characteristic length and V is a reference velocity. Expressing the pressure in terms of outer velocity, the final dimensionless boundary layer equations become

$$\frac{\partial u}{\partial x} + \frac{\partial v}{\partial y} = 0$$

$$u\frac{\partial u}{\partial x} + v\frac{\partial u}{\partial y} = U\frac{dU}{dx} + \varepsilon_1\frac{\partial^2 u}{\partial y^2} + \varepsilon_2\left(\frac{\partial u}{\partial y}\right)^{n-1}\frac{\partial^2 u}{\partial y^2}. \tag{2.6}$$

The classical boundary conditions for the problem are

$$u(x,0) = 0, \quad v(x,0) = 0, \quad u(x,\infty) = U(x). \tag{2.7}$$

For $\varepsilon_2 = 0$ or n = 1, the equations reduce qualitatively to those of Newtonian fluid.

LIE GROUP THEORY AND SYMMETRY REDUCTIONS

Lie group Theory is employed in search of symmetries of the equations. Details of the theory can be found in [28, 29]. The infinitesimal generator for the problem is

$$X = \xi_1(x,y,u,v)\frac{\partial}{\partial x} + \xi_2(x,y,u,v)\frac{\partial}{\partial y} + \eta_1(x,y,u,v)\frac{\partial}{\partial u} + \eta_2(x,y,u,v)\frac{\partial}{\partial v}. \tag{3.1}$$

A straightforward and tedious calculation (see [28, 29] for details) yields

$$\xi_1 = 3ax + b, \quad \xi_2 = ay + c(x), \quad \eta_1 = au, \quad \eta_2 = c'(x)u - av. \tag{3.2}$$

The classifying relation for the outer velocity is

$$(3ax + b)\frac{d}{dx}(UU') + a(UU') = 0. \tag{3.3}$$

Symbolic packages developed to calculate symmetries fail to produce the above results due to the arbitrary outer velocity function, and hand calculation becomes inevitable for precise results. There are two finite parameter Lie group symmetries represented by parameters "a" and "b," the former corresponding to scaling symmetry and the latter to translational symmetry in the x coordinate. There is an additional infinite parameter Lie group symmetry represented by c(x) function. For a general stress tensor which is an arbitrary function of the velocity gradient, the symmetries were calculated previously [18]. Our results for the special case of Sisko fluid confirm the previous calculations (see principal Lie Algebra presented in [18]).

Usually boundary conditions put much restriction on the symmetries which may lead to removal of all the symmetries. In our case, however, some of the symmetries remain stable after imposing the boundary conditions. For nonlinear equations, the generators should be applied to the boundaries and boundary conditions also [28]. Applying the generator to the boundary y = 0 yields c(x)= 0. Applying the generator to the boundary conditions does not impose further restrictions and hence the symmetries valid for the equations and boundary conditions reduce to

$$\xi_1 = 3ax + b, \qquad \xi_2 = ay, \qquad \eta_1 = au, \qquad \eta_2 = -av \tag{3.4}$$

with the classifying relation (3.3) remaining unchanged.

Selecting parameter "a" in the symmetries, the associated equations which define similarity variables are

$$\frac{dx}{3x} = \frac{dy}{y} = \frac{du}{u} = \frac{dv}{-v}. \tag{3.5}$$

Solving the system yields the similarity variables

$$\xi = \frac{y}{x^{1/3}}, \qquad u = x^{1/3} f(\xi), \qquad v = \frac{g(\xi)}{x^{1/3}}. \tag{3.6}$$

From (3.3) with $b = 0, U(x) = x^{1/3}$. Substituting all into the boundary layer equations yields the ordinary differential system

$$f - \xi f' + 3g' = 0$$

$$f^2 - \xi f f' + 3gf' = 1 + 3\varepsilon_1 f'' + 3\varepsilon_2 (f')^{n-1} f''. \tag{3.7}$$

The boundary conditions also transform to

$$f(0) = 0, \qquad g(0) = 0, \qquad f(\infty) = 1. \tag{3.8}$$

NUMERICAL RESULTS

Equation (3.7) with the boundary conditions (3.8) are numerically integrated using a finite difference code. The finite difference code implements the three-stage Lobatto IIIA formula that provides continuous solutions of fourth order accuracy in the problem domain. Description of the numerical method is given by Shampine et al. [30] and Kierzenka and Shampine [31]. In Figure f function and in Figure 2, g function related to the x and y components of the velocities are drawn for different ε_1 parameters. Boundary layer becomes thicker for an increase in ε_1.

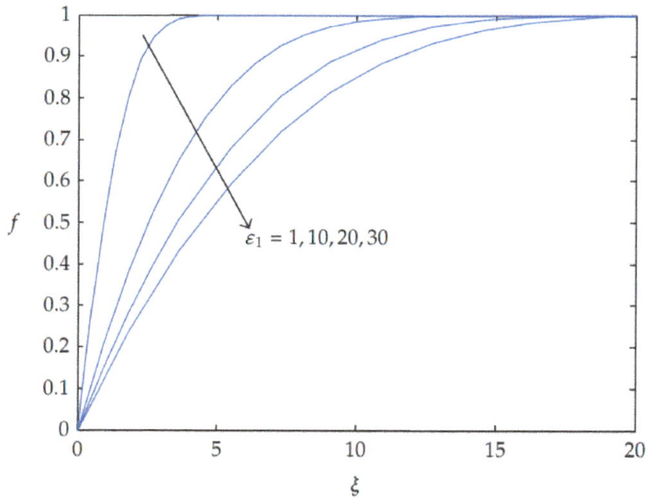

Figure 1: Effect of ε_1 parameter on the similarity function f related to the x component of velocity ($\varepsilon_2 = 1$, n = 2).

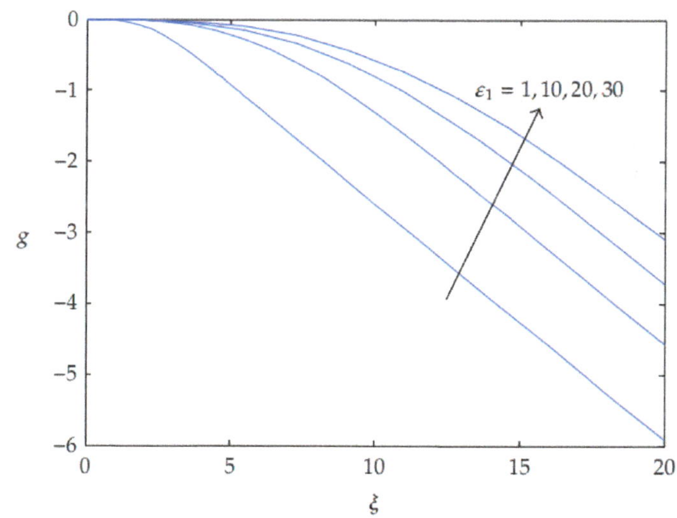

Figure 2: Effect of ε_1 parameter on the similarity function g related to the y component of velocity ($\varepsilon_2 = 1$, n = 2).

There is an increase in y component of velocity which is negative over the whole domain excluding the surface for an increase in ε_1 as can be seen from Figure 2. A similar trend is observed for parameter ε_2, although not as influential as in the previous case (see Figures 3 and 4).

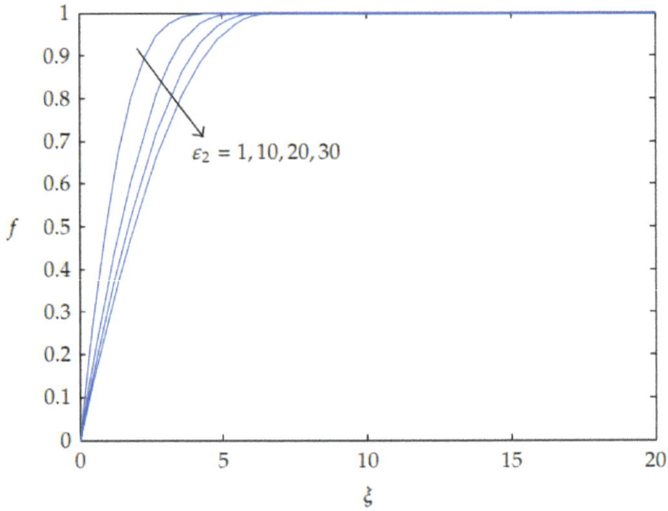

Figure 3: Effect of ε_2 parameter on the similarity function f related to the x component of velocity ($\varepsilon_2 = 1$, n = 2).

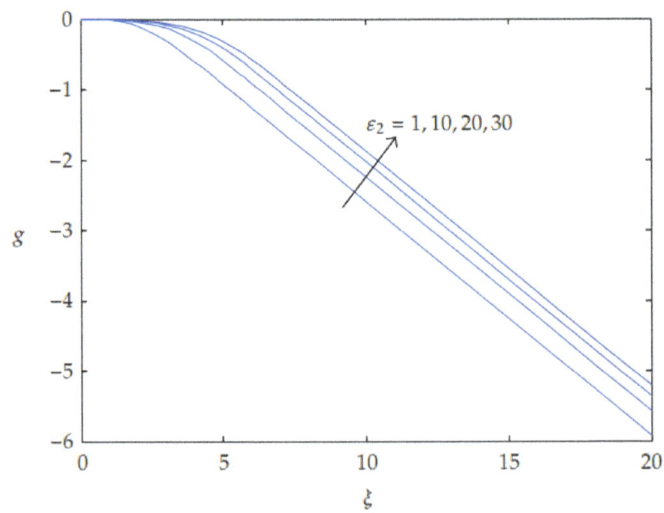

Figure 4: Effect of ε_2 parameter on the similarity function g related to the y component of velocity ($\varepsilon_2 = 1$, n = 2).

On the contrary, a reverse effect is observed for power index n, that is, as n increases, boundary layer becomes narrower eventually converging to a fixed thickness (see Figure 5). For the term related to y component of velocity, higher values of power index n cause a decrease which eventually converges to a solution as can be seen from Figure 6.

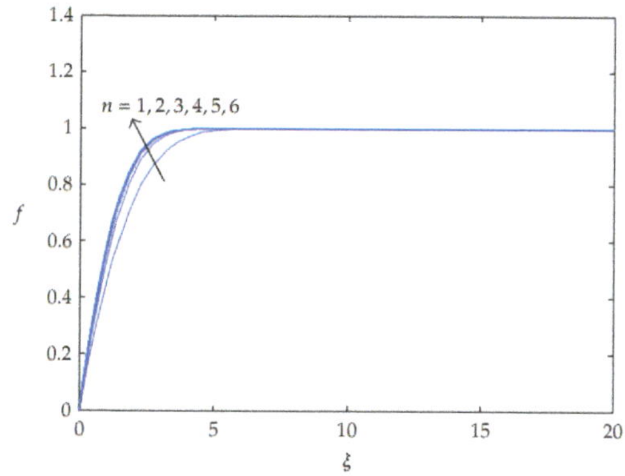

Figure 5: Effect of power index n on the similarity function f related to the x component of velocity ($\varepsilon_2 = 1$, $\varepsilon_2 = 1$).

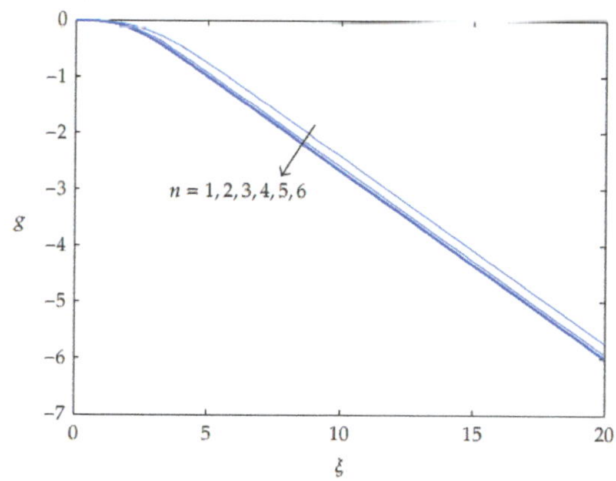

Figure 6: Effect of power index n on the similarity function g related to the y component of velocity ($\varepsilon_2 = 1$, $\varepsilon_2 = 1$).

CONCLUDING REMARKS

Boundary layer equations of Sisko fluid derived for the first time. Lie group theory is applied to the equations. Equations admit two finite parameter Lie group transformations and an infinite parameter Lie group transformation. The infinite parameter Lie group transformation is not stable with respect to usual boundary layer conditions. Using the scaling symmetry which is one of the finite parameter transformations, the partial differential system is transferred into an ordinary differential system. Resulting ordinary differential system is solved numerically using a finite difference scheme. Effects of non-Newtonian parameters on the boundary layers are discussed in detail. It is found that an increase in the parameters ε_1 and ε_2 causes the boundary layer to thicken whereas a reverse effect is observed for power index n.

There is an increase in y component of velocity which is negative over the whole domain excluding the surface for an increase in ε_1 as can be seen from Figure 2. A similar trend is observed for parameter ε_2, although not as influential as in the previous case (see Figures 3 and 4).

ACKNOWLEDGMENTS

This work is completed during mutual short visits of T. Hayat to Turkey and M. Pakdemirli to Pakistan. Funding supports of TUBITAK of Turkey and HEC of Pakistan are highly appreciated.

REFERENCES

1. Acrivos, M. J. Shahand, and E. E. Petersen, "Momentum and heat transfer in laminar boundary layer flows of non-Newtonian fluids past external surfaces," AIChE Journal, vol. 6, pp. 312–317, 1960.
2. W. R. Schowalter, "The application of boundary-layer theory to power-law pseudoplastic fluids: similarity solutions," AIChE Journal, vol. 6, pp. 25–28, 1960.
3. G. D. Bizzell and J. C. Slattery, "Non-newtonian boundary-layer flow," Chemical Engineering Science, vol. 17, no. 10, pp. 777–782, 1962.
4. C. Srivastava, "The flow of a non-Newtonian liquid near a stagnation point," Zeitschrift für Angewandte Mathematik und Physik, vol. 9, no. 1, pp. 80–84, 1958.
5. D. W. Beard and K. Walters, "Elastico-viscous boundary-layer flows," Proceedings of Cambridge Philosophical Society, vol. 60, pp. 667–674, 1964.
6. J. Astin, R. S. Jones, and P. Lockyer, "Boundary layers in non-Newtonian fluids," Journal Mechanique, vol. 12, no. 3, pp. 527–539, 1973.
7. K. R. Rajagopal, A. S. Gupta, and A. S. Wineman, "On a boundary layer theory for non-Newtonian fluids," International Journal of Engineering Science, vol. 18, no. 6, pp. 875–883, 1980.
8. K. R. Rajagopal, A. S. Gupta, and T. Y. Na, "A note on the Falkner-Skan flows of a non-Newtonian fluid,"International Journal of Non-Linear Mechanics, vol. 18, no. 4, pp. 313–320, 1983.
9. V. K. Garg and K. R. Rajagopal, "Flow of a non-Newtonian fluid past a wedge," Acta Mechanica, vol. 88, no. 1-2, pp. 113–123, 1991.
10. M. Massoudi and M. Ramezan, "Effect of injection or suction on the Falkner-Skan flows of second grade fluids," International Journal of Non-Linear Mechanics, vol. 24, no. 3, pp. 221–227, 1989.
11. V. K. Garg and K. R. Rajagopal, "Stagnation point flow of a non-Newtonian fluid," Mechanics Research Communications, vol. 17, no. 6, pp. 415–421, 1990.
12. M. Pakdemirli and E. S. Şuhubi, "Boundary layer theory for second order fluids," International Journal of Engineering Science, vol. 30, no. 4, pp. 523–532, 1992.

13. M. Pakdemirli, "The boundary layer equations of third-grade fluids," International Journal of Non-Linear Mechanics, vol. 27, no. 5, pp. 785–793, 1992.

14. M. Pakdemirli, "Boundary layer flow of power-law fluids past arbitrary profiles," IMA Journal of Applied Mathematics, vol. 50, no. 2, pp. 133–148, 1993.

15. R. K. Bhatnagar, G. Gupta, and K. R. Rajagopal, "Flow of an Oldroyd-B fluid due to a stretching sheet in the presence of a free stream velocity," International Journal of Non-Linear Mechanics, vol. 30, no. 3, pp. 391–405, 1995.

16. T. Hagen and M. Renardy, "Boundary layer analysis of the Phan-Thien-Tanner and Giesekus model in high Weissenberg number flow," Journal of Non-Newtonian Fluid Mechanics, vol. 73, no. 1-2, pp. 181–189, 1997.

17. M. Yürüsoy and M. Pakdemirli, "Symmetry reductions of unsteady three-dimensional boundary layers of some non-Newtonian fluids," International Journal of Engineering Science, vol. 35, no. 8, pp. 731–740, 1997.

18. M. Yürüsoy and M. Pakdemirli, "Group classification of a non-Newtonian fluid model using classical approach and equivalence transformations," International Journal of Non-Linear Mechanics, vol. 34, no. 2, pp. 341–346, 1999.

19. Y. Aksoy, M. Pakdemirli, and C. M. Khalique, "Boundary layer equations and stretching sheet solutions for the modified second grade fluid," International Journal of Engineering Science, vol. 45, no. 10, pp. 829–841, 2007.

20. W. Sisko, "The flow of lubricating greases," Industrial and Engineering Chemistry, vol. 50, no. 12, pp. 1789–1792, 1958.

21. M. Khan, Z. Abbas, and T. Hayat, "Analytic solution for flow of Sisko fluid through a porous medium,"Transport in Porous Media, vol. 71, no. 1, pp. 23–37, 2008.

22. M. Sajid and T. Hayat, "Wire coating analysis by withdrawal from a bath of Sisko fluid," Applied Mathematics and Computation, vol. 199, no. 1, pp. 13–22, 2008.

23. F. T. Akyıldız, K. Vajravelu, R. N. Mohapatra, E. Sweet, and R. A. Van Gorder, "Implicit differential equation arising in the steady flow of a Sisko fluid," Applied Mathematics and Computation, vol. 210, no. 1, pp. 189–196, 2009.

24. M. Molati, T. Hayat, and F. Mahomed, "Rayleigh problem for a MHD Sisko fluid," Nonlinear Analysis. Real World Applications, vol. 10, no. 6, pp. 3428–3434, 2009.

25. T. Hayat, R. J. Moitsheki, and S. Abelman, "Stokes' first problem for Sisko fluid over a porous wall,"Applied Mathematics and Computation, vol. 217, no. 2, pp. 622–628, 2010.

26. M. Khan, Q. Abbas, and K. Duru, "Magnetohydrodynamic flow of a Sisko fluid in annular pipe: a numerical study," International Journal for Numerical Methods in Fluids, vol. 62, no. 10, pp. 1169–1180, 2010.

27. M. Khan, S. Munawar, and S. Abbasbandy, "Steady flow and heat transfer of a Sisko fluid in annular pipe," International Journal of Heat and Mass Transfer, vol. 53, no. 7-8, pp. 1290–1297, 2010.

28. G. W. Bluman and S. Kumei, Symmetries and Differential Equations, vol. 81, Springer, New York, NY, USA, 1989.

29. H. Stephani, Differential Equations: Their Solution Using Symmetries, Cambridge University Press, New York, NY, USA, 1989.

30. L. F. Shampine, I. Gladwell, and S. Thompson, Solving ODEs with MATLAB, chapter 3, Cambridge University Press, Cambridge, UK, 2003.

31. J. Kierzenka and L. F. Shampine, "A BVP solver based on residual control and the MATLAB PSE,"Transactions on Mathematical Software, vol. 27, no. 3, pp. 299–316, 2001.

7
CHAPTER

LIE GROUP ANALYSIS OF UNSTEADY FLOW AND HEAT TRANSFER OVER A POROUS SURFACE FOR A VISCOUS FLUID

M. B. Akgül, G. Sarı, and M. Pakdemirli

INTRODUCTION

Boundary layer flow and heat transfer for an incompressible viscous fluid past a heated porous surface have attracted a great interest because of the importance in engineering applications such as cooling of nuclear reactors and turbine blades, food processing, crystal growth, flow control on airfoils, and electronic cooling. Blasius [1] first presented a similarity solution for velocity distribution within the boundary layer for a viscous incompressible fluid over a flat plate. Khaled and Vafai [2] studied hydromagnetic effects on dynamical and thermal boundary layer characteristics. This analysis is concerned with a family of squeezed flows. They displayed effects of the magnetic parameter, Prandtl number, suction/injection parameter, and squeezing parameter on velocity and temperature profiles graphically. Elbashbeshy and Bazid [3] studied the heat transfer over an unsteady stretching surface. They obtained a new similarity solution for the temperature field. They investigated the effect of the Prandtl number and the unsteadiness parameter which determine the velocity and temperature profiles and heat transfer coefficient. Mahmood et al. [4] studied flow and heat transfer over a permeable sensor surface placed in a squeezing channel. They displayed effects of the squeezing parameter on values of velocity and temperature and the variation of skin friction coefficient and heat transfer coefficient against the transpiration parameter with the graphics. Bataller [5] investigated the flow influenced by nonlinearly stretching of the sheet with heat transfer. They analyzed the effects of various physical parameters on heat transfer phenomena in a viscous flow over a nonlinearly stretching sheet. Mukhopadhyay [6] investigated the effects of thermal radiation on unsteady boundary

Citation: M. B. Akgül, G. Sarı, and M. Pakdemirli, "Lie Group Analysis of Unsteady Flow and Heat Transfer over a Porous Surface for a Viscous Fluid," Journal of Applied Mathematics, vol. 2012, Article ID 675287, 17 pages, 2012. doi:10.1155/2012/675287

layer mixed convection heat transfer problem from a vertical porous stretching surface embedded in a porous medium. Their study indicated that the flow and temperature field are significantly influenced by the unsteadiness parameter, buoyancy force, and suction parameter in both porous and nonporous media. Aziz [7] studied thermal boundary layer flow over a flat plate considering convective surface heat flux at the lower surface of the plate and determined the condition which the convection heat transfer coefficient must meet for a similarity solution to exist. Aziz [8] studied the boundary layer flow over a flat plate with slip flow and constant heat flux surface condition velocity and shear stress distributions were presented for a range of values of the parameter characterizing the slip flow. Ishak [9] considered the problem of steady laminar boundary layer flow and heat transfer over a stationary flat plate immersed in a uniform free stream with convective boundary condition. They found that suction increases the surface shear stress and as a consequence increases the heat transfer rate at the surface. Yao et al. [10] investigated heat transfer flow over a stretching/shrinking sheet with convective boundary conditions. They found that temperature at the surface of the plate increases with the increase of the surface convection parameter. Rahman [11] studied locally similar solutions for the hydromagnetic thermal boundary layer flow of a viscous incompressible electrically conducting fluid over a flat plate with a partial slip at the surface of boundary in the presence of the convective surface boundary condition numerically and displayed results for the velocity, temperature, and Prandtl number within the boundary layer delineating the effect of various parameters characterizing the flow.

In most of the published studies in the literature [12–16], the partial differential equations were converted to ordinary differential equations via a similarity transformation found by ad hoc methods. The resulting ordinary differential equations were then solved by numerical methods. Lie group methods are performed to investigate similarity transformation of the boundary layer equations in only a few studies [17–19]. However, a detailed investigation of the complete symmetries for unsteady boundary layer flow and heat transfer equations are lacking in the literature.

In this paper, Lie group analysis of unsteady flow and heat transfer over a porous surface in the presence of suction/injection velocity is analyzed. Group classification of the governing equations with respect to the variable free stream flow and suction/injection velocity is performed. The infinitesimal generators are obtained. The restrictions imposed by the boundary conditions on the generators are discussed. Using the resulting generators, the governing equations are transformed into ordinary differential equations and solved numerically. Numerical results are presented for the selected dimensionless problem parameters and discussed in detail. Present study reveals that injection decreases temperature gradient whereas suction increases the temperature gradient. The increase in the Prandtl number causes increase in the temperature gradient with increasing the slope.

MATHEMATICAL FORMULATION

Consider the unsteady two-dimensional, incompressible laminar flow over a porous surface. Under the boundary layer approximation, the continuity, momentum, and energy equations can be written as

$$\frac{\partial \bar{u}}{\partial \bar{x}} + \frac{\partial \bar{v}}{\partial \bar{y}} = 0,$$

$$\frac{\partial \bar{u}}{\partial \bar{t}} + \bar{u}\frac{\partial \bar{u}}{\partial \bar{x}} + \bar{v}\frac{\partial \bar{u}}{\partial \bar{y}} = \frac{\partial \bar{U}}{\partial \bar{t}} + \bar{U}\frac{\partial \bar{U}}{\partial \bar{x}} + \nu\frac{\partial^2 \bar{u}}{\partial \bar{y}^2},$$

$$\frac{\partial \bar{T}}{\partial \bar{t}} + \bar{u}\frac{\partial \bar{T}}{\partial \bar{x}} + \bar{v}\frac{\partial \bar{T}}{\partial \bar{y}} = \alpha\frac{\partial^2 \bar{T}}{\partial \bar{y}^2}, \tag{2.1}$$

where $\bar{x}$ is the coordinate along the surface and $\bar{y}$ is the coordinate vertical to $\bar{x}$ $\bar{t}$ is time, u and v are the velocity components in $\bar{x}$ and $\bar{y}$ coordinates. $\bar{U}$ is the free stream velocity outside the boundary layer, $\bar{T}$ is the temperature, ν and α are the kinematic viscosity and thermal diffusivity of the fluid, respectively

The boundary conditions are

$$\bar{u}(\bar{x},0,\bar{t}) = 0, \quad \bar{v}(\bar{x},0,\bar{t}) = \bar{V}(\bar{x},\bar{t}), \quad \bar{T}(\bar{x},0,\bar{t}) = T_s,$$

$$\bar{u}(\bar{x},\infty,\bar{t}) = \bar{U}(\bar{x},\bar{t}), \quad \bar{T}(\bar{x},\infty,\bar{t}) = T_\infty, \tag{2.2}$$

where $\bar{V}(\bar{x},\bar{t})$ is the suction/injection velocity of the permeable surface. T_s is the constant surface temperature and T_∞ is the temperature of the ambient fluid

The nondimensional form of (2.1) and boundary conditions are as follows:

$$\frac{\partial u}{\partial x} + \frac{\partial v}{\partial y} = 0, \tag{2.3}$$

$$\frac{\partial u}{\partial t} + u\frac{\partial u}{\partial x} + v\frac{\partial u}{\partial y} = F(x,t) + \frac{1}{\mathrm{Re}}\frac{\partial^2 u}{\partial y^2}, \tag{2.4}$$

$$\frac{\partial T}{\partial t} + u\frac{\partial T}{\partial x} + v\frac{\partial T}{\partial y} = \frac{1}{\mathrm{Re\,Pr}}\frac{\partial^2 T}{\partial y^2}, \tag{2.5}$$

$$u(x,0,t) = 0, \quad v(x,0,t) = V(x,t), \quad T(x,0,t) = 1,$$

$$u(x,\infty,t) = U(x,t), \quad T(x,\infty,t) = 0. \tag{2.6}$$

The nondimensional parameters are related to the dimensional ones through the following relations:

$$u = \frac{\bar{u}}{U_0}, \quad v = \frac{\bar{v}}{U_0}, \quad U = \frac{\bar{U}}{U_0}, \quad T = \frac{(\bar{T} - T_\infty)}{(T_s - T_\infty)}, \quad x = \frac{\bar{x}}{L}, \quad y = \frac{\bar{y}}{L}, \quad t = \frac{U_0\bar{t}}{L},$$

$$\mathrm{Re} = \frac{U_0 L}{\nu}, \quad \mathrm{Pr} = \frac{\nu}{\alpha}, \quad F(x,t) = \frac{\partial U}{\partial t} + U\frac{\partial U}{\partial x}, \tag{2.7}$$

where U_0, L, Re, and Pr are the average free stream velocity, length of the horizontal surface, Reynolds number, and Prandtl number, respectively. F(x, t) is used for simplicity in the equations.

SYMMETRY ANALYSIS

The symmetry groups of (2.3)–(2.5) are calculated by using classical Lie group approach. Details of the theory can be found in Bluman and Kumei [20] and Stephani

[21]. The one-parameter infinitesimals Lie group of transformations are defined as

$$x^* = x + \varepsilon\xi_1(x,y,t,u,v,T),$$

$$y^* = y + \varepsilon\xi_2(x,y,t,u,v,T),$$

$$t^* = t + \varepsilon\xi_3(x,y,t,u,v,T),$$

$$u^* = u + \varepsilon\eta_1(x,y,t,u,v,T),$$

$$v^* = v + \varepsilon\eta_2(x,y,t,u,v,T),$$

$$T^* = T + \varepsilon\eta_3(x,y,t,u,v,T). \tag{3.1}$$

The prolonged infinitesimal generator which includes higher order derivatives is

$$X = \xi_1\frac{\partial}{\partial x} + \xi_2\frac{\partial}{\partial y} + \xi_3\frac{\partial}{\partial t} + \eta_1\frac{\partial}{\partial u} + \eta_2\frac{\partial}{\partial v} + \eta_3\frac{\partial}{\partial T} + \eta_{1x}\frac{\partial}{\partial u_x} + \eta_{1y}\frac{\partial}{\partial u_y} + \eta_{1t}\frac{\partial}{\partial u_t}$$
$$+ \eta_{2y}\frac{\partial}{\partial v_y} + \eta_{3x}\frac{\partial}{\partial T_x} + \eta_{3y}\frac{\partial}{\partial T_y} + \eta_{3t}\frac{\partial}{\partial T_t} + \eta_{1yy}\frac{\partial}{\partial u_{yy}} + \eta_{3yy}\frac{\partial}{\partial T_{yy}}. \tag{3.2}$$

The invariance conditions for (2.3)–(2.5) are obtained by applying the generator to the equations:

$$\eta_{1x} + \eta_{2y} = 0,$$

$$\eta_{1t} + \eta_1 u_x + u\eta_{1x} + \eta_2 u_y + v\eta_{1y} = \xi_1 F_x + \xi_3 F_t + \frac{1}{\mathrm{Re}}\eta_{1yy},$$

$$\eta_{3t} + \eta_1 T_x + u\eta_{3x} + \eta_2 T_y + v\eta_{3y} = \frac{1}{\mathrm{Re\,Pr}}\eta_{3yy}. \tag{3.3}$$

MathLie program [22] is used to calculate higher order infinitesimals and then equations are separated with respect to higher order variables. The initial restrictions on the infinitesimals are

$$\xi_1 = \xi_1(x,t), \qquad \xi_2 = \xi_2(x,y,t), \qquad \xi_3 = \xi_3(t),$$

$$\eta_1 = \eta_1(x,y,t,u), \qquad \eta_2 = \eta_2(x,y,t,u,v), \qquad \eta_3 = \eta_3(x,y,t,T), \tag{3.4}$$

and the remaining determining equations are

$$\frac{\partial^2 \eta_1}{\partial u^2} = 0,$$

$$\frac{\partial^2 \eta_3}{\partial T^2} = 0,$$

$$-2\frac{\partial \xi_2}{\partial y} + \frac{\partial \xi_3}{\partial t} = 0,$$

$$\frac{\partial \eta_1}{\partial x} + \frac{\partial \eta_2}{\partial y} = 0,$$

$$-\frac{\partial \xi_2}{\partial x} + \frac{\partial \eta_2}{\partial u} = 0,$$

$$\eta_1 - \frac{\partial \xi_1}{\partial t} - u\frac{\partial \xi_1}{\partial x} + 2u\frac{\partial \xi_2}{\partial y} = 0,$$

$$-\frac{\partial \xi_1}{\partial x} + \frac{\partial \xi_2}{\partial y} + \frac{\partial \eta_1}{\partial u} - \frac{\partial \eta_2}{\partial v} = 0,$$

$$\Pr \mathrm{Re}\,\eta_2 - \Pr \mathrm{Re}\frac{\partial \xi_2}{\partial t} - \Pr \mathrm{Re}\,u\frac{\partial \xi_2}{\partial x} + \Pr \mathrm{Re}\,v\frac{\partial \xi_2}{\partial y} + \frac{\partial^2 \xi_2}{\partial y^2} - 2\frac{\partial^2 \eta_3}{\partial y \partial T} = 0,$$

$$\mathrm{Re}\,\eta_2 - \mathrm{Re}\frac{\partial \xi_2}{\partial t} - \mathrm{Re}\,u\frac{\partial \xi_2}{\partial x} + \mathrm{Re}\,v\frac{\partial \xi_2}{\partial y} + \frac{\partial^2 \xi_2}{\partial y^2} - 2\frac{\partial^2 \eta_1}{\partial y \partial u} = 0,$$

$$\Pr \mathrm{Re}\frac{\partial \eta_3}{\partial t} + \Pr \mathrm{Re}\,u\frac{\partial \eta_3}{\partial x} + \Pr \mathrm{Re}\,v\frac{\partial \eta_3}{\partial y} - \frac{\partial^2 \eta_3}{\partial y^2} = 0,$$

$$-\mathrm{Re}\,\xi_1\frac{\partial F}{\partial x} - \mathrm{Re}\,\xi_3\frac{\partial F}{\partial t} - 2F\,\mathrm{Re}\frac{\partial \xi_2}{\partial y} + \mathrm{Re}\frac{\partial \eta_1}{\partial t} + F\,\mathrm{Re}\frac{\partial \eta_1}{\partial u} + \mathrm{Re}\,u\frac{\partial \eta_1}{\partial x} + \mathrm{Re}\,v\frac{\partial \eta_1}{\partial y} - \frac{\partial^2 \eta_1}{\partial y^2} = 0. \tag{3.5}$$

Equations (3.5) are an overdetermined partial differential system. Solving the system, the infinitesimals are finally obtained:

$$\xi_1 = ax + b(t),$$

$$\xi_2 = \frac{c}{2}y + d(x,t),$$

$$\xi_3 = ct + e,$$

$$\eta_1 = (a-c)u + \frac{db}{dt},$$

$$\eta_2 = -\frac{c}{2}v + \frac{\partial d}{\partial t} + u\frac{\partial d}{\partial x},$$

$$\eta_3 = gT + h. \tag{3.6}$$

F(x, t) is to be determined from the following equation:

$$\frac{d^2b}{dt^2} + F(a-2c) - \xi_1\frac{\partial F}{\partial x} - \xi_3\frac{\partial F}{\partial t} = 0. \tag{3.7}$$

The generator is applied to the boundary conditions which should also be invariant under the transformation leading to restrictions in (3.6):

$$\xi_1 = ax + b,$$

$$\xi_2 = \frac{c}{2}y,$$

$$\xi_3 = ct + e,$$

$$\eta_1 = (a - c)u,$$

$$\eta_2 = -\frac{c}{2}v,$$

$$\eta_3 = 0,$$

$$F(a - 2c) - \xi_1 \frac{\partial F}{\partial x} - \xi_3 \frac{\partial F}{\partial t} = 0 \tag{3.8}$$

or

$$F = (ct + e)^{(a-2c)/c} f(\mu), \qquad \mu = \frac{(ax + b)^{1/a}}{(ct + e)^{1/c}}. \tag{3.9}$$

V (x, t) and U(x, t) functions should satisfy the following equations:

$$-\frac{c}{2}V = (ax + b)\frac{\partial V}{\partial x} + (ct + e)\frac{\partial V}{\partial t},$$

$$(a - c)U = (ax + b)\frac{\partial U}{\partial x} + (ct + e)\frac{\partial U}{\partial t}. \tag{3.10}$$

Different functions can be obtained from (3.10) for V (x, t) and U(x, t) via setting the parameters as follows.

(I) Parameter a: selecting parameter a = 1 without a loss of generality and all other parameters zero yields

$$x\frac{\partial V}{\partial x} = 0 \Longrightarrow V = s(t),$$

$$U = x\frac{\partial U}{\partial x} \Longrightarrow U = xf(t). \tag{3.11}$$

In this case, suction or injection velocity is an arbitrary function of time. Solving (3.11) for U(x, t) yields U xf(t) , where f(t) is an arbitrary function of time for this special choice

(II) Parameter c: selecting parameter c - 1 and all other parameters zero yields

$$-\frac{V}{2} = t\frac{\partial V}{\partial t} = 0 \Longrightarrow V = \frac{s(x)}{\sqrt{t}},$$

$$-U = t\frac{\partial U}{\partial t} \Longrightarrow U = \frac{1}{t}f(x). \tag{3.12}$$

For this choice,

$V(x,t)$ and $U(x,t)$ yield $V = s(x)/\sqrt{t}$ and $U = (1/t)f(x)$ from (3.10).

(III) Parameter b: selecting parameter b - 1 and all other parameters zero yields

$$\frac{\partial V}{\partial x} = 0 \Longrightarrow V = s(t),$$
$$\frac{\partial U}{\partial x} = 0 \Longrightarrow U = f(t). \tag{3.13}$$

In this case, suction or injection velocity and free stream velocity are arbitrary functions of the spatial variable x.

SIMILARITY TRANSFORMATIONS AND REDUCTIONS

A Similarity Transformation

In this section, we choose the first case (case I) for obtaining similarity transformations as follows:

$$\frac{dx}{x} = \frac{dy}{0} = \frac{dt}{0} = \frac{du}{u} = \frac{dv}{0} = \frac{dT}{0}. \tag{4.1}$$

Similarity variables and functions are obtained from (4.1):

$$\mu_1 = y, \quad \mu_2 = t, \quad u = xF_1(\mu_1,\mu_2), \quad v = F_2(\mu_1,\mu_2), \quad T = F_3(\mu_1,\mu_2),$$
$$V = s(\mu_2), \quad U = xf(\mu_2). \tag{4.2}$$

The substitution of these variables and functions into the original equations and boundary conditions yields

$$F_1 + \frac{\partial F_2}{\partial \mu_1} = 0,$$
$$\frac{\partial F_1}{\partial \mu_2} + F_1^2 + F_2\frac{\partial F_1}{\partial \mu_1} = \frac{df(\mu_2)}{d\mu_2} + f(\mu_2)^2 + \frac{1}{\mathrm{Re}}\frac{\partial^2 F_1}{\partial {\mu_1}^2}, \tag{4.3}$$
$$\frac{\partial F_3}{\partial \mu_2} + F_2\frac{\partial F_3}{\partial \mu_1} = \frac{1}{\mathrm{Re\,Pr}}\frac{\partial^2 F_3}{\partial {\mu_1}^2},$$

$$F_1(0,\mu_2) = 0, \quad F_2(0,\mu_2) = s(\mu_2), \quad F_3(0,\mu_2) = 1, \quad F_1(\infty,\mu_2) = f(\mu_2),$$
$$F_3(\infty,\mu_2) = 0. \tag{4.4}$$

The partial differential system with three independent variables is reduced to a system with two independent variables

The Second Reduction

To reduce (4.3) and (4.4) into an ordinary differential system, one-parameter infinitesimal Lie group of transformations is defined as

$$\mu_1^* = \mu_1 + \varepsilon\gamma_1(\mu_1, \mu_2, F_1, F_2, F_3),$$

$$\mu_2^* = \mu_2 + \varepsilon\gamma_2(\mu_1, \mu_2, F_1, F_2, F_3),$$

$$F_1^* = F_1 + \varepsilon\varphi_1(\mu_1, \mu_2, F_1, F_2, F_3),$$

$$F_2^* = F_2 + \varepsilon\varphi_2(\mu_1, \mu_2, F_1, F_2, F_3),$$

$$F_3^* = F_3 + \varepsilon\varphi_3(\mu_1, \mu_2, F_1, F_2, F_3). \tag{4.5}$$

The prolonged infinitesimal generator is

$$X = \gamma_1\frac{\partial}{\partial\mu_1} + \gamma_2\frac{\partial}{\partial\mu_2} + \varphi_1\frac{\partial}{\partial F_1} + \varphi_2\frac{\partial}{\partial F_2} + \varphi_3\frac{\partial}{\partial F_3} + \varphi_{1_{\mu_1}}\frac{\partial}{\partial F_{1_{\mu_1}}} + \varphi_{2_{\mu_1}}\frac{\partial}{\partial F_{2_{\mu_1}}}$$

$$+ \varphi_{3_{\mu_1}}\frac{\partial}{\partial F_{3_{\mu_1}}} + \varphi_{1_{\mu_2}}\frac{\partial}{\partial F_{1_{\mu_2}}} + \varphi_{3_{\mu_2}}\frac{\partial}{\partial F_{3_{\mu_2}}} + \varphi_{1_{\mu_1\mu_1}}\frac{\partial}{\partial F_{1_{\mu_1\mu_1}}} + \varphi_{3_{\mu_1\mu_1}}\frac{\partial}{\partial F_{3_{\mu_1\mu_1}}}. \tag{4.6}$$

The invariance conditions for (4.3) are

$$\varphi_1 + \varphi_{2_{\mu_1}} = 0, \tag{4.7}$$

$$\varphi_{1_{\mu_2}} + 2F_1\varphi_1 + \varphi_2 F_{1_{\mu_1}} + F_2\varphi_{1_{\mu_1}} = \gamma_2\frac{d^2 f}{d\mu_2^2} + 2\gamma_2 f\frac{df}{d\mu_2} + \frac{1}{\mathrm{Re}}\varphi_{1_{\mu_1\mu_1}}, \tag{4.8}$$

$$\varphi_{3_{\mu_2}} + \varphi_2 F_{3_{\mu_1}} + F_2\varphi_{3_{\mu_1}} = \frac{1}{\mathrm{Re\,Pr}}\varphi_{3_{\mu_1\mu_1}}. \tag{4.9}$$

Substituting the higher order infinitesimals, separating the equations with respect to higher order variables yields

$$\frac{\partial^2\varphi_1}{\partial {F_1}^2} = 0,$$

$$\frac{\partial^2\varphi_3}{\partial {F_3}^2} = 0,$$

$$-2\frac{\partial\gamma_1}{\partial\mu_1} + \frac{\partial\gamma_2}{\partial\mu_2} = 0,$$

$$\varphi_2 - \frac{\partial\gamma_1}{\partial\mu_2} + \frac{1}{\mathrm{Re}}\frac{\partial^2\gamma_1}{\partial\mu_1^2} - 2\frac{1}{\mathrm{Re}}\frac{\partial^2\varphi_1}{\partial\mu_1\partial F_1} + F_2\frac{\partial\gamma_1}{\partial\mu_1} = 0,$$

$$\varphi_2 - \frac{\partial\gamma_1}{\partial\mu_2} + \frac{1}{\mathrm{RePr}}\frac{\partial^2\gamma_1}{\partial\mu_1^2} - 2\frac{1}{\mathrm{RePr}}\frac{\partial^2\varphi_3}{\partial\mu_1\partial F_3} + F_2\frac{\partial\gamma_1}{\partial\mu_1} = 0,$$

$$\frac{1}{\mathrm{Re}}\frac{\partial\varphi_3}{\partial\mu_2} + \frac{1}{\mathrm{RePr}}f^2\frac{\partial\varphi_3}{\partial F_1} + \frac{1}{\mathrm{RePr}}f'\frac{\partial\varphi_3}{\partial F_1} - \frac{1}{\mathrm{Re^2Pr}}\frac{\partial^2\varphi_3}{\partial{\mu_1}^2} - \frac{1}{\mathrm{RePr}}{F_1}^2\frac{\partial\varphi_3}{\partial F_1} + \frac{1}{\mathrm{Re}}\frac{\partial^2\varphi_3}{\partial\mu_1\partial F_2} = 0,$$

$$-2ff'\gamma_2 - 2f^2\frac{\partial\gamma_1}{\partial\mu_1} - 2f'\frac{\partial\gamma_1}{\partial\mu_1} + \frac{\partial\varphi_1}{\partial\mu_2} + f^2\frac{\partial\varphi_1}{\partial F_1} + f'\frac{\partial\varphi_1}{\partial F_1} - f''\gamma_2 - \frac{1}{\mathrm{Re}}\frac{\partial^2\varphi_1}{\partial{\mu_1}^2} + 2F_1\varphi_1$$

$$+ 2F_1^2\frac{\partial\gamma_1}{\partial\mu_1} - F_1^2\frac{\partial\varphi_1}{\partial F_1} + F_2\frac{\partial\varphi_1}{\partial\mu_1} = 0,$$

$$-\frac{2}{\mathrm{Re^2Pr}}\frac{\partial^2\varphi_3}{\partial\mu_1\partial F_1} - \frac{1}{\mathrm{RePr}}F_2\frac{\partial\varphi_3}{\partial F_1} + \frac{1}{\mathrm{Re}}F_2\frac{\partial\varphi_3}{\partial F_1} = 0,$$

$$-\frac{2}{\mathrm{Re^2Pr}}\frac{\partial^2\varphi_1}{\partial\mu_1\partial F_3} + \frac{1}{\mathrm{RePr}}F_2\frac{\partial\varphi_1}{\partial F_3} - \frac{1}{\mathrm{Re}}F_2\frac{\partial\varphi_1}{\partial F_3} = 0,$$

$$\varphi_1 + \frac{\partial\varphi_2}{\partial\mu_1} + F_1\frac{\partial\gamma_1}{\partial\mu_1} - F_1\frac{\partial\varphi_2}{\partial F_2} = 0, \tag{3.10}$$

from which the above equations are solved:

$$\gamma_1 = -\frac{c_1}{2}\mu_1 + c_2(\mu_2),$$

$$\gamma_2 = -c_1\mu_2 + c_3,$$

$$\varphi_1 = c_1 F_1,$$

$$\varphi_2 = \frac{c_1}{2}F_2 + \frac{dc_2}{d\mu_2},$$

$$\varphi_3 = c_4 F_3 + c_5. \tag{3.11}$$

The determining equation for $f(\mu_2)$ which is selected to U(x, t) , outer velocity, is obtained as follows:

$$-2f\frac{df}{d\mu_2}\gamma_2 - \frac{d^2 f}{d\mu_2^2}\gamma_2 - f^2\gamma_2' - \frac{df}{d\mu_2}\gamma_2' + \left(f^2 + \frac{df}{d\mu_2}\right)\frac{\partial\varphi_1}{\partial F_1} = 0. \tag{3.12}$$

The boundaries and boundary conditions should be invariant under the transformations also (Bluman and Kumei [20]):

$$\gamma_1 = -\frac{c_1}{2}\mu_1,$$

$$\gamma_2 = -c_1\mu_2 + c_3,$$

$$\varphi_1 = c_1 F_1,$$

$$\varphi_2 = \frac{c_1}{2}F_2,$$

$$\varphi_3 = 0,$$

$$-2f\frac{df}{d\mu_2}\gamma_2 - \frac{d^2 f}{d\mu_2^2}\gamma_2 - f^2\gamma_2' - \frac{df}{d\mu_2}\gamma_2' + \left(f^2 + \frac{df}{d\mu_2}\right)\frac{\partial\varphi_1}{\partial F_1} = 0. \tag{3.13}$$

Two parameters are involved in Lie group Transformations.

$S(\mu_2)$ and $f(\mu_2)$ functions should satisfy the following equations:

$$\frac{c_1}{2}s = (-c_1\mu_2 + c_3)\frac{ds}{d\mu_2},$$

$$c_1 f = (-c_1\mu_2 + c_3)\frac{df}{d\mu_2}. \tag{3.14}$$

Different functions can be obtained from the above equations for $s(\mu_2)$ and $f(\mu_2)$ via setting the parameters. The general case in which both parameters are involved will be considered.

Solving (4.14) results

$$s = \frac{\lambda_1}{\sqrt{c_1\mu_2 - c_3}}, \qquad f = \frac{1}{c_1\mu_2 - c_3}. \tag{3.15}$$

The equations determining the similarity transformations are

$$\frac{d\mu_1}{-(c_1/2)\mu_1} = \frac{d\mu_2}{-c_1\mu_2 + c_3} = \frac{dF_1}{c_1F_1} = \frac{dF_2}{(c_1/2)F_2} = \frac{dF_3}{0}, \tag{3.16}$$

from which the similarity variables and functions are derived:

$$\beta = \mu_1(c_1\mu_2 - c_3)^{-1/2}, \qquad \theta_1(\beta) = (c_1\mu_2 - c_3)F_1, \qquad \theta_2^*(\beta) = (c_1\mu_2 - c_3)^{1/2}F_2,$$
$$\theta_3(\beta) = F_3. \tag{3.17}$$

Inserting these variables and functions into (4.3) and (4.4) yields the following ordinary differential system:

$$\theta_1 + \theta^*_{2\beta} = 0,$$
$$\theta_1^2 - c_1\theta_1 + \theta_2^*\theta_{1\beta} - \frac{c_1\beta}{2}\theta_{1\beta} = (1 - c_1) + \frac{1}{\text{Re}}\theta_{1\beta\beta}$$
$$-\frac{c_1\beta}{2}\theta_{3\beta} + \theta_2\theta_{3\beta} = \frac{1}{\text{Re Pr}}\theta_{3\beta\beta}. \tag{3.18}$$

Further, we scale variables θ_2^* and β as

$$\theta_2^* = \frac{1}{\sqrt{\text{Re}}}\theta_2, \qquad \beta = \frac{1}{\sqrt{\text{Re}}}Y. \tag{3.19}$$

The substitution of these variables into (4.18) yields

$$\theta_{2Y}^2 + c_1\theta_{2Y} - \theta_2\theta_{2YY} + \frac{c_1Y}{2}\theta_{2YY} = (1 - c_1) - \theta_{2YYY},$$
$$-c_1Y\theta_{3Y} + \theta_2\theta_{3Y} = \frac{1}{\text{Pr}}\theta_{3YY}. \tag{3.20}$$

The associated boundary conditions are

$$\theta_1(0) = 0, \qquad \theta_2(0) = \lambda_1, \qquad \theta_3(0) = 1, \qquad \theta_1(\infty) = 1, \qquad \theta_3(\infty) = 0, \tag{3.21}$$

where λ_1 is the suction/injection parameter; c_1 and c_3 are the unsteadiness parameters.

RESULT AND DISCUSSION

The nonlinear reduced ordinary differential equation system (4.20) with the boundary conditions in (4.21) is solved numerically by applying a finite difference code that implements the three-stage Lobatto IIIA formula that provides continuous solutions of fourth order accuracy in the problem domain. Description of the numerical method is given by Shampine et al. [23] and Kierzenka and Shampine [24]. Results are displayed graphically for various values of problem parameters to see developments in the velocity and temperature fields.

The variation of the dimensionless x-component of velocity is shown for various values of unsteadiness parameter c_1 in the absence of suction/injection in Figure 1. It is clear that an increase in the c_1 causes a reduction in the velocity until a certain point. After this point, reverse effect in velocity is observed for increasing values of c_1.

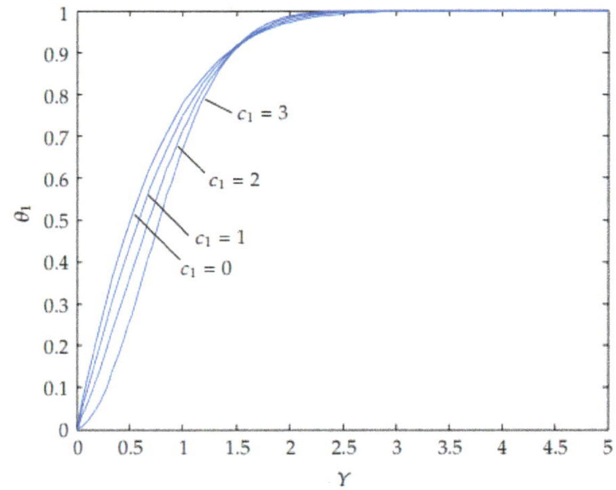

Figure 1: Variation of θ_1 for different values of unsteadiness parameter c_1 ($Pr = 10, \lambda_1 = 0$) .

The effect of suction/injection parameter on the velocity profile is presented in Figure 2. It is noted that suction decreases the boundary layer thickness whereas injection increases.

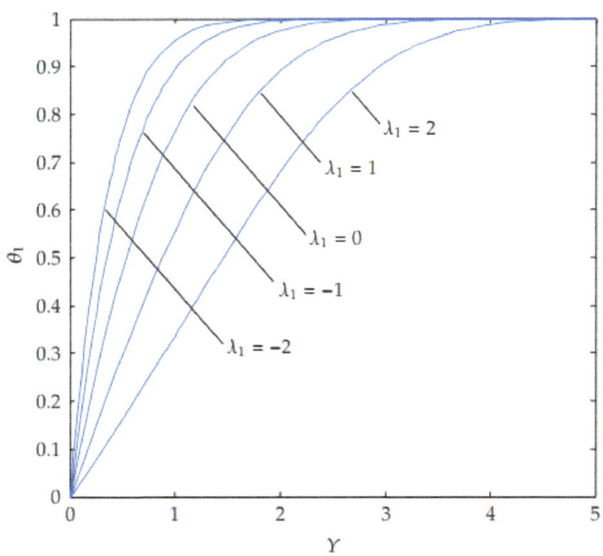

Figure 2: Variation of θ_1 for different values of suction/injection ($Pr = 10, c_1 = 0.5$) .

Effects of Prandtl number on the temperature profiles are given in Figure 3. An increase in the Prandtl number decreases the temperature profiles. Effects of unsteadiness parameter c_1 on the temperature profiles are depicted in Figure 4. An increase in c_1 decreases the temperature profiles. Thermal boundary layer is narrower for higher unsteadiness parameter. In Figure 5 effects of suction-injection parameter on the temperature profiles are depicted. $\lambda_1 = 0$ corresponds to no suction/injection case. Suction decreases the thickness of thermal boundary layer whereas injection increases the thickness of thermal boundary layer.

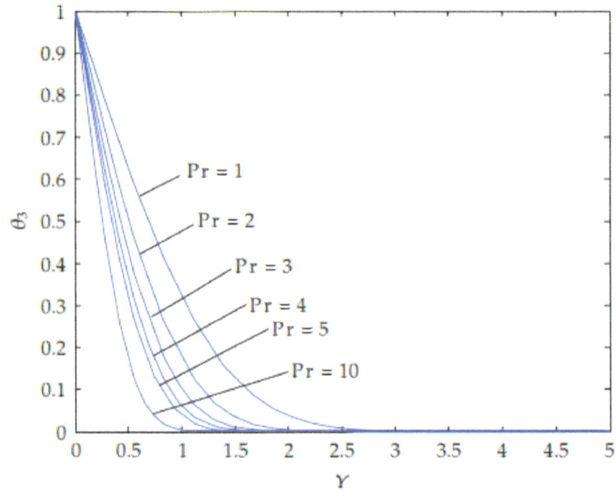

Figure 3: Variation of θ_3 for different values of the Prandtl number ($\lambda_1 = 0$, $c_1 = 0.5$).

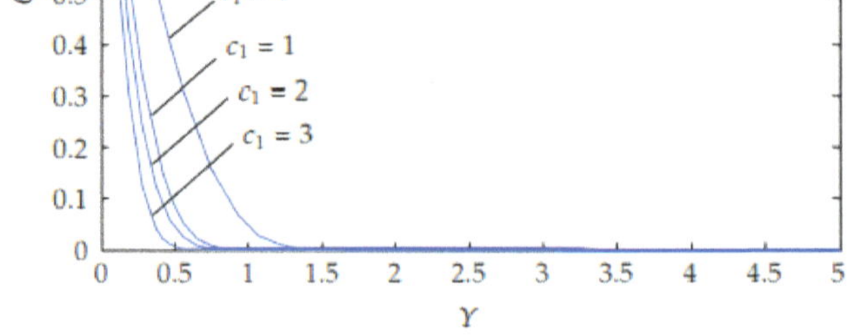

Figure 4: Variation of θ_3 for different values of c_1 ($Pr = 10$, $\lambda_1 = 0$).

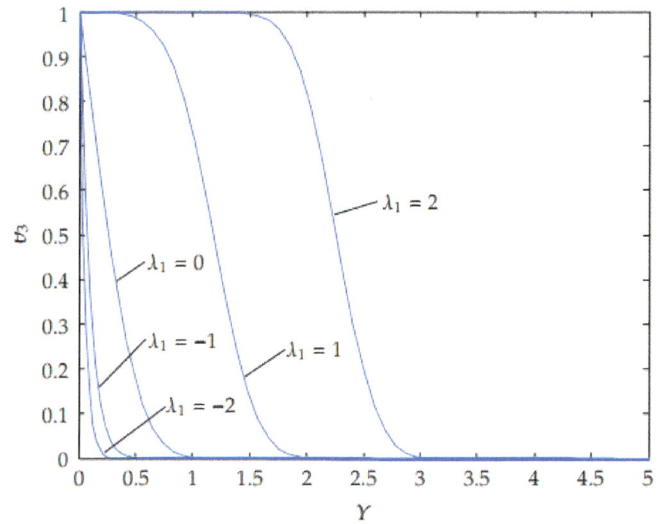

Figure 5: Variation of θ_3 for different values of suction/injection ($\lambda_1 = 0$, $c_1 = 0.5$) .

In Figure 6, we displayed the velocity gradient against suction/injection parameter for the case of flow along a horizontal porous surface for different values of the unsteadiness parameter, c_1. Injection decreases the velocity gradient whereas suction increases the velocity gradient. The reduction in c_1 causes an increase in velocity gradient with decreasing the slope.

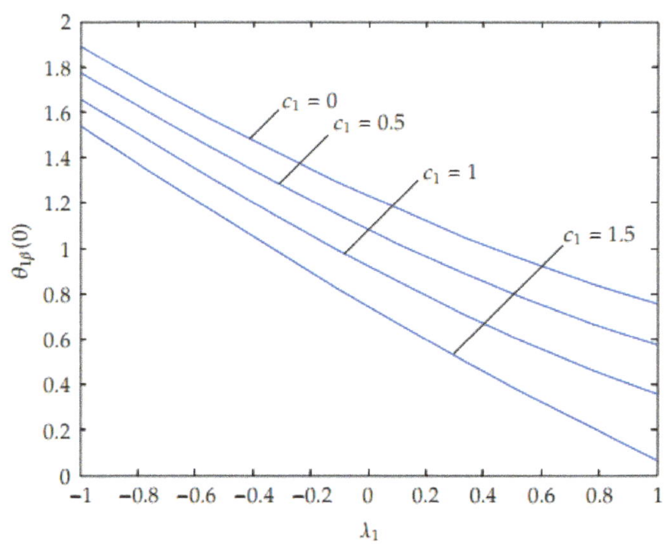

Figure 6: Variation of the velocity gradient with c_1 (Pr = 1)

In Figure 7, we show the temperature gradient against suction/injection parameter for different c_1-unsteadiness parameter. Injection decreases the temperature gradient whereas suction increases the temperature gradient. In Figure 8, we displayed the temperature gradient against unsteadiness parameter for different Prandtl numbers. The increase in the Prandtl number causes increase in temperature gradient.

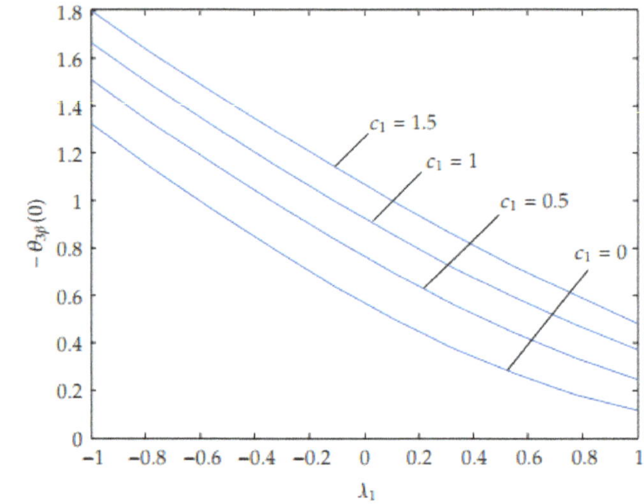

Figure 7: Variation of the temperature gradient with λ_1 (Pr = 1) .

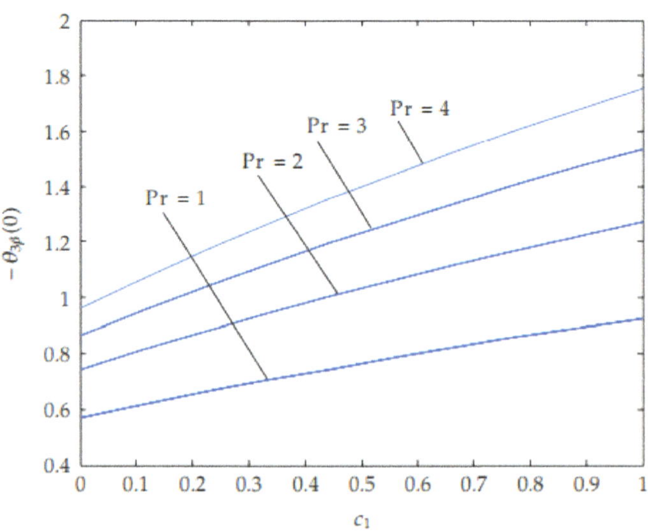

Figure 8: Variation of the temperature gradient with c_1 ($\lambda_1 = 0$) .

CONCLUSIONS

In this paper, we have studied the problem of two-dimensional, unsteady flow and heat transfer of a viscous fluid past a surface in the presence of variable suction/injection. The unsteadiness is due to time dependent free stream flow.

The governing equations are derived with the usual boundary layer approximation. Using Lie group theory, a group classification of the equations with respect to the variable free stream flow and suction/injection velocity is performed. Restrictions imposed by the boundary conditions on the symmetries are discussed. Adopting the obtained symmetry groups, governing partial differential equations are converted into ordinary differential equations and then solved numerically.

Effects of dimensionless problem parameters on the velocity and temperature profiles are outlined in the figures. From the present investigation the following results are obtained.

(i) An increase in the unsteadiness parameter c1 causes a reduction in the velocity.

(ii) Suction decreases the boundary layer thickness whereas injection increases.

(iii) An increase in the Prandtl number and unsteadiness parameter decreases the temperature profiles.

(iv) Suction decreases the thickness of thermal boundary layer whereas injection increases the thickness of boundary layer.

(v) Injection decreases the velocity gradient whereas suction increases. Moreover the reduction of c1causes an increase in the velocity gradient.

(vi) Injection decreases the temperature gradient whereas suction increases the temperature gradient. The increase in the Prandtl number causes increase in the temperature gradient with increasing the slope.

REFERENCES

1. H. Blasius, "Grenzschichten in Flussigkeiten mit kleiner Reibung," Zeitschrift für Mathematik und Physik, vol. 56, no. 1, pp. 1–37, 1908.

2. R. A. Khaled and K. Vafai, "Heat transfer and hydromagnetic control of flow exit conditions inside oscillatory squeezed thin films," Numerical Heat Transfer A, vol. 43, no. 3, pp. 239–258, 2003.

3. E. M. A. Elbashbeshy and M. A. A. Bazid, "Heat transfer over an unsteady stretching surface," Heat and Mass Transfer, vol. 41, no. 1, pp. 1–4, 2004.

4. M. Mahmood, S. Asghar, and M. A. Hossain, "Squeezed flow and heat transfer over a porous surface for viscous fluid," Heat and Mass Transfer, vol. 44, no. 2, pp. 165–173, 2007.

5. R. C. Bataller, "Similarity solutions for flow and heat transfer of a quiescent fluid over a nonlinearly stretching surface," Journal of Material Processing Technology, vol. 203, no. 1–3, pp. 176–183, 2008.

6. S. Mukhopadhyay, “Effect of thermal radiation on unsteady mixed convection flow and heat transfer over a porous stretching surface in porous medium,” Heat and Mass Transfer, vol. 52, no. 13-14, pp. 3261–3265, 2009.
7. Aziz, “A similarity solution for laminar thermal boundary layer over a flat plate with a convective surface boundary condition,” Communications in Nonlinear Science and Numerical Simulation, vol. 14, no. 4, pp. 1064–1068, 2009.
8. Aziz, “Hydrodynamic and thermal slip flow boundary layers over a flat plate with constant heat flux boundary condition,” Communications in Nonlinear Science and Numerical Simulation, vol. 15, no. 3, pp. 573–580, 2010.
9. Ishak, “Similarity solutions for flow and heat transfer over a permeable surface with convective boundary condition,” Applied Mathematics and Computation, vol. 217, no. 2, pp. 837–842, 2010.
10. S. Yao, T. Fang, and Y. Zhong, “Heat transfer of a generalized stretching/shrinking wall problem with convective boundary conditions,” Communications in Nonlinear Science and Numerical Simulation, vol. 16, no. 2, pp. 752–760, 2011.
11. M. M. Rahman, “Locally similar solutions for hydromagnetic and thermal slip flow boundary layers over a flat plate with variable fluid properties and convective surface boundary condition,” Meccanica, vol. 46, no. 5, pp. 1127–1143, 2011.
12. Ishak, R. Nazar, and I. Pop, “Heat transfer over an unsteady stretching permeable surface with prescribed wall temperature,” Nonlinear Analysis. Real World Applications, vol. 10, no. 5, pp. 2909–2913, 2009.
13. G. C. Layek, S. Mukhopadhyay, and S. A. Samad, “Heat and mass transfer analysis for boundary layer stagnation-point flow towards a heated porous stretching sheet with heat absorption/generation and suction/blowing,” International Communications in Heat and Mass Transfer, vol. 34, pp. 347–356, 2007.
14. R. Tsai, K. H. Huang, and J. S. Huang, “Flow and heat transfer over an unsteady stretching surface with non-uniform heat source,” International Communications in Heat and Mass Transfer, vol. 35, no. 10, pp. 1340–1343, 2008.
15. S. Sivasankaran, M. Bhuvaneswari, P. Kandaswamy, and E. K. Ramasami, “Lie group analysis of natural convection heat and mass transfer in an inclined surface,” Lithuanian Association of Nonlinear Analysis, vol. 11, no. 2, pp. 201–212, 2006.
16. S. Sivasankaran, M. Bhuvaneswari, P. Kandaswamy, and E. K. Ramasami, “Lie group analysis of natural convection heat and mass transfer in an inclined porous surface with heat generation,” International Journal of Applied Mathematics and Mechanics, vol. 2, no. 1, pp. 34–40, 2006.
17. M. Pakdemirli, “Similarity analysis of boundary layer equations of class of non-Newtonian fluids,”International Journal of Nonlinear Mechanics, vol. 29, no. 2, pp. 187–196, 1994.
18. M. Pakdemirli, M. Yürüsoy, and A. Küçükbursa, “Symmetry groups of boundary layer equations of class of non-Newtonian fluids,” International Journal of Nonlinear Mechanics,, vol. 31, no. 3, pp. 267–276, 1996.

19. M. Yürüsoy and M. Pakdemirli, "Symmetry reductions of unsteady three-dimensional boundary layers of some non-Newtonian fluids," International Journal of Engineering Science, vol. 35, no. 8, pp. 731–740, 1997.

20. G. W. Bluman and S. Kumei, Symmetries and Differential Equations, Springer, New York, NY, USA, 1989.

21. H. Stephani, Differential Equations: Their Solution Using Symmetries, Cambridge University Press, New York, NY, USA, 1989.

22. Wolfram Mathematica, MathLie 3.0, Dr. Gerd Baumann, 1992.

23. L. F. Shampine, I. Gladwell, and S. Thompson, Solving ODEs with MATLAB, chapter 3, Cambridge University Press, 2003.

24. J. Kierzenka and L. F. Shampine, "A BVP solver based on residual control and the MATLAB PSE," ACM Transactions on Mathematical Software, vol. 27, no. 3, pp. 299–316, 2001

8
CHAPTER

LIE GROUP METHOD OF THE DIFFUSION EQUATIONS

Jian-Qiang Sun, Rong-Fang Huang, Xiao-Yan Gu, and Ling Yu

INTRODUCTION

The subject of solving stiff system has been well studied in the literature, including linearly implicit methods, semi-implicit methods, time-splitting methods, projection methods, multiscale methods, integrating factor methods, the exponential time difference (ETD) methods, and Lie group methods [1– 12]. Lie group methods have been developed by MuntheKaas, Iserles, Nørsett, and their collaborators. In general, if a Lie group G acts freely and transitively on a manifold M, then a differential equation on M uniquely determines a differential equation on G. Thus, replacing the original differential on M by the equivalent differential equation on Lie algebra g allows one to apply appropriate Lie group integrators. There are many Lie group methods, such as Mangus method, RKMK method, and Fer method. Lie group methods can preserve the numerical solutions of the differential equations on the same manifolds, which have good stability and the same accuracy as the classical numerical methods [1, 2, 11, 13, 14]. In the paper, the exponential integrator method of the general nonlinear differential equations is proposed based on the idea of Lie group methods on manifolds and applied to the diffusion equations

The paper is organized as follows: in Section 2, the Lie group method and the exponential time difference method are introduced. The exponential integrator method for the general nonlinear differential equations is proposed. In Section 3, the quasilinear diffusion equation is solved by the explicit exponential integrator method and the explicit Runge-Kutta methods. In Section 4, the Allen-Cahn equation is solved by the explicit exponential integrator method and the explicit Runge-Kutta methods. At last, some conclusions are obtained.

Citation: Jian-Qiang Sun, Rong-Fang Huang, Xiao-Yan Gu, and Ling Yu, "Lie Group Method of the Diffusion Equations," Advances in Mathematical Physics, vol. 2014, Article ID 641918, 6 pages, 2014. doi:10.1155/2014/641918

EXPONENTIAL INTEGRATOR

Suppose that M is a manifold. There exist a Lie algebra g with a Lie bracket $[\cdot, \cdot]$, a (left) Lie algebra action : $g \times M \rightarrow M$, and a function $f: R \times M \rightarrow g$ such that the equation for $y(t) \in M$ can be written as

$$y' = (\lambda_* f(t, y))(y), \quad y(0) = p, \tag{1}$$

where

$$\lambda_* : g \longrightarrow \aleph(M) \tag{2}$$

is a homomorphism. The left algebra action $\lambda: g \times M \rightarrow M$ assigns to the arguments V $\in$ g and $y_0(x) \in M$ the solution of

$$y_t = (\lambda_* v)(y) \tag{3}$$

at time t=1with the initial condition $y0(x)$ and some boundary conditions. There is freedom to choose the function f, the corresponding differential equation in the Lie algebra

$$\frac{dv}{dt} = \mathrm{dexpinv}\left(v, \mathrm{f}\left(\mathrm{t}, \lambda\left(v, \mathrm{y}_0\right)\right)\right), \quad v(0) = \mathrm{O}. \tag{4}$$

Lie algebra is chosen as functions corresponding to simpler vector fields on M which are easy to exponentiate. The correspondence between the Lie algebra and the vector fields on M will be one to one. The Runge-Kutta method is applied to solve (4). It is the well-known RKMK method [1].

Let $a_{i,}$ and b_i be the coefficients of an s-stage, q order classical Runge-Kutta method, and let $c_i = \sum_{j=1}^{s} a_{i,j}$. The algorithm integrates (1) from t=0 to t=h:

Set $y_0 = p$.

For i = 1, 2, . . . , s

$$u_i = h\sum_{j=1}^{s} a_{i,j}\tilde{k}_j, \qquad k_i = f\left(hc_i, \lambda\left(u_i, y_0\right)\right)$$

$$\tilde{k}_i = \mathrm{dexpinv}\left(\mathrm{u_i}, \mathrm{k_i}, \mathrm{q}\right), \qquad \mathrm{q} = \text{the order of RK method.} \tag{5}$$

End

$$v = h\sum_{j=1}^{s} b_j\tilde{k}_j,$$

$$y_1 = \lambda\left(v, y_0\right), \tag{6}$$

where

$$\mathrm{dexpinv}\left(\mathrm{u}, v, \mathrm{q}\right) = \sum_{\mathrm{k!}}^{\mathrm{q}-1} \frac{\mathrm{B_k}}{\mathrm{k!}} \mathrm{ad}_{\mathrm{u}}^{\mathrm{k}}(v), \tag{7}$$

$ad_u^0(v) = v$, $ad_u^1(v) = [u, v]$ and $ad_u^k(v) = ad_u(ad_u^{k-1}(v))$ The number q denotes the order of the underlying Runge-Kutta method and B_k is the kth Bernoulli number [2].

A system of ordinary differential equations is often obtained

$$y_t = Ly + N(y,t), \quad y \in R^n, \tag{8}$$

where $L \in R^{n\times n}$. The ETD method can be described in the context of solving (8). Integrating the equation over a single time step from $t=t_n$ to $t_{n+1} = t_n + h$, we can get

$$y(t_{n+1}) = e^{Lh} y(t_n) + e^{Lh} \int_0^h e^{-Ls} N(y(t_n + s), t_n + s)\, ds. \tag{9}$$

Equation (9) is exact, and the various ETD schemes come from the approximation to the integral. Suppose y_n is the approximation to (t_n), and the first order ETD method is given by

$$y_{n+1} = e^{Lh} y_n + L^{-1}\left(e^{Lh} - I\right) N(y_n, t_n). \tag{10}$$

Higher order ETD schemes can also be found [7, 9, 10]. Some preliminary numerical studies conducted by a group of material scientists at the Penn State University have indicated that the higher order ETD schemes can be several orders of magnitudes faster than the low-order semi-implicit methods in some simulations of microstructure evolution [8].

In essence, for nonlinear time dependent equations, the ETD scheme provides a systematic coupling of the explicit treatment of nonlinearities and the implicit and possibly exact integration of the stiff linear part of the equation, while achieving high accuracy and maintaining good stability. The exponential integrator of (8) is constructed based on the idea of Lie group methods on manifolds. In general, (8) can be written as

$$y' = A(t, y)\, y, \quad y(0) = y_0, \quad y \in R^n, \tag{11}$$

where $A: R\times R^n \to R^{n\times n}$. In addition, A should be locally Lipschitz function (to ensure existence and uniqueness). The classical fourth order explicit RK scheme of (11) is

$$\begin{aligned}
V_1 &= Y_n, & K_1 &= A(t_n, V_1) V_1, \\
V_2 &= Y_n + \frac{1}{2} h K_1, & K_2 &= A\left(t_n + \frac{1}{2}h, V_2\right) V_2, \\
V_3 &= Y_n + \frac{1}{2} h K_2, & K_3 &= A\left(t_n + \frac{1}{2}h, V_3\right) V_3, \\
V_4 &= Y_n + h K_3, & K_4 &= A(t_n + h, V_4) V_4, \\
Y_{n+1} &= Y_n + h\left(\frac{1}{6}K_1 + \frac{1}{3}K_2 + \frac{1}{3}K_3 + \frac{1}{6}K_4\right). & &
\end{aligned} \tag{12}$$

According to a classical result of Hausdorff, the solution of (11) is

$$y(t) = e^{X(t)} y(0), \tag{13}$$

$(t) : R \to R^{n\times n}$ is the solution of the initial value problem

$$X(t)' = d\exp^{-1}_{X(t)} A(t, Y), \quad X(t_0) = O, \tag{14}$$

where

$$= \sum_{j=0}^{\infty} \frac{B_p}{p!} ad^p_X A, \tag{15}$$

$ad^k_X A = [X, ad^{k-1}_X A]$, and $[X, A] = XA - AX$. Taking p as the corresponding number and applying the Runge-Kutta method to (14), we can get the numerical solution of (14). So the numerical solution of (11) is obtained. The corresponding p order exponential integrator of (11) is

$$X_k = \sum_{l=1}^{p} a_{k,l} F_l,$$

$$F_k = d\exp^{-1}_X (A_k, p),$$

$$A_k = hA(t_n + c_k h, \exp(X_k) y_n),$$

$$X = \sum_{k=1}^{p} F_k,$$

$$y_{n+1} = \exp(X) y_n \tag{16}$$

for $n \in N$ and it is explicit provided that the underlying RK scheme is explicit. $d\exp^{-1}_X(A_k, p)$ is the p order truncation error of the $d\exp^{-1}_X(A_k)$. An explicit fourth order exponential integrator of (11) is as follows:

$$V_1 = O, \qquad K_1 = A\left(t_n, y_n\right),$$

$$V_2 = \frac{1}{2}hK_1,$$

$$K_2 = A\left(t_n + \frac{1}{2}h, e^{V_2}y_n\right) - \frac{1}{2}\left[V_2, A\left(t_n + \frac{1}{2}h, e^{V_2}y_n\right)\right],$$

$$V_3 = \frac{1}{2}hK_2,$$

$$K_3 = A\left(t_n + \frac{1}{2}h, e^{V_3}y_n\right) - \frac{1}{2}\left[V_3, A\left(t_n + \frac{1}{2}h, e^{V_3}y_n\right)\right],$$

$$V_4 = hK_3,$$

$$K_4 = A\left(t_n + h, e^{V_4}y_n\right) - \frac{1}{2}\left[V_4, A\left(t_n + h, e^{V_4}y_n\right)\right],$$

$$\Delta = h\left(\frac{1}{6}K_1 + \frac{1}{3}K_2 + \frac{1}{3}K_3 + \frac{1}{6}K_4\right),$$

$$y_{n+1} = e^{\Delta}y_n. \tag{17}$$

As the Lie group method on manifolds, the exponential integrator method of the general nonlinear differential equation has the same accuracy as the corresponding Runge-Kutta methods. In the subsequent sections, the fourth order explicit exponential integrator (EEI) method and the corresponding explicit Runge-Kutta (ERK) method are applied to the diffusion equations.

NUMERICAL EXPERIMENTS I: THE QUASILINEAR CONVECTION DIFFUSION EQUATION

To the quasilinear convection diffusion equation

$$\frac{\partial u}{\partial t} + u\frac{\partial u}{\partial x} = \frac{1}{\mathrm{Re}}\frac{\partial^2 u}{\partial x}, \tag{18}$$

the equation is the one-dimensional quasilinear parabolic differential equation, which is known as Burgers' equation, where $\mathrm{Re} > 0$ is a constant representing the kinematic viscosity of the fluid. Burgers' equation first appeared in a paper by Bateman and he gave a special solution of (18). In remarkable series of papers from 1939 to 1965, Burgers' investigated various aspects of turbulence and he used it as a model in studies of turbulence. Cole studied the general properties of Burgers' equation and outlined some of its various applications [15–17]. With the initial condition $(x, 0) = \sin(\pi x)$, $0 \le x \le 1$, and $t > 0$, the exact solution of (18) is

$$u(x,t) = \frac{4\pi}{\mathrm{Re}}\left(\left(\sum_{n=1}^{\infty} e^{-n^2\pi^2 t/\mathrm{Re}} n I_n\left(\frac{\mathrm{Re}}{2\pi}\right)\sin(n\pi x)\right)\right.$$
$$\times\left(I_0\left(\frac{\mathrm{Re}}{2\pi}\right)+2\sum_{n=1}^{\infty} e^{-n^2\pi^2 t/\mathrm{Re}}\right.$$
$$\left.\left.\times\, I_n\left(\frac{\mathrm{Re}}{2\pi}\right)\cos(n\pi x)\right)^{-1}\right), \tag{19}$$

and $I_n(x)$ is the Bessel function. Recently some numerical methods of Burgers' equation have been proposed [15–17].

Burgers' equation can be transformed into the nonlinear ordinary differential equation by discretization in the spacial direction. The spacial derivative of Burger' equation can be discretized as follows: $(\partial u/\partial x)|_{x=x_i} = ((u_{i+1} - u_{i-1})/2\Delta x)$. $(\partial^2 u/\partial^2 x)|_{x=x_i} = ((u_{i+1} - 2u_i + u_{i-1})/(\Delta x)^2)$. So the variable $u_i = (x_i, t)$ can satisfy

$$\frac{\partial u_i}{dt} + u_i\frac{(u_{i+1} - u_{i-1})}{2\Delta x} = \frac{1}{\mathrm{Re}}\frac{u_{i+1} - 2u_i + u_{i-1}}{(\Delta x)^2}. \tag{20}$$

Supposing that $U = (u_1, u_2, \ldots,_{-1})$, (20) can be written as

$$\frac{dU}{dt} = B(U)U, \tag{21}$$

...

$$B(U) = \frac{\eta}{(\Delta x)^2}\begin{pmatrix} -2 & 1 & 0 & \cdots & 0 & 0 \\ 1 & -2 & 1 & \cdots & 0 & 0 \\ \vdots & \vdots & \vdots & \ddots & \vdots & \vdots \\ 0 & 0 & 0 & \cdots & -2 & 1 \\ 0 & 0 & 0 & \cdots & 1 & -2 \end{pmatrix}$$
$$+\mu\begin{pmatrix} u_2 & 0 & 0 & \cdots & 0 & 0 \\ 0 & c_2 & 0 & \cdots & 0 & 0 \\ \vdots & \vdots & \vdots & \ddots & \vdots & \vdots \\ 0 & 0 & 0 & \cdots & c_{N-2} & 0 \\ 0 & 0 & 0 & \cdots & 0 & c_{N-1} \end{pmatrix}, \tag{22}$$

where $\mu = (1/2\Delta x)$, $\eta = (1/\mathrm{Re})$, $c_k = u_{k+1} - u_{k-1}$, and $k = 2, \ldots, N-2$, $c_{N-2} = -u_{N-2}$. EEI method and classical ERK method are applied to (21).

In Table 1, we compare the stability property of ERK method and EEI methods with Re = 100. The convergent steps with different spacial step are shown in Table 1. The convergent step lengths are obtained by computing three hundred steps to decide whether the step length is convergent. From Table 1, we can see that EEI method has better stability than ERK method. In Table 2, we compare the global errors of ERK method and EEI method with $N = 100$ and $h = 0.002$. The global error is defined as $E_{rr} = \max |Y^n - (n\Delta t)|$, where $(n\Delta t)$ is the exact solution at $t = n\Delta t$. They have the same accuracy. In Figure 1, the numerical solutions of Burgers' equation at $t \in [0, 3]$ are obtained by ERK method and the EEI method with $N = 100$, Re = 100, and $h = 0.005$. Figure 1(a) was obtained by ERK method and Figure 1(b) was obtained by EEI method. From Figure 1, we can get that they have the same accuracy.

Table 1: The stability comparison of ERK method and EEI method.

	ERK	EEI
$N = 50$	0.028	0.10
$N = 100$	0.007	0.080
$N = 200$	0.0016	0.025
$N = 400$	0.0005	0.0075

Table 2: The errors of the numerical solutions and the exact solutions at different times by ERK method and EEI method.

	ERK	EEI
$t = 1$	5.041×10^{-5}	5.044×10^{-5}
$t = 5$	8.765×10^{-4}	8.764×10^{-4}
$t = 10$	4.831×10^{-3}	4.831×10^{-3}
$t = 15$	1.265×10^{-2}	1.265×10^{-2}

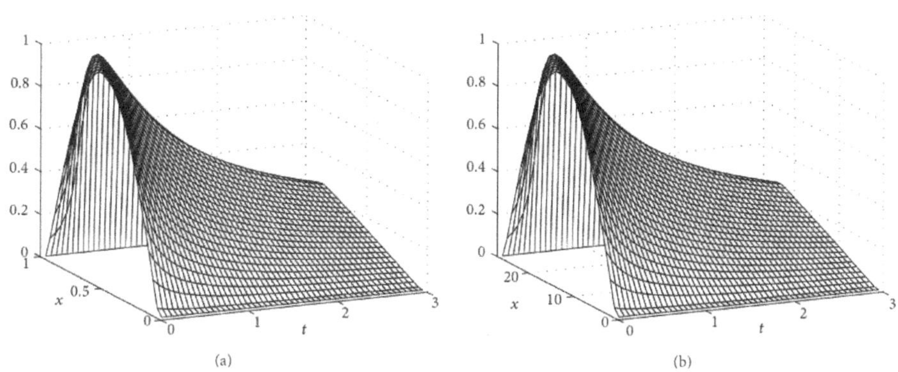

Figure 1: The numerical solutions of Burgers' equation at $t \in [0, 3]$ obtained by ERK method and EEI method

NUMERICAL EXPERIMENTS II: THE ALLEN-CAHN EQUATION

To the Allen-Cahn equation

$$\frac{\partial u}{\partial t} = \varepsilon u_{xx} + u - u^3, \tag{23}$$

with $\varepsilon = 0.01$ and initial condition $u(x, 0) = 0.53x + 0.47 \sin(-1.5\pi x)$. Suppose that $U = (u_1, u_2, \ldots)$, and (23) can be discretized into

$$\frac{dU}{dt} = B(U)U, \tag{24}$$

$$B(U) = \frac{\varepsilon}{(\Delta x)^2} \begin{pmatrix} -2 & 1 & 0 & \cdots & 0 & 0 \\ 1 & -2 & 1 & \cdots & 0 & 0 \\ \vdots & \vdots & \vdots & \ddots & \vdots & \vdots \\ 0 & 0 & 0 & \cdots & -2 & 1 \\ 0 & 0 & 0 & \cdots & 1 & -2 \end{pmatrix} + \begin{pmatrix} c_1 & 0 & 0 & \cdots & 0 & 0 \\ 0 & c_2 & 0 & \cdots & 0 & 0 \\ \vdots & \vdots & \vdots & \ddots & \vdots & \vdots \\ 0 & 0 & 0 & \cdots & c_{N-1} & 0 \\ 0 & 0 & 0 & \cdots & 0 & c_N \end{pmatrix}, \tag{25}$$

where $c_k = 1 - u_k^2$ and $k = 1, \ldots, N$. EEI method and classical ERK method are applied to (24).

In Table 3, we compare the stability property of ERR method and EEI method by solving (24). When N is different, the convergent step with different spacial step are shown in Table 3. The convergent step lengths are obtained by computing three hundred steps to decide whether the step length is convergent. From Table 3, we can see that the EEI method has better stability than ERK method. In Table 4, we compare the errors of the numerical solution at $x = -0.4$ with $h = 0.01$ obtained by ERK method and the numerical solution obtained by EEI method. The error is up to 10^{-8} at different times; the error is (h^4). We can conclude that they have the same accuracy. In Figure 2, the numerical solutions of the Allen-Cahn equation in $t \in [0, 200]$ were obtained by ERK method and EEI method with $N = 100$, $h = 0.01$, and $x = -0.4$. Figure 2(a) was obtained by ERK method. Figure 2(b) was obtained by EEI method. From Figure 2, we can get that the numerical results are the same. We can also conclude that the two methods have the same accuracy.

Table 3: The stability comparison of ERK method and EEI method.

	ERK	EEI
$N = 50$	0.1	0.7
$N = 100$	0.02	0.25
$N = 200$	0.006	0.1
$N = 400$	0.0015	0.06

Table 4: The errors of the numerical solutions at different times obtained by ERK method and EEI method.

	ERK	EEI	Err
$t = 5$	0.50893803918253	0.50893804639290	10^{-8}
$t = 10$	0.49120333210012	0.49120334154949	10^{-8}
$t = 15$	0.46413587595252	0.46413588834075	10^{-8}
$t = 20$	0.42975156048078	0.42975157699815	10^{-8}

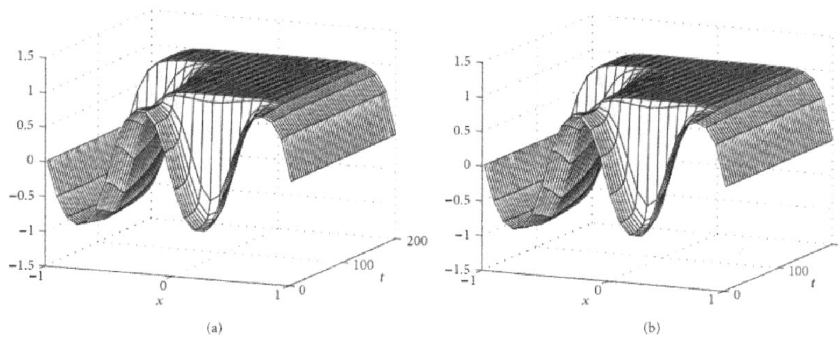

Figure 2: The numerical solutions of the Allen-Cahn equation at $t \in [0, 200]$ obtained by the two methods with $N = 100$ and $\varepsilon = 0.01$.

CONCLUSIONS

In this paper, the EEI method based on the idea of the Lie group method was proposed. The EEI method and the corresponding ERK method were applied to Burgers' equation and Allen-Cahn equation. Numerical results showed that the EEI method has better stability than the corresponding ERK methods. The two methods have the same accuracy. It is obvious that the explicit EER method is better than the ERK method in computing some stiff ordinary differential equations.

CONFLICT OF INTERESTS

The authors declare that there is no conflict of interests regarding the publication of this paper.

ACKNOWLEDGMENT

This work was supported by the National Natural Science Foundation of China (11161017 and 10905022).

REFERENCES

1. H. Munthe-Kaas, "Runge-Kutta methods on Lie groups," BIT Numerical Mathematics, vol. 38, no. 1, pp. 92–111, 1998.
2. Iserles, H. Munthe-Kaas, S. P. Nørsett, and A. Zanna, "Lie group methods," Acta Numerica, vol. 9, pp. 215–365, 2000.
3. G. J. Cooper, "Stability of Runge-Kutta methods for trajectory problems," IMA Journal of Numerical Analysis, vol. 7, no. 1, pp. 1–13, 1987.
4. P. E. Crouch and R. Grossman, "Numerical integration of ordinary differential equations on manifolds,"Journal of Nonlinear Science, vol. 3, no. 1, pp. 1–33, 1993.
5. Iserles, "Solving linear ordinary differential equations by exponentials of iterated commutators,"Numerische Mathematik, vol. 45, no. 2, pp. 183–199, 1984.
6. M. P. Calvo, A. Iserles, and A. Zanna, "Conservative methods for the Toda lattice equations," IMA Journal of Numerical Analysis, vol. 19, no. 4, pp. 509–523, 1999.
7. S. M. Cox and P. C. Matthews, "Exponential time differencing for stiff systems," Journal of Computational Physics, vol. 176, no. 2, pp. 430–455, 2002.
8. L. Q. Chen, C. Wolverton, V. Vaithyanathan, and Z. Liu, "Modeling solid-state phase transformations and microstructure evolution," MRS Bulletion, vol. 26, pp. 197–292, 2001.
9. G. Akrivis, M. Crouzeix, and C. Makridakis, "Implicit-explicit multistep methods for quasilinear parabolic equations," Numerische Mathematik, vol. 82, no. 4, pp. 521–541, 1999.
10. E. Weinan, "Analysis of the heterogeneous multiscale method for ordinary differential equations,"Communications in Mathematical Sciences, vol. 3, pp. 423–436, 2003.
11. Zanna, "Collocation and relaxed collocation for the Fer and the Magnus expansions," SIAM Journal on Numerical Analysis, vol. 36, no. 4, pp. 1145–1182, 1999.
12. L. R. Petzold, L. O. Jay, and J. Yen, "Numerical solution of highly oscillatory ordinary differential equations," Acta Numerica, vol. 6, pp. 437–483, 1997.
13. E. Hairer and G. Wanner, Solving Ordinary Differential Equations II: Stiff and Differential Algebraic Problems, Springer, New York, NY, USA, 1999.

14. M. Z. Qin and Y. S. Wang, Structure-Preserving Algorithm for Partial Differential Equation, Zhejiang Science and Technology Publishing, 2012.

15. R. C. Mittal and P. Singhal, "Numerical solution of Burger's equation," Communications in Numerical Methods in Engineering, vol. 9, pp. 163–175, 1993.

16. S. Kutluay, A. R. Bahadir, and A. Özdeş, "Numerical solution of one-dimensional Burgers› equation: explicit and exact-explicit finite difference methods," Journal of Computational and Applied Mathematics, vol. 103, no. 2, pp. 251–261, 1999.

17. E. R. Benton and G. W. Platzman, "A table of solution of the one-deimensional Burgers› equation,"Quarterly of Applied Mathematics, vol. 30, pp. 195–212, 1972.

9
CHAPTER

DYNAMICAL EQUATIONS OF MULTIBODY SYSTEMS ON LIE GROUPS

Wenjie Yu and Zhenkuan Pan

Keywords: Multibody system dynamics, Euler–Lagrange equations, Lie group, Euler–Poinare´ equations, reconstruction equations, constraint equations

INTRODUCTION

Multibody systems are mechanical systems that include interconnected bodies through different joints driven by some types of forces; their dynamical equations are usually denoted by nonlinear highly coupled ordinary differential equations or differential/algebraic equations, which must be solved numerically. The computational efficiency and stability of these dynamical equations depend strongly on the configuration description of multibody systems [1–9] and especially on the expressions of the orientations of bodies.

The classical Euler angles are usually used to describe the rotations of bodies in multibody systems,1 but the resulting dynamical equations of motion are nonlinear inertia coupled with singular problems. If the Euler angles are replaced by the Euler parameters,2[–6] the singular problems can be overcome. However, the nonlinear inertia coupling still exists. In addition, we must introduce new constraints in order to satisfy the definition of Euler parameters. If the three directors from direction cosine matrix of everybody are used to express the orientation,10 the generalized mass matrices of the dynamical equations are constant, which contributes to high computational efficiency, but six new constraint equations must be introduced for each body which enlarges the computational scales. If we use the orientation matrices as special orthogonal groups to describe the rotations,11[-14] the motions of bodies in multibody systems can be directly

Citation: Wenjie Yu and Zhenkuan Pan, Dynamical equations of multibody systems on Lie groups, DOI: 10.1177/1687814015575959

denoted by the Lie groups and their corresponding Lie algebras. Meanwhile, we do not need any additional constraints for the rotation description. Noticeably, the resulting generalized mass matrices are constant. This Lie group modeling method has become one of foundations for structure-preserving algorithms.

Technically, different descriptions of system configuration lead to different types of dynamical equations even though we use same modeling principle. In vectorized parameterization space, the Euler–Lagrange equations or Hamilton's equations of multibody system dynamics can be derived under the Hamilton's principle. While under the framework of Lie group and Lie algebra, these equations are the reduced Euler–Poinaré and Lie–Poisson equations.15[–17] By making use of the integrators of Lie group,18[–21] the Euler–Poinaré and Lie–Poisson equations can be easily applied to design the structure-preserving integrators of Lie group. These integrators have been implemented for long-term simulation of free rigid body as well as full rigid body.22Additionally, by combining Lie group methods and the variational principle of discrete mechanics,23·24 Lee et al.25 and Bou-Rabee and Marsden26 proposed the more stable algorithms of symplecticity–momentum–energy-Lie group preserving. Mizzi,27 Liu,28and Boyer and Primault29 used Lie group methods to construct equations for the motion of multibody systems. Unfortunately, the motions of rotation are parameterized finally. Park and Chung,30 Brüls et al.,31 and Sonneville and Brüls32·33 focused on Lie group integrators of multibody system dynamics, but they did not present the systematic derivations of dynamic equations under Lie group framework. In this article, based on Euler–Poinaré theory, we focus on how to derive the dynamical equations of rigid multibody systems by using Lie group and Lie algebra expressions of rotations.

The rest of this article is as follows. The classic Euler–Poinaré equations of free rigid body are introduced in section "The classical Euler–Poinaré equations of a rigid body." In section "The Euler–Poinaré equations of rigid multibody system dynamics," we derive the Euler–Poinaré equations of rigid multibody systems with joint constraints. In section "Generalized external forces in Lie groups and Lie algebras," we analyze some generalized external forces under Lie group and Lie algebra. Section "Examples" presents two examples using the proposed method. In the last section, the contributions of this article are summarized and some future research topics are also presented.

THE CLASSICAL EULER–POINARÉ EQUATIONS OF A RIGID BODY

Lie group is a smooth differentiable manifold on which composition operations such as matrix product can be defined. The Euler–Poinaré principle is based on matrix Lie group. If Lie group g is used to describe the configuration space of a dynamic system, its corresponding Lie algebras ξ and η can be denoted as15·16

$$\xi = g^{-1}\dot{g} \tag{1}$$

$$\eta = g^{-1}\delta g \tag{2}$$

where ξ and η are skew-symmetric matrices in vector space and δg is the variation in g. For the problem of fixed time terminals, $\delta g(0) = \delta g(T) = 0$, $\eta(0) = \eta(T) = 0$, and $t \in [0, T]$. ξ and η satisfy the following relationship of constrained variation

$$\delta\xi = \dot{\eta} + ad_{\xi}\eta \tag{3}$$

where $ad_{\xi}\eta = [\xi, \eta] = \xi\eta - \eta\xi$ is the commutation operator between two Lie algebras ξ and η. Using Lie group and Lie algebra, the reduced Hamilton action is given as

$$\begin{aligned} S &= \int_0^T L(g, \dot{g})dt = \int_0^T L(g^{-1}g, g^{-1}\dot{g})dt \\ &= \int_0^T l(e, \xi)dt = \int_0^T l(\xi)dt \end{aligned} \quad (4)$$

(4)

where e is the unit element of Lie group, and L and l are the traditional Lagrangian and reduced Lagrangian, respectively. Hamilton's principle based on Lie group and Lie algebra is

$$\begin{aligned} \delta S = \delta\int_0^T l(\xi)dt &= \int_0^T \left\langle \frac{\partial l}{\partial \xi}, \delta\xi \right\rangle dt = \int_0^T \left\langle \frac{\partial l}{\partial \xi}, \dot{\eta} + ad_{\xi}\eta \right\rangle dt \\ &= \int_0^T \left\langle -\frac{d}{dt}\left(\frac{\partial l}{\partial \xi}\right) + ad_{\xi}^{*}\left(\frac{\partial l}{\partial \xi}\right), \eta \right\rangle dt = 0 \end{aligned}$$

which is the reduced Euler–Poinaré principle, and its classic Euler–Poinaré equation reads

$$\frac{d}{dt}\left(\frac{\partial l}{\partial \xi}\right) - ad_{\xi}^{*}\left(\frac{\partial l}{\partial \xi}\right) = 0 \tag{6}$$

where $\mathrm{ad}_{\xi}^{*}\eta = [\eta, \xi] = \eta\xi - \xi\eta$ is the conjugate of $ad_{\xi}\eta$. After ξ is solved fromequation (6), the corresponding Lie group g can be recovered by the following reconstruction equation which is the classic Poisson equation

$$\dot{g} = g\xi \tag{7}$$

For a free rigid body, its orientation matrix can be selected as the special orthogonal group, that is, $g = R \in S(3)$, $R^{-1} = R^{T}$, and $det(R) = 1$. Its corresponding Lie algebras $\xi, \eta \in \mathfrak{s}(o)$ are the skew-symmetric matrices of the angular velocity and the variation in quasi rotation angles, respectively, that is, $\xi = \tilde{\Omega}$, $\eta = \delta\tilde{\Theta}$. Thus,equations (1)–(3) can be rewritten as

$$\tilde{\Omega} = R^{T}\dot{R} \tag{8}$$

$$\delta\tilde{\Theta} = R^T \delta R \tag{9}$$

$$\delta\tilde{\Omega} = \delta\dot{\tilde{\Theta}} + ad_{\tilde{\Omega}}\delta\tilde{\Theta} \tag{10}$$

So, we can rewrite equations (6) and (7) as follows

$$\frac{d}{dt}\left(\frac{\partial l}{\partial\tilde{\Omega}}\right) - ad^*_{\tilde{\Omega}}\left(\frac{\partial l}{\partial\tilde{\Omega}}\right) = 0 \tag{11}$$

$$\dot{R} = R\tilde{\Omega} \tag{12}$$

The equivalent vector form of equation (10) is

$$\delta\Omega = \delta\dot{\Theta} + \Omega \times \delta\Theta \tag{13}$$

Based on equation (13), we rewrite the Euler–Poinaré principle (5) using vector expression as follows

$$\begin{aligned}\delta S &= \delta\int_0^T l(\tilde{\Omega})dt = \delta\int_0^T l(\Omega)dt = \int_0^T \left\langle \frac{\partial l}{\partial\Omega}, \delta\Omega \right\rangle dt \\ &= \int_0^T \left\langle \frac{\partial l}{\partial\Omega}, \delta\dot{\Theta} + \Omega \times \delta\Theta \right\rangle dt \\ &= \int_0^T \left\langle -\frac{d}{dt}\left(\frac{\partial l}{\partial\Omega}\right) + \frac{\partial l}{\partial\Omega} \times \Omega, \delta\Theta \right\rangle dt\end{aligned} \tag{14}$$

The corresponding Euler–Poinaré equation of a free rigid body is in the following vector form

$$\frac{d}{dt}\left(\frac{\partial l}{\partial\Omega}\right) + \Omega \times \frac{\partial l}{\partial\Omega} = 0 \tag{15}$$

The Lagrangian of a free rigid body is

$$l = \frac{1}{2}\Omega \cdot J\Omega \tag{16}$$

Thus, equation (15) can be transformed into the following classical Euler equation of rotation

$$J\dot{\Omega} + \Omega \times J\Omega = 0 \tag{17}$$

THE EULER–POINARÉ EQUATIONS OF RIGID MULTIBODY SYSTEM DYNAMICS

A typical multibody system is depicted in Figure 1, which contains n bodies and m constraints. For every body $B_i(i = 1, 2, \ldots, n)$ in the system, we set up the body-fixed frame at its mass center C_i and its axes coincide with its three inertia principle axes. The position of a typical point P on B_i in the inertia space is

$$r_i^P = r_i + R_i s_i' \tag{18}$$

where r_i is the position coordinate of the mass center of body B_i in inertia space, and R_i is the direction cosine matrix of B_i . s'_i is the relative position coordinate of point P with respect to the mass center of B_i expressed in the body-fixed frame. The derivative of equation (18) with respect to time t is

$$\dot{r}_i^P = \dot{r}_i + \dot{R}_i s' = \dot{r}_i + R_i R_i^T \dot{R}_i s_i' = \dot{r}_i + R_i \tilde{\Omega}_i s_i' \tag{19}$$

where

$$\tilde{\Omega}_i = R_i^T \dot{R}_i \tag{20}$$

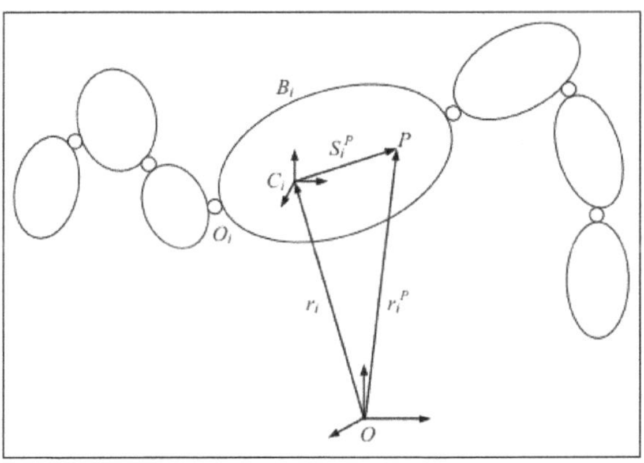

Figure 1: A typical multibody system.

The kinetic energy of B_i is

$$T_i = \frac{1}{2} \int_{V_i} \rho \dot{r}_i^{PT} \dot{r}_i^P dV = \frac{1}{2} m_i \dot{r}_i^T \dot{r} + \frac{1}{2} tr(\dot{R}_i \hat{J}_i \dot{R}_i^T) \tag{21}$$

where $\hat{J}_i$ is the non-standard moment of inertia of B_i , which is

$$\hat{J}_i = \int_{V_i} \rho s'_i s'^T_i dV = E \int_{V_i} \|s'_i\|^2 dV - J_i$$

(22.1)

where J_i is the standard moment of inertia of B_i

$$J_i = \int_{V_i} \rho \tilde{s}'^T_i \tilde{s}'_i dV$$

(22.2)

Equation (21) can be rewritten using angular velocity as

$$T_i = \frac{1}{2} m_i \|\dot{r}_i\|^2 + \frac{1}{2} \int_{V_i} \rho \left\| \tilde{\Omega}_i s'_i \right\|^2 dV = \frac{1}{2} m_i \|\dot{r}_i\|^2 + \frac{1}{2} tr\left(\tilde{\Omega}_i \hat{J}_i \tilde{\Omega}_i^T\right)$$

(23.1)

or

$$T_i = \frac{1}{2} m_i \|\dot{r}_i\|^2 + \frac{1}{2} \Omega_i \cdot J_i \Omega_i$$

(23.2)

Here, we assume that the potential energy of the system is the following form

$$U = U(r, R) = U(r_1, r_2, \ldots, r_n, R_1, R_2, \ldots, R_n)$$

(24)

and express the holonomic constraints of the system in the following compact form

$$\phi = \phi(r, R) = \begin{pmatrix} \phi_1(r_1, r_2, \ldots, r_n, R_1, R_2, \ldots, R_n) \\ \phi_2(r_1, r_2, \ldots, r_n, R_1, R_2, , R_n) \\ \ldots \\ \phi_m(r_1, r_2, \ldots, r_n, R_1, R_2, \ldots, R_n) \end{pmatrix} = 0$$

(25)

where $r = [\, r_1^T \quad r_2^T \quad \ldots \quad r_n^T \,]^T$ and $R = (R_1, R_2, \ldots, R_n)$.

Next, we separately denote the translation velocities, angular velocities, and their skew-symmetric matrices as $v = \dot{r}$, $v = [\, v_1^T \quad v_2^T \quad \ldots \quad v_n^T \,]^T$, $\Omega = [\, \Omega_1^T \quad \Omega_2^T \quad \ldots \quad \Omega_n^T \,]^T$, and $\tilde{\Omega} = \mathrm{dig}[\, \tilde{\Omega}_1 \quad \tilde{\Omega}_2 \quad \ldots \quad \tilde{\Omega}_n \,]$. Thus, the reduced Lagrangian of the system is

$$L = T - U = L(r, \dot{r}, R, \dot{R}) = l(r, v, R, \tilde{\Omega}) = T(r, v, \tilde{\Omega}) - U(r, R)$$

(26)

where $T = \sum_{i=1}^{n} T_i$ is the kinetic energy. Based on these notations, the Hamilton principle can be stated as the following reduced form

$$\delta S = \delta\int_0^T \left(L(r,v,R,\dot{R}) - \lambda^T\phi(r,R)\right)dt = \delta\int_0^T \left(l(r,v,R,\tilde{\Omega}) - \lambda^T\phi(r,R)\right)dt = 0 \tag{27}$$

According to equation (10), equation (27) can be manipulated further as

$$\begin{aligned}\delta S = &\int_0^T \left(\left\langle\frac{\partial l}{\partial r},\delta r\right\rangle + \left\langle\frac{\partial l}{\partial v},\delta\dot{r}\right\rangle + \left\langle\lambda^T\frac{\partial\phi}{\partial r},\delta r\right\rangle\right)dt\\ &+ \sum_{i=1}^{n}\int_0^T \left(\left\langle\frac{\partial l}{\partial R_i},\delta R_i\right\rangle + \left\langle -\frac{d}{dt}\left(\frac{\partial l}{\partial\tilde{\Omega}_i}\right) + ad^*_{\tilde{\Omega}_i}\left(\frac{\partial l}{\partial\tilde{\Omega}_i}\right),\delta\tilde{\Theta}_i\right\rangle\right)dt\\ &- \int_0^T \left(\langle\Phi,\delta\lambda\rangle + \left\langle\frac{\partial(\lambda^T\phi(r,R))}{\partial R_i},\delta R_i\right\rangle\right)dt = 0\end{aligned} \tag{28}$$

For any two matrices $A_{n\times n}$ and $B_{n\times n}$, the following properties are applicable

$$\langle A,B\rangle := \sum_{p,q=1}^{n} A_{p,q}B_{p,q} = tr(AB^T) \tag{29.1}$$

$$A = \frac{1}{2}(A + A^T) + \frac{1}{2}(A - A^T) = A_{Sym} + A_{ssy} \tag{29.2}$$

$$tr(AB) = tr(BA) = tr(B^TA^T) = tr(A^TB^T) \tag{29.3}$$

where $A_{Sym} = 1/2(A + A^T)$ and $A_{ssy} = 1/2(A - A^T)$ are, respectively, symmetric matrix and skew-symmetric matrix. So, we have

$$\begin{aligned}\left\langle\frac{\partial l}{\partial R_i},\delta R_i\right\rangle &= \left\langle\frac{\partial l}{\partial R_i},R_iR_i^T\delta R_i\right\rangle = \left\langle\frac{\partial l}{\partial R_i},R_i\delta\tilde{\Theta}_i\right\rangle\\ &= tr\left(\frac{\partial l}{\partial R_i}\left(R_i\delta\tilde{\Theta}_i\right)^T\right) = tr\left(\frac{\partial l}{\partial R_i}\delta\tilde{\Theta}_i^TR_i^T\right)\\ &= tr\left(R_i^T\frac{\partial l}{\partial R_i}\delta\tilde{\Theta}_i^T\right) = \left\langle R_i^T\frac{\partial l}{\partial R_i},\delta\tilde{\Theta}_i\right\rangle\\ &= \left\langle\frac{1}{2}\left(R_i^T\frac{\partial l}{\partial R_i} + \left(\frac{\partial l}{\partial R_i}\right)^TR_i\right)\right.\\ &\quad\left. + \frac{1}{2}\left(R_i^T\frac{\partial l}{\partial R_i} - \left(\frac{\partial l}{\partial R_i}\right)^TR_i\right),\delta\tilde{\Theta}_i\right\rangle \quad (30)\end{aligned} \tag{30}$$

As $1/2(R_i^T(\partial l/\partial R_i)+(\partial l/\partial R_i)^T R_i)$ is a symmetric matrix and $\delta\tilde{\Theta}_i$ is a skew-symmetric matrix, we have

$$\left\langle \frac{1}{2}\left(R_i^T\frac{\partial l}{\partial R_i}+\left(\frac{\partial l}{\partial R_i}\right)^T R_i\right),\delta\tilde{\Theta}_i\right\rangle = 0 \quad (31.1)$$

(31.1)

Because $1/2(R_i^T(\partial l/\partial R_i)+(\partial l/\partial R_i)^T R_i)$ is a skew-symmetric matrix, we can define

$$\tilde{\Xi}_i = \frac{1}{2}\left(R_i^T\frac{\partial l}{\partial R_i}-\left(\frac{\partial l}{\partial R_i}\right)^T R_i\right) = -\frac{1}{2}\tilde{\Xi}_i$$

(31.2)

For any vectors x, y, $-1/2(\tilde{x},\tilde{y}) = (x,y) = x^T y$, equation (30) can be written as

$$\left\langle \frac{\partial l}{\partial R_i},\delta R_i\right\rangle = \left\langle \tilde{\Xi}_i,\delta\tilde{\Theta}_i\right\rangle = -\frac{1}{2}\langle\tilde{\Xi}_i,\delta\tilde{\Theta}_i\rangle = \langle\Xi_i,\delta\Theta_i\rangle \quad (31.3)$$

(31.3)

where

$$\frac{\partial l}{\partial R_i} = \frac{\partial U}{\partial R_i}$$

(31.4)

Based on the identity $(\widetilde{x\times y}) = \tilde{x}\tilde{y}-\tilde{y}\tilde{x} = yx^T - xy^T$, we can give the physical explanation of $\tilde{\Xi}_i$ and Ξ_i as follows

$$\begin{aligned} R_i^T\frac{\partial U}{\partial R_i}-\left(\frac{\partial U}{\partial R_i}\right)^T R_i &= \begin{bmatrix} R_{i1}^T & R_{i2}^T & R_{i3}^T\end{bmatrix}\begin{bmatrix} U_{R_{i1}} \\ U_{R_{i2}} \\ U_{R_{i3}}\end{bmatrix} \\ &- \begin{bmatrix} U_{R_{i1}}^T & U_{R_{i2}}^T & U_{R_{i3}}^T\end{bmatrix}\begin{bmatrix} R_{i1} \\ R_{i2} \\ R_{i3}\end{bmatrix} \\ &= \left(R_{i1}^T U_{R_{i1}} - U_{R_{i1}}^T R_{i1}\right) + \left(R_{i2}^T U_{R_{i2}} - U_{R_{i2}}^T R_{i2}\right) \\ &+ \left(R_{i3}^T U_{R_{i3}} - U_{R_{i3}}^T R_{i3}\right) = -\left(\widetilde{R_{i1}^T\times U_{R_{i1}}^T}\right) \\ &- \left(\widetilde{R_{i2}^T\times U_{R_{i2}}^T}\right) - \left(\widetilde{R_{i3}^T\times U_{R_{i3}}^T}\right) \end{aligned} \quad (31.5)$$

So, we have

$$-\tilde{\Xi}_i = -\left(\widetilde{R_{i1}^T\times U_{R_{i1}}^T}\right) - \left(\widetilde{R_{i2}^T\times U_{R_{i2}}^T}\right) - \left(\widetilde{R_{i3}^T\times U_{R_{i3}}^T}\right)$$

(31.6)

$$-\Xi_i = -\left(R_{i1}^T\times U_{R_{i1}}^T\right) - \left(R_{i2}^T\times U_{R_{i2}}^T\right) - \left(R_{i3}^T\times U_{R_{i3}}^T\right)$$

(31.7)

Compared with equation (31.3), we can obtain

$$\Xi_i = \frac{\partial U}{\partial \Theta_i} \tag{31.8}$$

Ξ_i is the partial derivative of potential energy with respect to quasi-angles, while $\tilde{\Xi}_i$ is its skew-symmetric matrix. Using similar method, we can derive the variation in constraint terms in equation (28) with respect to orientation matrix as follows

$$\begin{aligned}
&\left\langle \frac{\partial(\lambda^T \phi(r,R))}{\partial R_i}, \delta R_i \right\rangle \\
&= \frac{1}{2}\left\langle \left(R_i^T \frac{\partial(\lambda^T \phi(r,R))}{\partial R_i} - \left(\frac{\partial(\lambda^T \phi(r,R))}{\partial R_i}\right)^T R_i \right), \delta\tilde{\Theta}_i \right\rangle \\
&= \left\langle \frac{1}{2}\left(R_i^T \frac{\partial\left(\sum_{j=1}^{m} \lambda_j \phi_j(r,R)\right)}{\partial R_i} - \left(\frac{\partial\left(\sum_{j=1}^{m} \lambda_j \phi_j(r,R)\right)}{\partial R_i}\right)^T R_i \right), \delta\tilde{\Theta}_i \right\rangle \\
&= \sum_{j=1}^{m} \left\langle \lambda_j, \left\langle \frac{1}{2}\left(R_i^T \frac{\partial \phi_j(r,R)}{\partial R_i} - \left(\frac{\partial \phi_j(r,R)}{\partial R_i}\right)^T R_i \right), \delta\tilde{\Theta}_i \right\rangle \right\rangle \\
&= \sum_{j=1}^{m} \left\langle \lambda_j, \left\langle \tilde{\Phi}_{j,i}, \delta\tilde{\Theta}_i \right\rangle \right\rangle \\
&= -\sum_{j=1}^{m} \left\langle \lambda_j, \left\langle \frac{1}{2}\tilde{\Phi}_{j,i}, \delta\tilde{\Theta}_i \right\rangle \right\rangle = \sum_{j=1}^{m} \left\langle \lambda_j, \left\langle \Phi_{j,i}, \delta\Theta_i \right\rangle \right\rangle \\
&= \sum_{j=1}^{m} \left\langle \lambda_j, \left\langle \phi_{j,\Theta_i}, \delta\Theta_i \right\rangle \right\rangle = \lambda^T \phi_\Theta \delta\Theta
\end{aligned} \tag{31.9}$$

where

$$\Phi_{j,i} = \phi_{j,\Theta_i} = \frac{\partial \phi_j}{\partial \Theta_i}$$

and its corresponding skew-symmetric matrix is

$$-\tilde{\Phi}_{j,i} = R_i^T \frac{\partial \phi_j(r,R)}{\partial R_i} - \left(\frac{\partial \phi_j(r,R)}{\partial R_i}\right)^T R_i$$

Substituting equations (31.3) and (31.9) into equation (28), we obtain

$$\delta S = \int_0^T \left(\left\langle \frac{\partial l}{\partial r}, \delta r \right\rangle + \left\langle -\frac{d}{dt}\left(\frac{\partial l}{\partial v}\right), \delta r \right\rangle - \left\langle \lambda^T \frac{\partial \phi}{\partial r}, \delta r \right\rangle \right) dt$$

$$- \int_0^T \langle \Phi, \delta\lambda \rangle dt + \sum_{i=1}^{n} \int_0^T \left(\left\langle \Xi_i, \delta\tilde{\Theta}_i \right\rangle \right.$$

$$\left. + \left\langle -\frac{d}{dt}\left(\frac{\partial l}{\partial \tilde{\Omega}_i}\right) + ad^*_{\tilde{\Omega}_i}\left(\frac{\partial l}{\partial \tilde{\Omega}_i}\right), \delta\tilde{\Theta}_i \right\rangle - \left\langle \lambda^T \Phi_i, \delta\tilde{\Theta}_i \right\rangle \right) dt = 0 \tag{32}$$

where the dynamical equations of multibody systems can be derived as

$$\begin{cases} \frac{d}{dt}\left(\frac{\partial l}{\partial v}\right) - \frac{\partial l}{\partial r} + \phi_r^T \lambda = 0 \\ \frac{d}{dt}\left(\frac{\partial l}{\partial \tilde{\Omega}_i}\right) - ad^*_{\tilde{\Omega}_i}\left(\frac{\partial l}{\partial \tilde{\Omega}_i}\right) - \Xi_i + \lambda^T \Phi_i = 0 \quad (i = 1, 2, \ldots, n) \\ \phi = 0 \\ \dot{r} = v \\ \dot{R}_i = R_i \tilde{\Omega}_i \quad (i = 1, 2, \ldots, n) \end{cases} \tag{33}$$

where the first equation is the traditional Euler–Lagrange equation on translational motion of mass center. The second one is the Euler–Poinaré equation on rotational motion of every body. The third one is constraint equation. Note that the last two equations are reconstruction equations.

By using equation (13), we can rewrite the first two equations of equation (33) as the following vector form

$$m\dot{v} - \frac{\partial U}{\partial r} + \phi_r^T \lambda = 0 \tag{34.1}$$

$$\frac{d}{dt}\left(\frac{\partial l}{\partial \Omega}\right) + \Omega \times \left(\frac{\partial l}{\partial \Omega}\right) - U_\Theta + \phi_\Theta^T \lambda = 0 \tag{34.2}$$

For rigid bodies, we have

$$\frac{\partial l}{\partial \Omega} = J\Omega \tag{35}$$

With equation (35), equation (34.2) can be rewritten as the following simple form

$$J\dot{\Omega} + \Omega \times J\Omega - U_\Theta + \phi_\Theta^T \lambda = 0 \tag{36}$$

The constraints for the interconnected bodies are important features of multimode systems. In this section, we employ the method proposed in Haug5·6 to construct constraint library for the multibody systems according to the definition of basic constraints. The constraint equations can be expressed as the following general form

$$\phi(r_i, R_i, r_j, R_j) = 0 \tag{37.1}$$

Its variations, first-order derivatives, and second order-derivatives are, respectively

$$\delta\phi = \phi_{,r_i}\delta r_i + \phi_{\Theta_i}\delta\Theta_i + \phi_{\Theta_j}\delta\Theta_j + \phi_{,r_j}\delta r_j \tag{37.2}$$

$$\dot{\phi} = \phi_{,r_i}\dot{r}_i + \phi_{\Theta_i}\Omega_i + \phi_{\Theta_j}\Omega_j + \phi_{,r_j}\dot{r}_j = 0 \tag{37.3}$$

$$\begin{aligned}\ddot{\phi} &= \phi_{,r_i}\ddot{r}_i + \phi_{\Theta_i}\dot{\Omega}_i + \phi_{\Theta_j}\dot{\Omega}_j + \phi_{,r_j}\ddot{r}_j + \dot{\phi}_{,r_i}\dot{r}_i \\ &\quad + \dot{\phi}_{\Theta_i}\Omega_i + \dot{\phi}_{\Theta_j}\Omega_j + \dot{\phi}_{,r_j}\dot{r}_j = 0\end{aligned} \tag{37.4}$$

For the velocity level constraint equation (37.3), we present its derivation in detail as follows

$$\begin{aligned}\dot{\phi} &= \dot{\phi}(r_i, R_i, r_j, R_j) \\ &= \phi_{,r_i}\dot{r}_i + \phi_{,r_j}\dot{r}_j + \left\langle \frac{\partial\phi}{\partial R_i}, \dot{R}_i \right\rangle + \left\langle \frac{\partial\phi}{\partial R_j}, \dot{R}_j \right\rangle \\ &= \phi_{,r_i}\dot{r}_i + \phi_{,r_j}\dot{r}_j + \left\langle \frac{\partial\phi}{\partial R_i}, R_i\tilde{\Omega}_i \right\rangle + \left\langle \frac{\partial\phi}{\partial R_j}, R_j\tilde{\Omega}_j \right\rangle \\ &= \phi_{,r_i}\dot{r}_i + \phi_{,r_j}\dot{r}_j + \left\langle \frac{\partial\phi}{\partial R_i}, R_i\tilde{\Omega}_i \right\rangle + \left\langle \frac{\partial\phi}{\partial R_j}, R_j\tilde{\Omega}_j \right\rangle \\ &= \phi_{,r_i}\dot{r}_i + \phi_{,r_j}\dot{r}_j - tr\left(\tilde{\Omega}_i R_i^T \frac{\partial\phi}{\partial R_i}\right) - tr\left(\tilde{\Omega}_j R_j^T \frac{\partial\phi}{\partial R_j}\right) \\ &= \phi_{,r_i}\dot{r}_i + \phi_{,r_j}\dot{r}_j - tr\left(R_i^T \frac{\partial\phi}{\partial R_i}\tilde{\Omega}_i\right) - tr\left(R_j^T \frac{\partial\phi}{\partial R_j}\tilde{\Omega}_j\right) \\ &= \phi_{,r_i}\dot{r}_i + \phi_{,r_j}\dot{r}_j + \left\langle R_i^T \frac{\partial\phi}{\partial R_i}, \tilde{\Omega}_i \right\rangle + \left\langle R_j^T \frac{\partial\phi}{\partial R_j}, \tilde{\Omega}_j \right\rangle \\ &= \phi_{,r_i}\dot{r}_i + \phi_{,r_j}\dot{r}_j + \phi_{\Theta_i}\Omega_i + \phi_{\Theta_j}\Omega_j\end{aligned} \tag{37.5}$$

GENERALIZED EXTERNAL FORCES IN LIE GROUPS AND LIE ALGEBRAS

In this section, we focus on some conservative external forces, such as gravity, elastic force due to linear translational and twisted springs. The corresponding generalized forces are minus derivatives of energy with respect to quasi-coordinates.

For the generalized external forces due to gravity, we let $\overline{g} = \begin{bmatrix} 0 & 0 & g \end{bmatrix}^T$ denote gravity acceleration matrix. Based on the assumption that the origin of body-fixed frame is located at mass center and its axes are along the three inertia principle axes, we have the following three equations

$$U_i = r_i^T m_i \bar{g} \tag{38.1}$$

$$\delta U_i = \delta r_i^T m_i \bar{g} \tag{38.2}$$

$$\frac{\partial U_i}{\partial r_i} = m_i \bar{g}, \frac{\partial U_i}{\partial \Theta_i} = 0 \tag{38.3}$$

Equations (38.1), (38.2), and (38.3) are potential energy of every body, its variation, and partial derivative with quasi-coordinates, respectively.

By taking account into the linear translational spring between point P on B_i and point $q$on B_j, we denote d_{ij} and k as the relative position vector between points P and q and the elastic modulus, respectively. The elastic potential energy and its variation are given as follows

$$U_{ij} = \frac{1}{2} k \left| |d_{ij}| - |d_{ij0}| \right|^2 \tag{39.1}$$

$$\delta U_{ij} = F_{ij} \frac{d_{ij}^T}{|d_{ij}|} \delta d_{ij} \tag{39.2}$$

where the variation in d_{ij} is

$$\begin{aligned} \delta d_{ij} &= \delta r_j + \delta R_j s_j^{\prime q} - \delta r_i - \delta R_i s_i^{\prime P} \\ &= \delta r_j + R_j \delta \tilde{\Theta}_j s_j^{\prime q} - \delta r_i - R_i \delta \tilde{\Theta}_i s_i^{\prime P} \\ &= \delta r_j - R_j \tilde{s}_j^{\prime q} \delta \Theta_j - \delta r_i + R_i \tilde{s}_i^{\prime P} \delta \Theta_i \end{aligned} \tag{39.3}$$

According to equation (39.3), we have $F_{ij} = k \left| |d_{ij}| - |d_{ij0}| \right|$ in equation (39.2).

From equation (39.2), we can derive the derivatives of energy with respect to quasi-coordinates as follows

$$\begin{aligned} &\frac{\partial U_{ij}}{\partial r_j} = F_{ij} \frac{d_{ij}^T}{|d_{ij}|}, \frac{\partial U_{ij}}{\partial r_i} = -F_{ij} \frac{d_{ij}^T}{|d_{ij}|}, \\ &\frac{\partial U_{ij}}{\partial \Theta_j} = -F_{ij} \frac{d_{ij}^T}{|d_{ij}|} R_j \tilde{s}_j^{\prime q}, \frac{\partial U_{ij}}{\partial \Theta_i} = F_{ij} \frac{d_{ij}^T}{|d_{ij}|} R_i \tilde{s}_i^{\prime P} \end{aligned} \tag{39.4}$$

For a twisted spring between B_i and B_j, the elastic energy and its variation are

$$U_{ij} = \frac{1}{2}k\left|\left|\Theta_{ij}\right| - \left|\Theta_{ij0}\right|\right|^2 \tag{40.1}$$

$$\delta U_{ij} = M_{ij}\frac{\Theta_{ij}^T}{\left|\Theta_{ij}\right|}\delta\Theta_{ij} \tag{40.2}$$

where $M_{ij} = k\left|\left|\Theta_{ij}\right| - \left|\Theta_{ij0}\right|\right|$. According to the following equations

$$R_j = R_i R_{ij}, R_{ij} = R_i^T R_j \tag{40.3}$$

$$\delta R_{ij} = \delta R_i^T R_j + R_i^T \delta R_j \tag{40.4}$$

$$R_{ij}R_{ij}^T \delta R_{ij} = -R_i^T \delta R_i R_i^T R_j + R_i^T R_j R_j^T \delta R_j \tag{40.5}$$

$$R_{ij}\delta\tilde{\Theta}_{ij} = -\delta\tilde{\Theta}_i R_{ij} + R_{ij}\delta\tilde{\Theta}_j \tag{40.6}$$

One can obtain

$$\delta\tilde{\Theta}_{ij} = \delta\tilde{\Theta}_j - R_{ji}\delta\tilde{\Theta}_i R_{ij} = \delta\tilde{\Theta}_j - Ad_{R_{ji}}\delta\tilde{\Theta}_i \tag{40.7}$$

$$\delta\Theta_{ij} = \delta\Theta_j - R_{ji}\delta\Theta_i \tag{40.8}$$

Finally, the partial derivatives of energy with respect to quasi-coordinates are written as follows

$$\frac{\partial U_{ij}}{\partial r_i} = 0, \frac{\partial U_{ij}}{\partial r_j} = 0, \frac{\partial U_{ij}}{\partial \Theta_j} = M_{ij}\frac{\Theta_{ij}^T}{\left|\Theta_{ij}\right|},$$
$$\frac{\partial U_{ij}}{\partial \Theta_j} = -M_{ij}\frac{\Theta_{ij}^T}{\left|\Theta_{ij}\right|}R_{ji} \tag{40.9}$$

EXAMPLES

A Heavy Top

For a heavy top presented in Figure 2, we set up its body-fixed frame according to section "The Euler–Poinaré equations of rigid multibody system dynamics." Its kinetic energy and potential energy are, respectively

$$T = \frac{1}{2}mv^T v + \frac{1}{2}\Omega^T J \Omega$$

$$U = r^T m\bar{g}$$

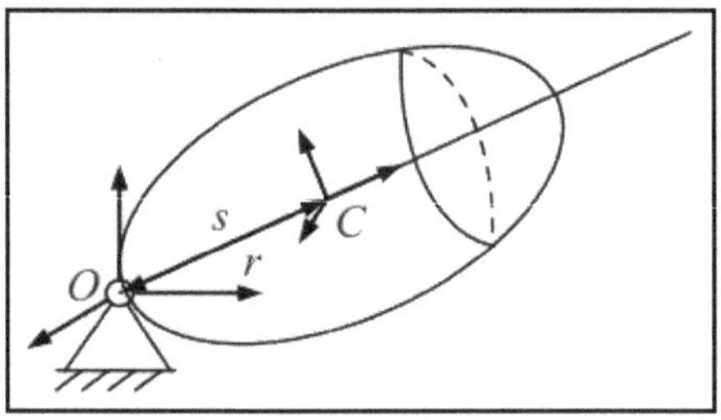

Figure 2: Heavy top.

The constraint equation at joint O and its variation are, respectively

$$\phi(r, R) = r + Rs' = 0$$

$$\begin{aligned}\delta\phi(r, R) &= \delta r + \delta R s' = \delta r + RR^T \delta R s' \\ &= \delta r + R\delta\tilde{\Theta}s' = \delta r - R\tilde{s}'\delta\Theta\end{aligned}$$

where s' is the coordinate of O in the body-fixed frame. The dynamical equations of the heavy top can be derived as follows according to the results in section "The Euler–Poinaré equations of rigid multibody system dynamics"

$$\begin{cases} m\dot{v} - m\bar{g} + \lambda = 0 \\ J\dot{\Omega} + \Omega \times J\Omega + Rs' \times \lambda = 0 \\ r + Rs' = 0 \\ \dot{r} = v \\ \dot{R} = R\tilde{\Omega} \end{cases}$$

A Spatial Pendulum with Three Rigid Bodies

A spatial pendulum shown in Figure 3 consists of three bodies interconnected through three spherical joints. The kinetic energy, potential energy, constraint equations, and their variations are, respectively

$$T = \frac{1}{2}\sum_{i=1}^{3} m_i v_i^T v_i + \frac{1}{2}\sum_{i=1}^{3} \Omega_i^T J_i \Omega_i$$

$$U = \sum_{i=1}^{3} r_i^T m_i \bar{g}$$

$$\phi(r,R) = \begin{bmatrix} \phi_1(r,R) \\ \phi_2(r,R) \\ \phi_3(r,R) \end{bmatrix} = \begin{bmatrix} r_1 + R_1 s_1'^1 \\ r_1 + R_1 s_1'^2 - r_2 - R_2 s_2'^2 \\ r_2 + R_2 s_2'^3 - r_3 - R_3 s_3'^3 \end{bmatrix} = 0$$

$$\begin{aligned} \delta\phi = & \begin{bmatrix} 1 & 0 & 0 \\ 1 & -1 & 0 \\ 0 & 1 & -1 \end{bmatrix} \begin{bmatrix} \delta r_1 \\ \delta r_2 \\ \delta r_3 \end{bmatrix} \\ & + \begin{bmatrix} -R_1 \tilde{s}_1'^1 & 0 & 0 \\ -R_1 \tilde{s}_1'^2 & R_2 \tilde{s}_2'^2 & 0 \\ 0 & -R_2 \tilde{s}_2'^3 & R_3 \tilde{s}_3'^3 \end{bmatrix} \begin{bmatrix} \delta\Theta_1 \\ \delta\Theta_2 \\ \delta\Theta_3 \end{bmatrix} \\ & = \phi_r \delta r + \phi_\Theta \delta\Theta = 0 \end{aligned}$$

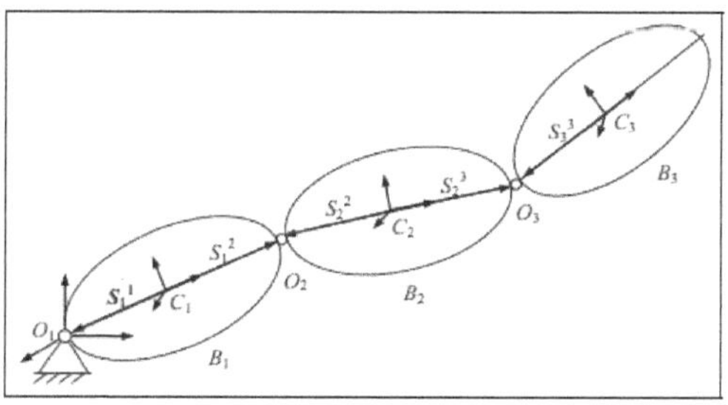

Figure 3: A spatial pendulum.

The dynamic equations of the system are derived as follows according to the results in section "The Euler–Poinaré equations of rigid multibody system dynamics"

$$\begin{cases} m\dot{v} - m\bar{g} + \phi_r^T \lambda = 0 \\ J\dot{\Omega} + \Omega \times J\Omega + \phi_\Theta^T \lambda = 0 \\ \phi(r,R) = 0 \\ \dot{r}_i = v_i \\ \dot{R}_i = R_i \tilde{\Omega}_i \\ (i = 1,2,3) \end{cases}$$

Obviously, numerical methods of differential/algebraic equations in Lie group appearing in this article will be a new research subject in multibody system dynamics, which are not discussed here due to scope of this research. Please refer to Uhlar and Betsch,10Brüls et al.,11 Brüls and Cardona,12 Celledoni et al.,13 Celledoni and Owren,14 and Müller and Terze34 for some new developments on this topic.

CONCLUSION

A new type of dynamical equations of rigid multibody systems is derived using Euler–Poinaré principle. The results include four typical equations: the Euler–Lagrange equations with constraints on translations of mass centers; the Euler–Poinaré equations with constraints on rotations of everybody; the reconstruction equations used to recover the orientation matrices from Lie algebras; and the constraint equations describing inter-connections of bodies. All these equations have simple forms which are easy for numerical implementation. The inertia matrices are constant, leading to high computational efficiency. In addition, the Lie group structure-preserving algorithms can be developed from the Lie group integrators, which is very important for long-term simulations of multibody systems.

This work can be extended to other new researches naturally, including the following: (1) the discrete Euler–Poinaré equations and Lie–Poisson equations of multibody systems based on the variational principles for discrete mechanics; (2) continuous and discrete Euler–Poinaré equations and Lie–Poisson equations of flexible multibody systems; and (3) Lie group variational integrators of multibody systems and their applications in the field of control and optimization and so forth.

DECLARATION OF CONFLICTING INTERESTS

The authors declare that there is no conflict of interest.

FUNDING

This research was supported by the Natural Science Foundation of China (Grant Nos 11472144, 11002075, 11272166, 11472143).

REFERENCES

1. Liu Y, Pan Z and Ge X. Dynamics of multibody systems. Beijing, China: Higher Education Press, 2014.
2. Hong J. Computational multibody system dynamics. Beijing, China: Higher Education Press, 1999.
3. Shabana AA. Dynamics of multibody systems. New York: Wiley, 1989.
4. Nikravesh PE. Computer-aided analysis of mechanical systems. Englewood Cliffs, NJ: Prentice Hall, 1988.
5. Haug EJ. Computer aided kinematics and dynamics of mechanical systems, volume 1: basic methods. Needham Heights, MA: Allyn & Bacon, 1989.
6. Haug EJ. Intermediate dynamics. Englewood Cliffs, NJ: Prentice Hall, 1992.
7. Ge´radin M and Cardona A. Flexible multibody dynamics: a finite element approach. Chichester: John Wiley & Sons, 2001.
8. Bauchau OA. Flexible multibody dynamics. Berlin: Springer, 2011.
9. Simeon B. Computational flexible multibody dynamics—a differential algebraic approach. Berlin: Springer, 2013.

10. Uhlar S and Betsch P. A rotationless formulation of multibody dynamics: modeling of screw joints and incorporation of control constraints. Multibody Syst Dyn 2009; 22(1): 69–95.
11. Bru¨ls O, Cardona A and Arnold M. Lie group generalized-a time integration of constrained flexible multibody systems. Mech Mach Theor 2012; 48: 121–137.
12. Bru¨ls O and Cardona A. On the use of Lie group time integrators in multibody dynamics. J Comput Nonlin Dyn 2010; 5: 1–13.
13. Celledoni E, Marthinsen H and Owren B. An introduction to Lie group integrators-basics, new developments and applications. J Comput Phys 2014; 257B: 1040–1061.
14. Celledoni E and Owren B. Lie group methods for rigid body dynamics and time integration on manifolds. Comput Meth Appl Mech Eng 2003; 192: 421–438.
15. Holm DD, Marsden JE and Ratiu TS. The Euler-Poinare′ equations and semidirect products with applications to continuum theories. Adv Math 1998; 137: 1–81.
16. Holm DD, Schmah T and Stoica C. Geometric mechanics and symmetry: from finite to infinite dimensions. Oxford: Oxford University Press, 2009.
17. Ma Z and Rowley CW. Lie-Poisson integrators: a Hamiltonian, variational approach. Int J Numer Meth Eng 2010; 82: 1609–1644.
18. Crouch PE and Grossman R. Numerical integration of ordinary differential equations on manifolds. J Nonlinear Sci 1993; 3: 1–33.
19. Lewis D and Simo JC. Conserving algorithms for the dynamics of Hamiltonian systems on Lie groups. J Nonlinear Sci 1994; 4: 253–299.
20. Munthe-Kaas H. High order Runge-Kutta methods on manifolds. Appl Numer Math 1999; 29: 115–127.
21. Iserles A, Munthe-Kaas HZ, Nørsett S, et al. Lie-group methods. Acta Numer 2000; 9: 215–365.
22. Hairer E, Lubich C and Wanner G. Geometric numerical integration: structure-preserving algorithms for ordinary differential equations. 2nd ed. Berlin: Springer-Verlag, 2006.
23. Marsden JE, Pekarsky S and Shkoller S. Discrete EulerPoinare′ and Lie-Poisson equations. Nonlinearity 1999; 12: 1647–1662.
24. Marsden JE and West M. Discrete mechanics and variational integrators. Acta Numer 2001; 10: 357–514.
25. Lee T, Leok M and McClamroch NH. Lie group variational integrators for the full body problem. Comput Meth Appl Mech Eng 2007; 196: 2907–2924.
26. Bou-Rabee N and Marsden JE. Hamilton-Pontryagin integrators on Lie groups. Part I: introduction and structure preserving properties. Found Comput Math 2009; 9: 197–219.
27. Mizzi JP. Equations of motion of a multibody system and Lie groups. Meccanica 1995; 30: 201–219.

28. Liu X. A Lie group formulation of Kane's equations for multibody systems. Multibody Syst Dyn 2008; 20: 29–49.

29. Boyer F and Primault D. The Poincare-Chetayev equations and flexible multibody systems. J Appl Math Mech 2005; 69: 925–942.

30. Park J and Chung W-K. Geometric integration on Euclidean group with application to articulated multibody systems. IEEE Trans Robot 2005; 21: 850–863.

31. Bru¨ls O, Arnold M and Cardona A. Two Lie group formulations for dynamic multibody systems with large rotations. In: Proceedings of the ASME 2011, international design engineering technical conferences and computers and information in engineering conference, IDETC/CIE 2011, Washington, DC, 28–31 August 2011, DETC2011/ MSNDC-48132.

32. Sonneville V and Bru¨ls O. Formulation of kinematic joints and rigidity constraints in multibody dynamics using a Lie group approach. In: The 2nd joint international conference on multibody system dynamics, Stuttgart, 29 May–1 June 2012.

33. Sonneville V and Bru¨ls O. A formulation on the special Euclidean group for dynamic analysis of multibody systems. J Comput Nonlin Dyn 2014; 9: 041002 (8 pp.).

34. Mu¨ller A and Terze Z. The significance of the configuration space Lie group for the satisfaction in numerical time integration of multibody systems. Mech Mach Theory 2014; 82: 173–202.

10
CHAPTER

QUANTUM MECHANICS AND PERMUTATION INVARIANTS OF FINITE GROUPS

V. V. Kornyak

INTRODUCTION

Quantum behavior is a natural consequence of symmetries of physical systems. In fact, it is a manifestation of the fundamental impossibility to trace the identity of indistinguishable objects in the process of their evolution. The only that can be extracted from any physical observation is information about invariant relations and quantities which describe ensembles of such objects. Thus, the fundamental problem is to study invariants of groups of symmetries of physical systems. Invariant bilinear forms, i.e., scalar products are particularly important for quantum physics. These forms in accordance with the Born rule provide the link between mathematical description and experimental observations

Physics, as an empirical science, is insensitive to whether finite or infinite concepts are used in formulating its laws. Moreover, effective modeling is possible only when the problem is formulated in finite terms. From the theory of quantum computing it is known that any unitary operator can be presented with arbitrary precision by a finite combination of matrix operators from a finite set of universal quantum gates. Such universal sets of gates generate an infinite group G_∞ which is everywhere dense in the full unitary group. By the theorem of Mal'cev [1], the group G_∞ — as being a finitely generated matrix group — is residually finite [2], i.e., there is a rich set of nontrivial homomorphisms from G_∞ to finite groups. Combining these facts, we can replace the full unitary group by some finite group G in any particular problem without distorting its physical content

It is well known that any linear representation of a finite group is unitary.1 Any unitary representation of a finite group is a sub representation of some permutation representation. Let U be a representation of the group G in a K-dimensional Hilbert

Citation: V. V. Kornyak, Quantum mechanics and permutation invariants of finite groups, doi:10.1088/1742-6596/442/1/012050.

space HK. Then U can be embedded into a permutation representation P of G in an N-dimensional Hilbert space HN, where $N \geq K$. The representation P is equivalent to an action of G on a set of things $\Omega = \{\omega_1, \ldots, \omega_N\}$ by permutations. In the proper case, i.e., when $N > K$, the embedding has the structure

$$T^{-1}PT = \begin{pmatrix} 1 & \\ & V \end{pmatrix}\!\Big\}\mathcal{H}_{N-K} \; \begin{matrix} \\ U\}\,\mathcal{H}_K \end{matrix}, \quad \mathcal{H}_N = \mathcal{H}_{N-K} \oplus \mathcal{H}_K,$$

where 1 is the trivial one-dimensional representation, mandatory for any permutation representation; V is a complementary sub representation which may or may not be. T is a matrix of transition from the basis of the permutation representation P to the basis in which the permutation space H_N is split into the invariant subspaces H_{N-K} and H_K. For brevity, we will refer to this basis as "quantum basis". The data in the spaces H_K and H_{N-}K are independent since both spaces are invariant subspaces of the space H_N. So we can consider the data in H_{N-K} as "hidden parameters" with respect to the data in H_K.

A trivial approach would be to set arbitrary (e.g., zero) data in the complementary subspace H_{N-K}. This approach is not interesting since it is not falsifiable by means of the standard quantum mechanics. In fact, it leads to the standard quantum mechanics modulo the empirically unobservable distinction between the "finite" and the "infinite". The only difference is technical: we can replace the linear algebra in the K-dimensional space H_K by permutations of N things

A more promising approach requires some changes in the concept of quantum amplitudes. We assume [3, 4] that they are projections onto invariant subspaces of the vectors of multiplicities ("occupation numbers") of elements of the set Ω on which the group G acts by permutations. The vectors of multiplicities

$$|n\rangle = \begin{pmatrix} n_1 \\ \vdots \\ n_N \end{pmatrix}$$

are elements of the module $H_N = \mathbb{N}^N$, where $\mathbb{N} = \{0, 1, 2, \ldots\}$is the semiring of natural numbers. We start with the natural permutation representation of G in the module H_N. In order to turn the module H_N into the Hilbert space H_N, it is sufficient just to add roots of unity — an algebraic incarnation of periodicity — to the natural numbers. Linear combinations of Pth roots of unity with natural coefficients form a semiring $\mathbb{N}_\mathcal{P}$. The period P (it is called also conductor) depends on the structure of the group G. Generally, it is a multiple of some divisor of the exponent of G, which is defined as the least common multiple of the orders of elements of G. We will always assume that $P \geq 2$. In this case, using the standard identities for the roots of unity, we can introduce the negative numbers, and thus, $\mathbb{N}_\mathcal{P}$ becomes a ring of cyclotomic integers. To complete the conversion of the module HN into the Hilbert space H_N, we introduce the cyclotomic field $\mathbb{Q}_\mathcal{P}$ as a field of fractions of the ring NP. If $P \geq 3$, then $\mathbb{Q}_\mathcal{P}$ is a dense subfield of the field of complex numbers $\mathbb{C}$. In particular, this gives a trivial explanation of why the complex numbers are so important in quantum mechanics. In fact, the algebraic properties of elements of $\mathbb{Q}_\mathcal{P}$ are quite sufficient for all our purposes, so we can forget the possibility to embed $\mathbb{Q}_\mathcal{P}$ into $\mathbb{C}$.

The connection between mathematical description and observation is provided by the born rule: the probability to register a particle described by the amplitude $|\phi\rangle$, by an apparatus tuned to the amplitude $|\phi\rangle$ is

$$\mathbf{P}(\phi,\psi) = \frac{|\langle\phi \mid \psi\rangle|^2}{\langle\phi \mid \phi\rangle \langle\psi \mid \psi\rangle}\ .$$

In the "finite" background the only reasonable interpretation of probability is the frequency interpretation: the probability is the ratio of the number of "favorable" combinations to the total number of combinations. So we expect that P(φ, ψ) must be a rational number if everything is arranged correctly.

To calculate via the Born rule quantum interferences of projections of natural amplitudes, it is sufficient to know the expressions for scalar products in invariant subspaces of the permutation representation. These expressions can be obtained by straightforward computation using the matrix of transition from the "permutation" to "quantum" basis. Unfortunately, there is no satisfactory algorithm for splitting a module over an associative algebra into irreducible submodules — and, hence, for computing the corresponding transition matrix — over important to us number systems of zero characteristic.2 However, we could try to cope with these difficulties if we could get the expressions for the scalar products directly, i.e., without the use of transition matrices. We describe here one of the possible approaches. This approach can be used for splitting invariant forms of representations for which one can easily calculate the centralizer ring. Permutation representations belong just to this type of representations.

BASIS OF PERMUTATION INVARIANT FORMS

For a start we construct the set of all invariant bilinear forms in the permutation basis. Suppose that the group $G \equiv G(\Omega)$, that acts on the set $\Omega = \{\omega_1, \ldots, \omega_N\}$, is generated by K elements $g_1, \ldots, g_k, \ldots, g_K$. Clearly, it suffices to verify invariance of the form with respect to these elements. For convenience, the set Ω is identified canonically with the set of indices of its elements: $\Omega = \vec{N} = \{1, \ldots, N\}$. Denote matrices of permutation representation of the generators by symbols $P_k = P(g_k)$. Obviously, $P_k^{\mathrm{T}} = P(g_k^{-1})$. The condition of invariance of the bilinear form $A = (a_{ij})$ under the group G can be written as a system of matrix equations

$$A = P_k A P_k^{\mathrm{T}}, \quad 1 \le k \le K.$$

It is easy to check that in terms of components these equations have the form

$$a_{ij} = a_{ig_k jg_k}. \tag{1}$$

Thus, the basis of all invariant bilinear forms is in one-to-one correspondence with the set of orbits $\Delta 1, \ldots, \Delta_R$ of the action of G on the Cartesian product $\Omega \times \Omega$. Orbits of the group on the product $\Omega \times \Omega$, called orbitals, play an important role in the theory of permutation groups and their representations [6, 7]. If the group is transitive, then there is a single orbital, which consists of all pairs of the form (i, i). Such an orbital is called trivial or diagonal. The number of orbitals R is called the rank of the group G. Each orbital Δr can be associated with

(i) a directed graph, whose vertices are the elements of Ω and edges are pairs (i, j) $\in \Delta_r$;

(ii) an $N \times N$ matrix $A(\Delta r)$, which is constructed according to the rule

$$A(\Delta_r)_{ij} = \begin{cases} 1, & \text{if } (i,j) \in \Delta_r, \\ 0, & \text{if } (i,j) \notin \Delta_r. \end{cases}$$

Properties of the graphs of orbitals reflect important properties of the groups. For example, a transitive permutation group is primitive if and only if the graphs of all its non-trivial orbitals are connected.

Matrices of orbitals form a basis of the centralizer ring of permutation representation of G. This ring plays an important role in the theory of group representations. We will denote it by $ZR \equiv ZR(G(\Omega))$. The multiplication table of basis elements of the ring ZR has the form

$$A(\Delta_p)\,A(\Delta_q) = \sum_{r=1}^{R} a_{pq}^{r} A(\Delta_r),$$

where all the structure constants a_{pq}^{r} are natural numbers

Note that the centralizer ring can be obtained via slightly different approach [8] which is originated from the works of Schur. The approach is based on the study of orbits of stabilizers of points in the set Ω. Recall that the stabilizer of the element $i \in \Omega$ is the maximal subgroup G_i of the group G that fixes this element, i.e., $g \in G_i \Leftrightarrow ig = i$. In the general case, the stabilizer G_i of a transitive group G acts on the set Ω intransitively, generating a set of orbits $\Lambda_1 = \{i\}, \Lambda_2, \ldots, \Lambda R$. There is a trivial one-to-one correspondence between the orbits of the stabilizer (sometimes they are called suborbits) and the orbitals (see, e.g., [6, 7]). Therefore the number R of orbits of the stabilizer Gi coincides with the rank of the permutation group $G(\Omega)$. Via the orbits of stabilizers the basis matrices of the centralizer ring are constructed as follows. We consider the orbits of the stabilizer G_1. Each orbit Λ of the stabilizer G_1 is associated with the matrix $A(\Lambda)$ by the rule

$$A(\Lambda)_{ij} = \begin{cases} 1, & \text{if there exist } g \in \mathrm{G} \text{ and } l \in \Lambda \text{ such that } 1g = i \text{ and } lg = j, \\ 0, & \text{otherwise.} \end{cases}$$

Technical advantage of this approach is the use of the isomorphism between the orbits of stabilizers of different points: initially, we construct the first row of the matrix using the orbits of the stabilizer of the point 1, and then, the remaining rows are created via the group translations.

We will call the matrices of orbitals $A_r \equiv A(\Delta_r)$ basic forms, since any permutation invariant bilinear form can be presented by their linear combination

Algorithm for computing the basic forms is reduced to construction of orbitals in accordance with (1). In a few words, it scans the elements of the set $\Omega \times \Omega$ in some, say, lexicographic, order and distributes these elements over the equivalence classes. The output of the algorithm is a complete basis $A_1, \ldots, A_R$ of permutation invariant bilinear forms. The algorithm is quite simple. Our implementation is just a few lines in C.

The following identity follows directly from the construction

$$\mathcal{A}_1 + \mathcal{A}_2 + \cdots + \mathcal{A}_R = L^T L = J_N \equiv \begin{pmatrix} 1 & 1 & \cdots & 1 \\ 1 & 1 & \cdots & 1 \\ \vdots & \vdots & \vdots & \vdots \\ 1 & 1 & \cdots & 1 \end{pmatrix}$$

where L is a covector of the form $\overbrace{(1, 1, \ldots, 1)}^{N}$, and J_N is an N × N "matrix of ones"

Let us illustrate the algorithm with the example of the group $\mathbb{Z}_3$ acting on the set $\Omega = \{1, 2, 3\}$. The group is generated by a single element, for example, $g1 = (1, 2, 3)$. We need to distribute the set of pairs of indices

$$\begin{matrix} (1,1) & (1,2) & (1,3) \\ (2,1) & (2,2) & (2,3) \\ (3,1) & (3,2) & (3,3) \end{matrix}$$

over the equivalence classes in accordance with (1). If we start from the top left corner of table (2) and will look for untreated pairs in the lexicographic order, the orbitals will be constructed in the following order:

$$\Delta_1 = \{(1,1), (2,2), (3,3)\}, \Delta_2 = \{(1,2), (2,3), (3,1)\}, \Delta_3 = \{(1,3), (2,1), (3,2)\}.$$

The corresponding complete basis of invariant forms reads

$$\mathcal{A}_1 = \begin{pmatrix} 1 & \cdot & \cdot \\ \cdot & 1 & \cdot \\ \cdot & \cdot & 1 \end{pmatrix}, \quad \mathcal{A}_2 = \begin{pmatrix} \cdot & 1 & \cdot \\ \cdot & \cdot & 1 \\ 1 & \cdot & \cdot \end{pmatrix}, \quad \mathcal{A}_3 = \begin{pmatrix} \cdot & \cdot & 1 \\ 1 & \cdot & \cdot \\ \cdot & 1 & \cdot \end{pmatrix}.$$

Note that in general, if we start the algorithm with the pair (1, 1) and G is transitive, then the first basis form will always be the identity matrix $A_1 = I_N$, corresponding to the trivial orbital.

PERMUTATION INVARIANT FORMS AND DECOMPOSITION INTO IRREDUCIBLE COMPONENTS

Despite the ease of obtaining the permutation invariant forms, they are sufficiently informative. Consider the decomposition of permutation representation into irreducible components using the transformation matrix T. This decomposition for transitive groups has the form

$$T^{-1}P(g)T = \begin{pmatrix} 1 & & & & \\ & I_{m_2} \otimes U_2(g) & & & \\ & & \ddots & & \\ & & & I_{m_k} \otimes U_k(g) & \\ & & & & \ddots \\ & & & & & I_{m_{N_{Irr}}} \otimes U_{N_{Irr}}(g) \end{pmatrix}, \quad g \in G.$$

Here N_{Irr} is the total number of different irreducible representations U_k ($U_1 \equiv 1$) of the group G, which are contained in the permutation representation P; m_k is the multiplicity of the subrepresentations Uk in the representation P; $\otimes$ denotes the Kronecker product of matrices; I_m is an $m \times m$ identity matrix.

The most general permutation invariant form is a linear combination of the basic forms

$$A = a_1\mathcal{A}_1 + a_2\mathcal{A}_2 + \cdots + a_R\mathcal{A}_R \ , \qquad (3)$$

here the coefficients a_i are elements of some abelian number field3 F which is defined concretely in the computations to be described below.

It is easy to show (see [6, 8]) that in a basis, which splits the permutation representation into irreducible components, the form in (3) takes the form

$$T^{-1}AT = \begin{pmatrix} \mathcal{B}_1 & & & & & \\ & \mathcal{B}_2 \otimes I_{d_2} & & & & \\ & & \ddots & & & \\ & & & \mathcal{B}_k \otimes I_{d_k} & & \\ & & & & \ddots & \\ & & & & & \mathcal{B}_{N_{Irr}} \otimes I_{d_{N_{Irr}}} \end{pmatrix} .$$

Here B_k is an $m_k \times m_k$ matrix, whose elements are linear combinations of the coefficients a_i from (3), while m_k is a multiplicity of the irreducible component U_k; d_k is the dimension of U_k. The structure of matrix (4) implies that the rank R of the group (i.e., the dimension of the centralizer ring) is equal to the sum of the squares of the multiplicities: $R = 1 + m_2^2 + \cdots + m_{N_{Irr}}^2$.

Let us now consider the determinant det (A). In terms of the variables a1, . . . , aR this determinant is a homogeneous polynomial of degree N. Since the determinant of a form does not depend on the choice of basis, and the determinant of a block diagonal matrix is the product of determinants of its blocks, from decomposition (4) it follows

$$\det(A) = \det(\mathcal{B}_1)\det(\mathcal{B}_2)^{d_2}\cdots\det(\mathcal{B}_k)^{d_k}\cdots\det(\mathcal{B}_{N_{Irr}})^{d_{N_{Irr}}} .$$

Here we have used the identity $\det(X \otimes Y) = \det(X)^m \det(Y)^n$ for the Kronecker product of an $n \times n$ matrix X and an $m \times m$ matrix Y .

It is clear that $\det(B_k)$ is a homogeneous polynomial of degree m_k in the variables $a_1, \ldots, a_R$: $\det(B_k) = E_k(a_1, \ldots, a_R)$. Thus, we have the following

Proposition. The determinant of linear combination (3) has the following decomposition into factors over a certain ring of cyclotomic integers:

$$\det\sum_{i=1}^{R} a_i\mathcal{A}_i = \prod_{k=1}^{N_{Irr}} E_k(a_1,\ldots,a_R)^{d_k}, \quad \deg E_k(a_1,\ldots,a_R) = m_k \ , \qquad (5)$$

where E_k is an irreducible polynomial corresponding to the irreducible component U_k in the representation P. Recall that N_{Irr} is the number of different irreducible components in P; d_k is dimension of U_k; m_k is the multiplicity of U_k in P.

From this proposition the idea of an algorithm to compute the invariant forms in irreducible subspaces of the permutation representation follows.

We must first calculate the polynomial det (A). This is a relatively simple task. In particular, algorithms based on the Gaussian elimination have a cubic complexity in the size of the matrix.

Then, the polynomial det(A) must be decomposed into a maximum number of irreducible factors. Note that advanced algorithms of polynomial factorization automatically determine an algebraic extension of rationals that guarantees maximal factorization. Such algorithms are called algorithms of "absolute factorization". There are many practical implementations of polynomial factorization algorithms with different estimates of the complexity

Thus, constructing decomposition (5) is an algorithmically realizable task. Solving it, we obtain the complete information about the dimensions and multiplicities of all irreducible subrepresentations.

The next natural step is to try to compute explicitly invariant forms B_k in the irreducible subspaces of the permutation representation. To do this, we exclude from consideration the factor E_k $(a_1, \ldots, a_R)$, related to the component B_k, and equate to zero the other factors. That is, we write the system of equations

$$E_1 = \cdots = \widehat{E_k} = \cdots = E_{N_{\mathrm{Irr}}} = 0. \qquad (6)$$

Example with irreducible components of multiplicity one.

If all the multiplicities mi = 1 (under this condition $R = N_{Irr}$ and the centralizer ring ZR is commutative), then all the polynomials E_i are linear. In this case, the computation of the scalar products in the invariant subspaces can easily be completed.

As an example, consider the group SL(2, 3) defined as a group of special linear transformations of two-dimensional space over the field of three elements $\mathbb{F}_3$. This group is used in particle physics where it is often denoted by T0 or 2T since it is a double cover of the symmetry group T ~= A_4 of a tetrahedron. We will consider its faithful permutation action of degree 8, which can be generated, for example, by the following two permutations

$$g_1 = (1, 5, 3, 2, 6, 4)\,(7, 8) \quad \text{and} \quad g_2 = (1, 3, 7, 2, 4, 8)\,(5, 6)\,.$$

The 8-dimensional permutation representation will be denoted by 8.

The following four matrices — obtained via constructing orbitals — form a basis of the ring $\mathrm{ZR} \equiv \mathrm{ZR}\left(\mathsf{SL}(2,3)\left(\vec{8}\right)\right)$ of permutation invariant forms:

$$A_1 = \begin{pmatrix} 1 & \cdot & \cdot & \cdot & \cdot & \cdot & \cdot & \cdot \\ \cdot & 1 & \cdot & \cdot & \cdot & \cdot & \cdot & \cdot \\ \cdot & \cdot & 1 & \cdot & \cdot & \cdot & \cdot & \cdot \\ \cdot & \cdot & \cdot & 1 & \cdot & \cdot & \cdot & \cdot \\ \cdot & \cdot & \cdot & \cdot & 1 & \cdot & \cdot & \cdot \\ \cdot & \cdot & \cdot & \cdot & \cdot & 1 & \cdot & \cdot \\ \cdot & \cdot & \cdot & \cdot & \cdot & \cdot & 1 & \cdot \\ \cdot & \cdot & \cdot & \cdot & \cdot & \cdot & \cdot & 1 \end{pmatrix}, \quad A_2 = \begin{pmatrix} \cdot & 1 & \cdot & \cdot & \cdot & \cdot & \cdot & \cdot \\ 1 & \cdot & \cdot & \cdot & \cdot & \cdot & \cdot & \cdot \\ \cdot & \cdot & \cdot & 1 & \cdot & \cdot & \cdot & \cdot \\ \cdot & \cdot & 1 & \cdot & \cdot & \cdot & \cdot & \cdot \\ \cdot & \cdot & \cdot & \cdot & \cdot & 1 & \cdot & \cdot \\ \cdot & \cdot & \cdot & \cdot & 1 & \cdot & \cdot & \cdot \\ \cdot & \cdot & \cdot & \cdot & \cdot & \cdot & \cdot & 1 \\ \cdot & \cdot & \cdot & \cdot & \cdot & \cdot & 1 & \cdot \end{pmatrix},$$

$$A_3 = \begin{pmatrix} \cdot & \cdot & 1 & \cdot & 1 & \cdot & 1 & \cdot \\ \cdot & \cdot & \cdot & 1 & \cdot & 1 & \cdot & 1 \\ \cdot & 1 & \cdot & \cdot & \cdot & 1 & 1 & \cdot \\ 1 & \cdot & \cdot & \cdot & 1 & \cdot & \cdot & 1 \\ \cdot & 1 & 1 & \cdot & \cdot & \cdot & \cdot & 1 \\ 1 & \cdot & \cdot & 1 & \cdot & \cdot & 1 & \cdot \\ \cdot & 1 & \cdot & 1 & 1 & \cdot & \cdot & \cdot \\ 1 & \cdot & 1 & \cdot & \cdot & 1 & \cdot & \cdot \end{pmatrix}, \quad A_4 = \begin{pmatrix} \cdot & \cdot & \cdot & 1 & \cdot & 1 & \cdot & 1 \\ \cdot & \cdot & 1 & \cdot & 1 & \cdot & 1 & \cdot \\ 1 & \cdot & \cdot & \cdot & 1 & \cdot & \cdot & 1 \\ \cdot & 1 & \cdot & \cdot & \cdot & 1 & 1 & \cdot \\ 1 & \cdot & \cdot & 1 & \cdot & \cdot & 1 & \cdot \\ \cdot & 1 & 1 & \cdot & \cdot & \cdot & \cdot & 1 \\ 1 & \cdot & 1 & \cdot & \cdot & 1 & \cdot & \cdot \\ \cdot & 1 & \cdot & 1 & 1 & \cdot & \cdot & \cdot \end{pmatrix}.$$

The determinant of their linear combination $A = {}_{a1}A_1 + {}_{a2}A_2 + {}_{a3}A_3 + {}_{a4}A_4$ is decomposed into linear factors over the ring of cyclotomic integers $\mathbb{N}_3$:

$$\det A = (a_1 + a_2 + 3a_3 + 3a_4)$$
$$\{a_1 - a_2 + (1 + 2\mathrm{r})\, a_3 - (1 + 2\mathrm{r})\, a_4\}^2$$
$$\{a_1 - a_2 - (1 + 2\mathrm{r})\, a_3 + (1 + 2\mathrm{r})\, a_4\}^2$$
$$(a_1 + a_2 - a_3 - a_4)^3,$$

where r is a third primitive root of unity. From this formula the structure of decomposition of the representation $\underline{8}$ into irreducible components is seen immediately:

$$\underline{8} = \mathbf{1} \oplus \mathbf{2} \oplus \mathbf{2}' \oplus \mathbf{3}.$$

Excluding sequentially the linear factors corresponding to the representations k = 1, 2, 2 0 , 3, and equating to zero the other factors, we obtain four systems of three linear equations.

Consider for example the subrepresentation 2. For this component, the system of equations (6) takes the form

$a_1 + a_2 + 3_{a3} + 3_{a4} = 0,$ (7)

$a_1 - a_2 - (1 + 2r)_{a3} + (1 + 2r)_{a4} = 0$ (8)

$a_1 + a_2 - a_3 - a_4 = 0$ (9)

The linear systems of this type consist of R − 1 equations, but contain R variables. Since the bilinear form (3) describes a non-degenerate scalar product, the coefficient a_1 at the diagonal basis form can not vanish. Thus, one can treat a_1 as a parameter and solve the system of equations for the remaining variables. In principle, the coefficient a_1 can be taken to be an arbitrary (nonzero) parameter, since Born's probability is independent of its value. However, it is reasonable to choose the value

$a_1 = d_k/N$ (10)

for each irreducible component. Under such a normalization the sum of scalar products in invariant subspaces will be equal to the standard scalar product in the space of permutation representation presentation. Solving linear system (7—9) and setting $a_1 = 2/8 = 1/4$, we obtain

$$a_2 = -\frac{1}{4},\ a_3 = -\frac{1+2r}{12},\ a_4 = \frac{1+2r}{12}.$$

Applying the same procedure to all other irreducible components we come to the following set of forms defining scalar products in the invariant subspaces

$$\begin{aligned}
\mathcal{B}_1 &= \frac{1}{8}\left(\mathcal{A}_1 + \mathcal{A}_2 + \mathcal{A}_3 + \mathcal{A}_4\right) \equiv \frac{1}{8}\mathrm{J}_8,\\
\mathcal{B}_2 &= \frac{1}{4}\left(\mathcal{A}_1 - \mathcal{A}_2 - \frac{1+2r}{3}\mathcal{A}_3 + \frac{1+2r}{3}\mathcal{A}_4\right),\\
\mathcal{B}_{2'} &= \frac{1}{4}\left(\mathcal{A}_1 - \mathcal{A}_2 + \frac{1+2r}{3}\mathcal{A}_3 - \frac{1+2r}{3}\mathcal{A}_4\right),\\
\mathcal{B}_3 &= \frac{3}{8}\left(\mathcal{A}_1 + \mathcal{A}_2 - \frac{1}{3}\mathcal{A}_3 - \frac{1}{3}\mathcal{A}_4\right).
\end{aligned}$$

It is easy to check that normalization (10) ensures the identity $B_1 + B_2 + B_{20} + B_3 = A_1 \equiv I_8$. Recall that in all such tasks B_1 does not require calculation, since the inner product in the subspace of the trivial representation always has the form $\frac{1}{N}J_N$.

Let $|n\rangle = (n_1, \cdots, n_8)^T$ and $|m\rangle = (m_1, \cdots, m_8)^T$ be natural vectors in the permutation basis, and $|\Psi_k\rangle$ and $|\Phi_k\rangle$ their projections onto the invariant subspaces, where k = 1, 2, 2 0 , 3. The scalar products of these projections in the invariant subspaces can be expressed in terms of the natural vectors as

$$\langle \Phi_k \,|\, \Psi_k \rangle = \langle m \,| \mathcal{B}_k |\, n \rangle\,.$$

For example, for the sub representation 1 we have

$$\langle \Phi_1 \,|\, \Psi_1 \rangle \ = \ \frac{1}{8}\left(m_1 + m_2 + \cdots + m_8\right)\left(n_1 + n_2 + \cdots + n_8\right).$$

It is clear that this expression never vanishes for natural vectors $|n\rangle$ and $|m\rangle$. In fact, in the general case the trivial one-dimensional sub representation contained in any permutation representation can be interpreted as the P "counter of particles", since the permutation invariant $\sum_{i=1}^{N} n_i$ corresponding to this sub representation is the total number of elements of the set Ω in the ensemble

As for the other subrepresentations, we can observe non-trivial "quantum behavior" in their invariant subspaces. In particular, the equations $\langle m \,| \mathcal{B}_k |\, n \rangle = 0$ — the conditions for destructive interference — have infinitely many solutions in natural vectors for k = 2, 2 0 , 3

Note also, that Born's probabilities computed via the forms B_2 and B_2 take irrational values for natural vectors. This contradicts the idea that any probability in the finite background must be rational. The contradiction is resolved by the fact that the permutation action, that we consider, is imprimitive, i.e., it moves some nontrivial subsets of Ω as single units. Such subsets are called blocks. In our example, the block system (called also the system of imprimitivity) is the following {[1, 2] , [3, 4] , [5, 6] , [7, 8]} . Thus, we cannot take the sub representations 2 and 2' separately and must consider their sum 2 ⊕ 2 0 instead. The invariant form for this sum $\mathcal{B}_{2\oplus 2'} = \frac{1}{2}(\mathcal{A}_1 - \mathcal{A}_2)$ does not contain irrationalities.

The case of multiple subrepresentations

In the case of multiple subrepresentations the situation becomes more complicated because of the nonlinearity of relations. The question of whether it is possible to develop a general algorithm for the case of multiple subrepresentations requires a deeper additional study. In many concrete examples, using analogy to the case of single subrepresentations and some ad hoc tricks, it is possible to carry out the complete decomposition of scalar products into irreducible components

The source of nonlinearity is an excess in the number of parameters, the definite values of which are inessential for computing the Born probabilities. As one can see from the structure of decomposition (4), any block of multiple components $B_k \otimes Id_k$ contains $m_k \times m_k$ such parameters. Our convention (10) allows to fix m_k diagonal elements of the matrix B_k. So we need to fix somehow the remaining $m_k^2 - m_k$ parameters. In some cases, the examination of the structure of the centralizer ring helps to do this

As an illustrative example, let us consider the Coxeter group A_2, which is also the Weyl group of the Lie group, say, SU(3). We will consider the natural action of A_2 on its root system $\Omega = \{1, 2, \cdots, 6\}$. The vectors of this root system form a two-dimensional regular hexagon. The generators of the action are, for example, $g_1 = (1, 4)(2, 3)(5, 6)$ and $g2 = (1, 3)(2, 5)(4, 6)$.

Computation of orbitals gives the following basis of the centralizer ring

$$\mathcal{A}_1 = \begin{pmatrix} 1 & \cdot & \cdot & \cdot & \cdot & \cdot \\ \cdot & 1 & \cdot & \cdot & \cdot & \cdot \\ \cdot & \cdot & 1 & \cdot & \cdot & \cdot \\ \cdot & \cdot & \cdot & 1 & \cdot & \cdot \\ \cdot & \cdot & \cdot & \cdot & 1 & \cdot \\ \cdot & \cdot & \cdot & \cdot & \cdot & 1 \end{pmatrix}, \quad \mathcal{A}_2 = \begin{pmatrix} \cdot & 1 & \cdot & \cdot & \cdot & \cdot \\ \cdot & \cdot & \cdot & \cdot & \cdot & 1 \\ \cdot & \cdot & \cdot & \cdot & 1 & \cdot \\ \cdot & \cdot & 1 & \cdot & \cdot & \cdot \\ \cdot & \cdot & \cdot & 1 & \cdot & \cdot \\ 1 & \cdot & \cdot & \cdot & \cdot & \cdot \end{pmatrix}, \quad \mathcal{A}_3 = \begin{pmatrix} \cdot & \cdot & 1 & \cdot & \cdot & \cdot \\ \cdot & \cdot & \cdot & 1 & \cdot & \cdot \\ 1 & \cdot & \cdot & \cdot & \cdot & \cdot \\ \cdot & 1 & \cdot & \cdot & \cdot & \cdot \\ \cdot & \cdot & \cdot & \cdot & \cdot & 1 \\ \cdot & \cdot & \cdot & \cdot & 1 & \cdot \end{pmatrix}$$

$$\mathcal{A}_4 = \begin{pmatrix} \cdot & \cdot & \cdot & 1 & \cdot & \cdot \\ \cdot & \cdot & \cdot & \cdot & 1 & \cdot \\ \cdot & \cdot & \cdot & \cdot & \cdot & 1 \\ 1 & \cdot & \cdot & \cdot & \cdot & \cdot \\ \cdot & 1 & \cdot & \cdot & \cdot & \cdot \\ \cdot & \cdot & 1 & \cdot & \cdot & \cdot \end{pmatrix}, \quad \mathcal{A}_5 = \begin{pmatrix} \cdot & \cdot & \cdot & \cdot & 1 & \cdot \\ \cdot & \cdot & 1 & \cdot & \cdot & \cdot \\ \cdot & 1 & \cdot & \cdot & \cdot & \cdot \\ \cdot & \cdot & \cdot & \cdot & \cdot & 1 \\ 1 & \cdot & \cdot & \cdot & \cdot & \cdot \\ \cdot & \cdot & \cdot & 1 & \cdot & \cdot \end{pmatrix}, \quad \mathcal{A}_6 = \begin{pmatrix} \cdot & \cdot & \cdot & \cdot & \cdot & 1 \\ 1 & \cdot & \cdot & \cdot & \cdot & \cdot \\ \cdot & \cdot & \cdot & 1 & \cdot & \cdot \\ \cdot & \cdot & \cdot & \cdot & 1 & \cdot \\ \cdot & \cdot & 1 & \cdot & \cdot & \cdot \\ \cdot & 1 & \cdot & \cdot & \cdot & \cdot \end{pmatrix}$$

Applying the algorithm of absolute factorization to the generic linear combination

$$A = a_1\mathcal{A}_1 + a_2\mathcal{A}_2 + a_3\mathcal{A}_3 + a_4\mathcal{A}_4 + a_5\mathcal{A}_5 + a_6\mathcal{A}_6 \ ,$$

we obtain the following decomposition:

$$\det A = (a_1 + a_2 - a_3 - a_4 - a_5 + a_6)\\ (a_1 + a_2 + a_3 + a_4 + a_5 + a_6)\\ \{a_1^2 + a_2^2 - a_3^2 - a_4^2 - a_5^2 + a_6^2 - a_1 a_2 - a_1 a_6 - a_2 a_6 + a_3 a_4 + a_3 a_5 + a_4 a_5\}^2 . \quad (11)$$

The structure of the action of the Weyl group A_2 on its roots in terms of the permutation representation follows immediately from (11): 6 = 1 ⊕ 1' ⊕ 2 ⊕ 2. Here we have four irreducible components and a six-dimensional centralizer ring. Thus, we need to eliminate the two extra degrees of freedom. Basis matrices derived from the orbitals are (0,1)-matrices with disjoint sets of unit entries. A result of addition of such matrices belongs to the same type of matrices. So, the natural idea is to sum up some of the matrices A_i in order to reduce their total number. Recall that if the centralizer ring is commutative, then all the multiplicities are equal to one. So we retain the commutative subset of matrices unchanged. In our case, such a subset consists of the matrices A_1, A_2 and A_6. "Killing noncommutativity" by replacing the remaining noncommuting matrices by their sum $A_\Sigma = A_3 + A_4 + A_5$, we write the linear combination of four matrices

$$A' = a_1\mathcal{A}_1 + a_2\mathcal{A}_2 + a_6\mathcal{A}_6 + a_\Sigma\mathcal{A}_\Sigma$$

Now the determinant is factorizable to the linear factors over the cyclotomic integer ring $\mathbb{N}_3$:

$$\det A' = (a_1 + a_2 + a_6 - 3a_\Sigma)\\ (a_1 + a_2 + a_6 + 3a_\Sigma)\\ \{a_1 + \mathrm{r}a_2 - (1+\mathrm{r})\, a_6\}^2\\ \{a_1 - (1+\mathrm{r})\, a_2 + \mathrm{r}a_6\}^2 ,$$

where r is a third root of unity. After the same manipulations as above, we come to the following set of scalar product forms in the invariant subspaces

$$\mathcal{B}_1 = \frac{1}{6}(\mathcal{A}_1 + \mathcal{A}_2 + \mathcal{A}_6 + \mathcal{A}_\Sigma) \equiv \frac{1}{6}\mathrm{J}_6,$$

$$\mathcal{B}_{1'} = \frac{1}{6}(\mathcal{A}_1 + \mathcal{A}_2 + \mathcal{A}_6 - \mathcal{A}_\Sigma),$$

$$\mathcal{B}_2 = \frac{1}{3}\{\mathcal{A}_1 - (1+\mathrm{r})\,\mathcal{A}_2 + \mathrm{r}\mathcal{A}_6\},$$

$$\widetilde{\mathcal{B}}_2 = \frac{1}{3}\{\mathcal{A}_1 + \mathrm{r}\mathcal{A}_2 - (1+\mathrm{r})\,\mathcal{A}_6\}.$$

Here B_2 and $\widetilde{\mathcal{B}}_2$ are different coordinate presentations of the same form, associated with the irreducible representation 2.

CONCLUSIONS

Elimination of actual infinities from the description of physical reality removes many technical difficulties. This allows to focus on the content aspects of physical problems.

We have considered some implications of the idea that quantum mechanics, as any reasonable physical theory, must allow effective modeling by finite means.

Using mathematical arguments of general nature we can show that any quantum problem can be reduced to permutations. If we assume also that the entities, which are subject to the permutations, have a physical meaning, we come to a very simple and self-consistent picture of the quantum behavior. To study the consequences of this assumption, we need to know the inner products in the invariant subspaces of permutation representations.

In this paper, we have considered a possible algorithmic approach to calculating these inner products. With this approach, we can for any permutation representation obtain the full information about the structure of its decomposition into irreducible components. If all irreducible components have unit multiplicities then the invariant inner products can be easily computed. There are many observations which indicate that in the case of multiple subrepresentations a reasonable algorithm is also possible.

ACKNOWLEDGMENTS

The work was partially supported by the grants 10-01-00200 from the Russian Foundation for Basic Research and 3802.2012.2 from the Ministry of Education and Science of the Russian Federation

REFERENCES

1. Mal'cev A 1940 On isomorphic matrix representations of infinite groups Mat. Sbor. 8(50)(3) 405–422 (Russian)
2. Magnus W 1969 Residually finite groups Bull. Amer. Math. Soc. 75(2) 305–316
3. Kornyak V V 2012 Permutation interpretation of quantum mechanics J. Phys.: Conf. Ser. 343 012059 doi:10.1088/1742-6596/343/1/012059
4. Kornyak V V 2013 Classical and quantum discrete dynamical systems Phys. Part. Nucl. 44(1) 47-91 (Preprint arXiv:1208.5734 [quant-ph])
5. Holt D F, Eick B and O'Brien E A 2005 Handbook of Computational Group Theory (Boca-Raton—London— New-York—Washington-D.C.: Chapman & Hall/CRC)
6. Cameron P J 1999 Permutation Groups (Cambridge: Cambridge University Press)
7. Dixon J D and Mortimer B 1996 Permutation Groups (Berlin: Springer)
8. Wielandt H 1964 Finite Permutation Groups (New-York—London: Academic Press)

11
CHAPTER

A GROUP THEORETICAL AND QUANTUM CHEMICAL STUDY OF ELECTRONIC ABSORPTION AND FLUORESCENCE, VIBRATIONAL SPECTRA, AND CONFORMATIONS OF TRIMETHINE CYANINE DYE MOLECULES

Vita Solomko, Petro Kondratenko and Yuriy Lopatkin

INTRODUCTION

The absorption spectra and fluorescence spectra of solutions of polymethine dyes have been studied quite extensively and described in scientific papers [1–6], since this class of dyes is widely used in practice [7, 8]. And yet, the energy structure of the molecules of these dyes was not enough studied, which does not allow us to definitely interpret the fluorescence spectra of the higher excited states that are typical of this class of dyes. The conformational structure and way of the transition of the molecule from one conformation to another which affect the quantum yield of fluorescence were not described in detail

Aims of this work are to describe the results of the investigation of excited states and conformational transformations, for example, model of the molecule, Dioksazoltrimethinecyanine cation (I) (see Scheme 1)

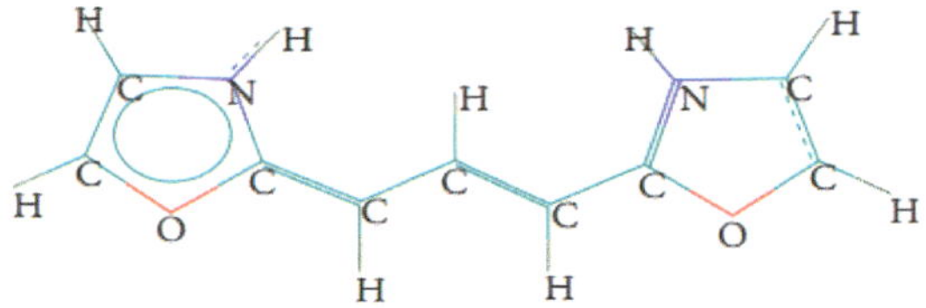

Scheme 1:

Citation: Vita Solomko, Petro Kondratenko, and Yuriy Lopatkin, "A Group Theoretical and Quantum Chemical Study of Electronic Absorption and Fluorescence, Vibrational Spectra, and Conformations of Trimethine Cyanine Dye Molecules," Advances in Physical Chemistry, vol. 2016, Article ID 6737494, 7 pages, 2016. doi:10.1155/2016/6737494

MATERIALS AND METHODS

For research, the software package with a powerful graphical user interface for molecular modeling and study of intermolecular interactions Hyper Chem (Release 8.0.6) was used, in particular the method AM1 [9], which has proved popular in the study of relaxation process of molecules with highly excited states [10–14].

Theoretical Group Analysis of the Electronic and Vibrational Structure of the Molecule (I)

In the ground state molecule (I) has a symmetric structure, which is described by the group of symmetry C_{2V}. For the understanding of the electronic and vibrational structure of this molecule, it was needed to analyze in the framework of groupC_{2V} symmetry of the molecular orbital (MO) and the vibrational modes of the molecule (I). The results of this analysis are shown in Table 1.

Table 1: Table characters of the group C_{2V} and assignment to him coordinate axes, broadcasts (T), rotation (R), the number of vibrational modes (n), and σ- and π-MO molecule (I).

C_{2v}	E	C_2	σ_h	σ_v		T	R	N	n	σ	π
A_1	1	1	1	1	y	1	0	22	21	25	0
A_2	1	1	−1	−1		0	1	10	9	0	6
B_1	1	−1	1	−1	x	1	1	22	20	23	0
B_2	1	−1	−1	1	z	1	1	12	10	0	7
X	1	−1	1	−1							
Y	1	1	1	1							
Z	1	−1	−1	1							
T	3	−1	1	1							
R	3	−1	−1	−1							
N	66	−2	22	2							
σ-MO	48	2	48	2							
π-MO	13	−1	−13	1							

It follows from Table 1 that the total number of degrees of freedom N distributed over representations is as follows: A_1, 22, A_2, 10, B_1, 22, and B_2, 12. In this work calculations of all vibrational frequencies were carried out. Some of them show themselves in the infrared spectra (polarizations x, y, and z) and Raman spectra (x^2 , y^2 , z^2 , xy, xz, and yz). But inasmuch only fully symmetric vibrations take part in the electronic absorption and fluorescence spectra; these vibrations are given in the table. In Table 2 fully symmetrical vibrations of molecules (I) are shown

Table 2: Symmetric vibrations of the molecule (I)

Number	Frequency, cm^{-1}	Localization of vibrations	Intensity
1	70	Two polynomial carried pendulum oscillations, changing the angle of the C-C-C-center	0.049
2	300	Folding: fragments move forward; methine bridge changes the angles	0.078
3	405	Methine bridge and ring in antiphase	0.011
4	829	Synchronous changing of angles O-C-N and central C-C-C	0.29
5	927	Vibrations angle O-C-C in rings	2.1
6	1067	Vibrations of methine angles bond and length O-C-N	0.084
7	1147	Pendulum oscillations of C-H groups in the rings	24
8	1196	Vibrations of the molecule along *y*-axis.	0.23
9	1280	Vibrations lengths S-O-S	2.5
10	1320	Pendulum oscillations of N-H are accompanied by vibrations of the bonds and the methine bridge	2.0
11	1325	Symmetric pendulum vibrations of C-H bonds bridge	2.4
12	1418	Vibrations rings along *x*; central heating C atoms along *y*.	0.63
13	1512	Horizontal vibrations C-N	0.096
14	1590	The length of the C-N and C-C-C	16
15	1686	C-C rings	3.8
16	1864	Bonds rings with the methine bridge	44
17	3129	All three groups C-H bridge in antiphase with the central C-H	17
18	3162	All three C-H groups in the bridge phase	70
19	3207	Two upper C-H bonds in the ring	150
20	3254	Two lower C-H bonds in the ring	53
21	3497	Two N-H- bonds	180

During the formation of the electronic structure of molecules 48 σ-MO and 13 π-MO are formed. In this case, the number of filled electrons MO is 33. 8 of which belong to π-MO, and the rest are in σ-MO. All the processes that we shall be interested in occur in the range from MO number 29 to MO number 43. Symmetry and nature of the MO are shown in Table 3

Table 3: Symmetry MO.

Number of MO	Symmetry and the type of MO
29	$\mathbf{A}_{\sigma-1}$ MO
30	B_2 π-MO
31	A_2 π-MO
32	B_2 π-MO
33	A_2 π-MO
34	B_2 π-MO
35	A_2 π-MO
36	B_2 π-MO
37	A_2 π-MO
38	$\mathbf{A}_{\sigma-1}$ MO
39	$\mathbf{B}_1$ σ-MO
40	$\mathbf{A}_{\sigma-1}$ MO
41	$\mathbf{B}_1$ σ-MO
42	B_2 π-MO
43	$\mathbf{A}_{\sigma-1}$ MO

Considering these data, we can find the polarization of the quantum transitions between separate MO and determine quantum transitions which are forbidden by symmetry (Table 4).

Table 4: The polarization of the quantum transitions between MO.

MO	34	35	36	37	38	39	40	41	42	43
29	*Z*	—	*z*	—	y	x	y	x	*z*	y
30	Y	x	y	x	*z*	—	*z*	—	y	*z*
31	x	y	x	y	—	*z*	—	*z*	x	—
32	y	x	y	x	*z*	—	*z*	—	y	*z*
33	x	y	x	y	—	*z*	—	*z*	x	—

It turned out that quantum transitions with polarization along *z*-axis correspond to $\pi \rightarrow \sigma^*$-quantum transitions. Moreover, since the integral of the overlap between π and σ^*- MO is close to zero, these quantum transitions significantly weakened.

The quantum transitions which are forbidden by symmetry are marked with the dashed line. We present the structure of four MO of those listed in Table 3 (see Scheme 2).

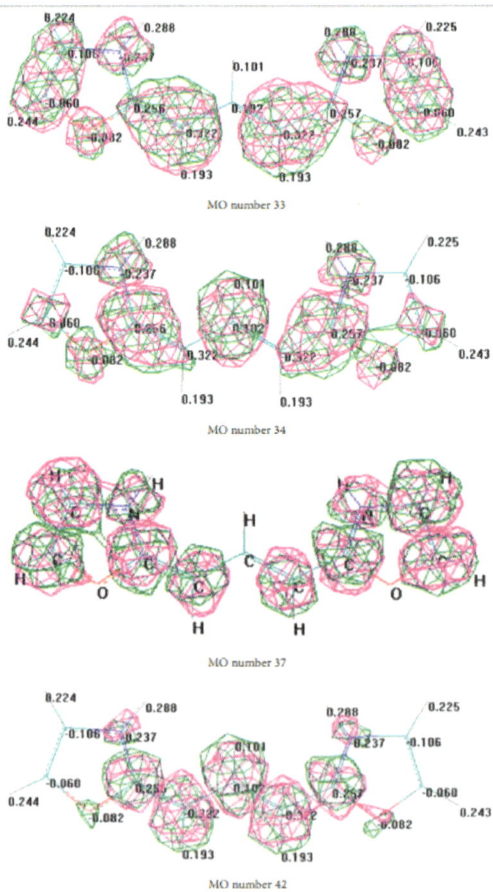

Scheme 2:

The Energy Structure and Conformation of Molecule.

Consideration of symmetry of MO (Table 3) shows that σ- MO may relate to representations $A1$ and $B1$; on the other hand, π-MO may relate to representations A_2 and B_2. Quantum transitions between σ-MO and π-MO can be permitted by symmetry but forbidden by space (the overlap integral between σ-MO and π-MO is close to zero, since the MO are oriented in mutually perpendicular directions).

As σ-MO and π-MO may be responsible for the absorption and emission of the molecule, only σ-MO is responsible for the formation of dissociative potential surfaces and photodissociation processes. On the other hand, only π-MO processes are responsible for cis-trans-isomerization

In this paper we consider the isomerization processes and their impact on radiative processes in the excited molecules.

In the case of trimethine dyes, three isomers can be isolated: trans (I), cis (II) and cis-cis (III) (see Scheme 3).

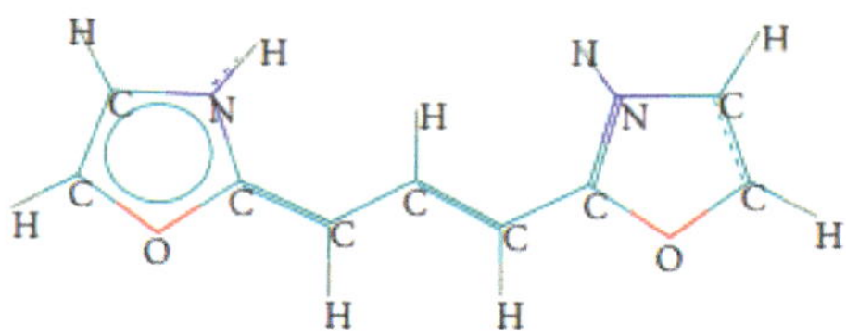

(I) *trans*-Isomer

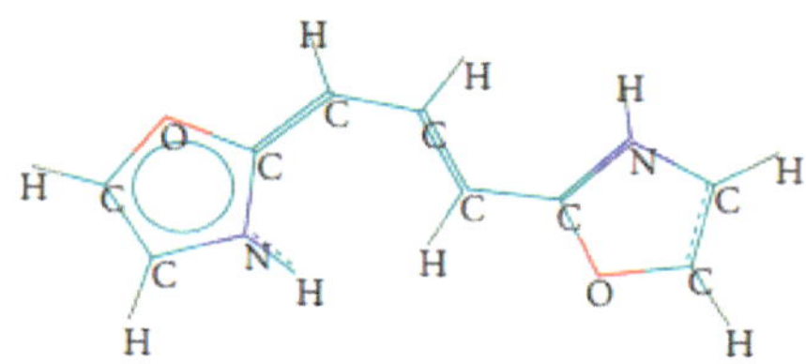

(II) *cis*-Isomer

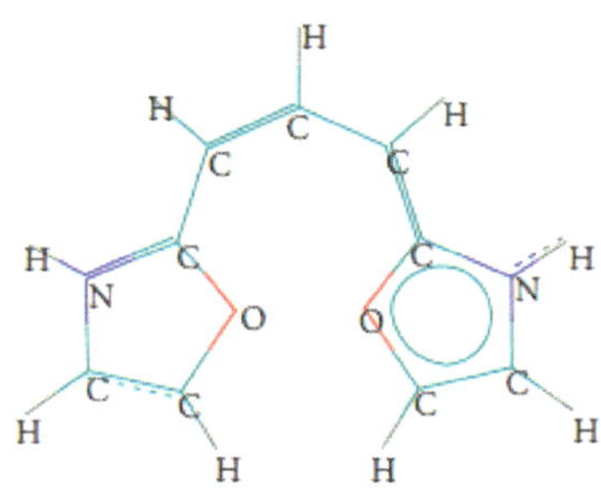

(III) *cis-cis*-Isomer

Scheme 3

The trans-isomer is a flat, highly symmetric molecule, while the other isomers are characterized by low symmetry. Furthermore, cis-isomer has a small angle (~8°) between the planes of the two rings and angles in the methine bridge increased (from left to right: about 128.653 and about 128.754, and the angle of trans is about 123.097). This provides a reduction in repulsion energy between hydrogen atoms, which would otherwise overlap

The angles in the thrimethine bridge of the cis-cis-isomer from left to right: about 129.345° , 133.993° , and 129.552° . Moreover, trimethine bridges are twisted, and the rings are almost not turned (see Scheme 4).

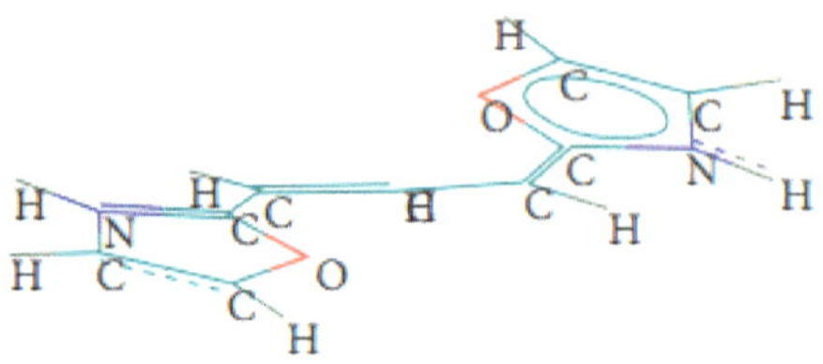

Scheme 4

The binding energy of the electron system in the molecule of I is −2.170.9248 kcal/mol = −94.14 eV, that of molecule II is −2.170.7688 kcal/mol = −94.1336 eV, and that of molecule III is −2.165.0604 kcal/mol = −93.8861 eV. If we expand cisisomer II bridge at about 90, we define the energy of the transition state to be equal to −2.148.3517 kcal/mol. So, we can assume that the first two isomers in thermodynamic equilibrium will be equally represented in the dye solution, whereas the third isomer, the binding energy of which is at least 5.3084 kcal/mol = 0.23 eV, shall be present in small amounts ($1 \cdot 10^{-4}$). However, this fact does not exclude the involvement of this isomer in the process of photoisomerization.

In the process of conversion of the isomer I into isomer II the magnitude of the molecule dipole moment varies, as shown in Figure 1.

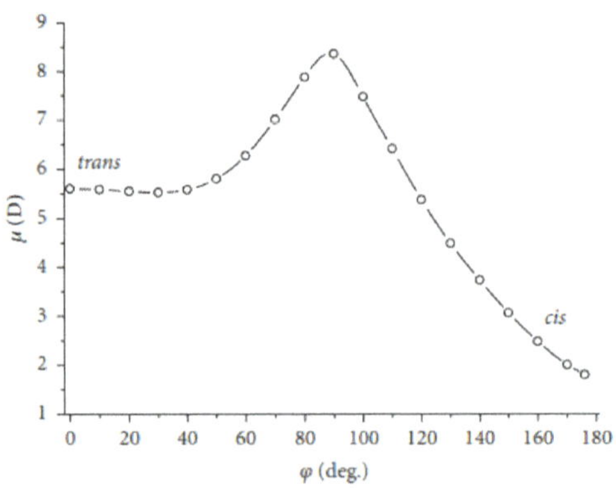

Figure 1: The dependence of the magnitude of dipole moment of the molecule (I) from the rotation angle of the molecule fragments around C-C in the methine group.

Since we are considering cation (I), then charge distribution in it, which determines the SSSSSSSSS of the dipole moment, will determine the structure of the solvation shell in the dye solution, and this in turn may affect the structure of the isomer and its interaction with the solvent, further stabilizing the isomer. In the case where isomer III dipole moment is 1.4135 D, during the transition from isomer I to isomer II, dependence of the dipole magnitude moment of the rotation angle around C-C bond is almost symmetric, reaching the maximum value at 90-degree orientation of fragments of the molecule. Enlarged magnitude of dipole moment in both cases of isomerization at 90-degree orientation of the molecule fragments is indicative of the possibility of increasing the energy of the transition state interactions with solvent molecules, thus reducing the height of the potential barrier to the dark transition between the isomers.

Investigation of the energy structure of the isomers in excited states (Figures 2 and 3) showed that in the ground state there is a barrier between the two isomers, with height of 20.5 kcal/mol = 0.89 eV. This value may be somewhat reduced due to the interaction with the solvent. However, this value indicates a very low rate of establishment of thermodynamic equilibrium between the isomers.

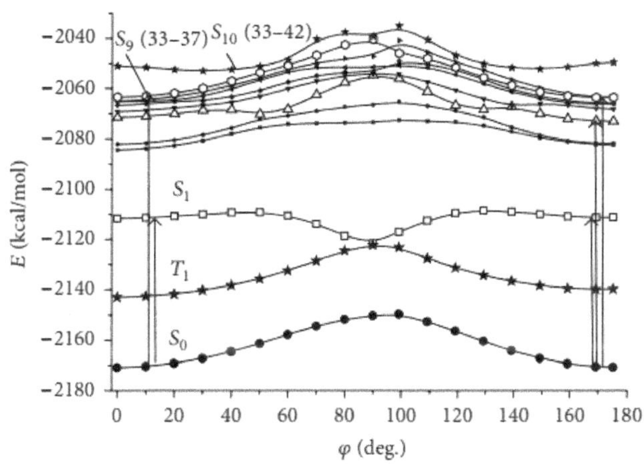

Figure 2: The angular dependence of the energy of the ground and excited singlet states of the molecule (I) isomer in the transition from I to II isomer.

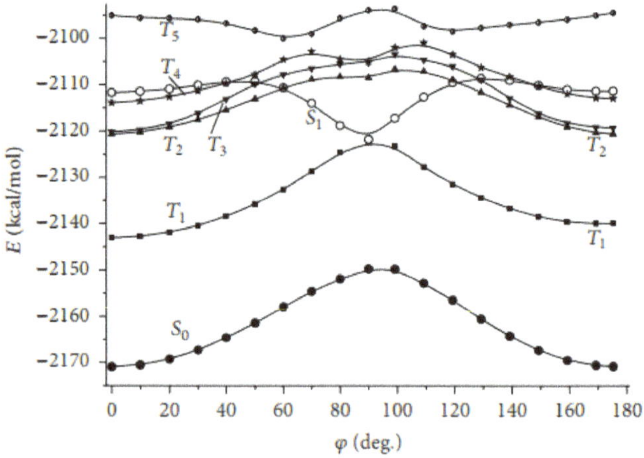

Figure 3: The angular dependence of the energy of the ground and triplet excited states of the molecule (I) isomer in the transition from I to II isomer.

Excitation of molecule (I) in S_1-state will lead to a decrease in the barrier height to 2.39 kcal/mol = 0.104 eV. With such magnitude of barrier, (Boltzmann factor is 0.016) photoisomerization will compete with radiating process, whereby the long-wavelength fluorescence quantum yield of polymethine dyes does not exceed 0.55 [15, 16].

Figure 2 shows the data for the quantum transitions into the singlet-excited state up to the highest π-MO number 42, because it has a disintegrating character for C-C bonds of trimethine fragment. However, this MO did not contribute to cis-trans-isomerization of molecule (I), as its disintegrating character stayed for the second isomer

As for the triplet states, excited T_5 state corresponding to the excitation of an electron from π-MO 33 on σ-MO 38 can cause isomerization of the molecule. Its intersection with state T_4 ($\pi \rightarrow \pi^*$ MO transition between number 33 $\rightarrow$ numbers 35 and 37), which led to the splitting of the energy levels, should not significantly affect the ability to flow isomerization from this state. Interaction between these triplet states has become possible because the rotation of one relative to the other moiety of the molecule greatly reduced the symmetry of the molecule and mixed π- and σ-MO.

Reference [9] noted that, under the influence of light, a photoisomer is created and long-wavelength band is shifted to lower energies with a short lifetime ($10^{-4} \div 10^{-3}$ sec); at that, photoisomer is formed with low quantum yield. However, the main channel of decrease of the fluorescence quantum yield is considered nonradiative dissipation of excitation energy.

The results of this work testify that the first two isomers are stable and have $S_0 \rightarrow S_1$-absorption bond in the same spectral region, and the third one may be unstable, and its corresponding absorption band is shifted to longer wavelengths

Photoisomerization processes between the first two isomers will enhance the probability of nonradiative transitions, since the 90-degree orientation of fragments of the molecule and the energy gap between S_0 and S_1 are significantly reduced. Nonradiative processes should behave analogously at the transition from the second isomer to the third one. In all the these processes, the probability of the third isomer will be insignificant, since the excitation of the first isomer as a result will lead to the creation of equal probability of the first and second isomers, and the excitation of the second isomer can lead to the creation of all three isomers. As a consequence, a third isomer is created with a small quantum yield.

In the case of the plane geometry of molecule (I) in the absorption spectrum bands corresponding to the transition between quantum MO numbers 33 and 34 (long-wavelength absorption band) as well as between MO numbers 33 and 37 should appear (Figure 2). Other quantum transitions that lie between these transitions are characterized by a small oscillator strength and therefore do not appear in the absorption spectra

In isomer II, the long-wavelength absorption band is in the same spectral region as that in isomer I, but instead of a single absorption band, noted above, there are two bands (Table 5), since in this case the oscillator strength $\pi \rightarrow \pi^*$ transition is between MO numbers 33 and 35 which increased by almost an order of magnitude.

Table 5: Singlet-singlet quantum transitions in isomers I and II.

trans-Isomer		*cis*-Isomer	
Type of transition	l, nm, (f)	Type of transition	l, nm, (f)
$S_0 \rightarrow S_1$ $\pi \rightarrow \pi^*$ (33 → 34)	482.7 (1.0418)	$\pi \rightarrow \pi^*$ (33 → 34)	481.0 (0.8252)
$S_0 \rightarrow S_2$ $\pi \rightarrow \sigma^*$ (33 → 38)	330.7 (0.0000)	$\pi \rightarrow \sigma^*$ (33 → 38 33 → 39 32 → 39)	324.0 (0.0011)
$S_0 \rightarrow S_3$ $\pi \rightarrow \sigma^*$ (33 → 39 32 → 38)	322.1 (0.0064)	$\pi \rightarrow \sigma^*$ (33 → 38 33 → 39 32 → 38)	321.5 (0.0059)
$S_0 \rightarrow S_4$ $\pi \rightarrow \pi^*$ (33 → 35 32 → 34)	**287.4 (0.0141)**	**$\pi \rightarrow \pi^*$ (33 → 35)**	**292.6 (0.1068)**
$S_0 \rightarrow S_5$ $\pi \rightarrow \sigma^*$ (33 → 40 32 → 41)	281.2 (0.0000)	$\pi \rightarrow \sigma^*$ (33 → 40 33 → 41)	279.0 (0.0031)
$S_0 \rightarrow S_6$ $3\pi \rightarrow \sigma^*$ (3 → 41 32 → 40)	274.3 (0.0000)	$\pi \rightarrow \pi^*$ (32 → 34)	273.5 (0.0228)
$S_0 \rightarrow S_7$ $\pi \rightarrow \pi^*$ (33 → 35 32 → 34)	270.5 (0.0121)	$\pi \rightarrow \sigma^*$ (33 → 41)	272.7 (0.0160)
$S_0 \rightarrow S_8$ $\pi \rightarrow \pi^*$ (33 → 36)	269.9 (0.0285)	**$\pi \rightarrow \pi^*$ (33 → 36 33 → 37)**	**266.7 (0.1628)**
$S_0 \rightarrow S_9$ **$\pi \rightarrow \pi^*$ (33 → 37)**	**266.0 (0.3301)**	$\pi \rightarrow \pi^*$ (33 → 37 32 → 36)	266.3 (0.0772)
$S_0 \rightarrow S_{10}$ $\pi \rightarrow \pi^*$ (33 → 42)	238.3 (0.0000)	$\pi \rightarrow \pi^*$ (31 → 34)	236.0 (0.0139)
$S_0 \rightarrow S_{11}$ $\pi \rightarrow \pi^*$ (31 → 34)	235.9 (0.0123)	$\pi \rightarrow \pi^*$ (33 → 42)	232.0 (0.0897)

For us it is important to determine which quantum transition may be responsible for the short-wave fluorescence band.

Analysis of the nature of the second (apparent) absorption band indicates that the corresponding $\pi \rightarrow \pi^*$ (33 → 37) quantum transition cannot provide the fluorescence of the highly excited states, as other $\pi \rightarrow \pi^*$ transitions are placed below

For fluorescence the transition $S_4 \rightarrow S_0$ might be responsible, but in this case the large Stokes shift (increased 2.800 cm^{-1}) must be observed. Relative of this transition following the transition $S_3 \rightarrow S_0$ (of $\pi\sigma^*$-state) is shifted to 3750 cm^{-1}, which, given the spatial band, will not contribute to rapid internal conversion. By analyzing the experimental results, it can be seen that high-energy band fluorescence is really characterized by a large Stokes shift [17, 18]. Consequently, we conclude that this is responsible for $\pi \rightarrow \pi^*$ (33 → 35) transition

We have considered above the totally symmetric vibrations of molecule (I), which will give the structure of longwavelength band in the absorption spectrum. The transition of the molecule to the other geometric structures (conformations) is accompanied by a significant lowering of the symmetry of the molecules. This will facilitate the possibility of manifestation of the electronic absorption spectrum of other vibrational frequencies of the molecule and, consequently, change the shape of the long-wavelength absorption band.

CONCLUSIONS

On the basis of the conducted researches of energetic structure and processes of photoisomerization of trimethine cyanine dye (I) we can conclude the following

(1) Molecule (I) may reside in three different conformational states, one of which is highly symmetric (symmetry C_{2V}) and the other two with low symmetry. In this case, two conformers (trans- and cis-conformers) have in the ground state almost the same energy of the electronic system and the same position longwavelength absorption band. The third conformer is characterized by lower 0.23 eV binding energy of the electronic system, and the long-wavelength absorption band is shifted to lower energies.

(2) The group theoretical analysis of molecule (I) allowed systematizing the vibrational and electronic quantum transitions and identifying the bands in the absorption spectra

(3) It is shown that the excitation of the molecule in S_1- state causes trans-cis-isomerization. The presence of the barrier height of ~0.1 eV allows the process fluorescence to compete with isomerization process, but isomerization causes a decrease in the fluorescence quantum yield of the dye. Since there are two stable isomers and one unstable isomer, photoisomerization processes with low quantum yield lead to the creation of unstable isomer. The probability of nonradiative processes on each channel increases, which reduces the fluorescence quantum yield of the dye in the isomerization process.

(4) Investigation of the absorption spectra of the isomers in the UV region of the spectrum showed that in the isomer of I a single absorption band manifested with a large value of the oscillator strength ($f = 0.33$), the excitation of which can fix the short-wavelength fluorescence with a significant Stokes shift. Calculation showed that close to this absorption band at lower energies there are other quantum transitions with small oscillator strength that pertain both to a $\pi \rightarrow \pi^*$ and to $\pi \rightarrow \sigma^*$ quantum transitions. In the second isomer, one of such $\pi \rightarrow \pi^*$ transitions have sufficient appreciable oscillator strength ($f = 0.1$), whereas in the first one, there is a less order of magnitude. Relaxation of excitation in this state may result in a large Stokes shift of high energy fluorescence.

CONFLICT OF INTERESTS

The authors declare that there is no conflict of interests regarding the publication of this paper.

REFERENCES

1. J. M. Hales, J. Matichak, S. Barlow et al., "Design of polymethine dyes with large third-order optical nonlinearities and loss figures of merit," Science, vol. 327, no. 5972, pp. 1485–1488, 2010.
2. A. Mishra, R. K. Behera, P. K. Behera, B. K. Mishra, and G. B. Behera, "Cyanines during the 1990s: a review," Chemical Reviews, vol. 100, no. 6, pp. 1973–2012, 2000.
3. N. Peyghambarian, L. Dalton, J. Alex et al., "Technological advances brighten horizons for organic nonlinear optics," Laser Focus World, vol. 42, no. 8, pp. 85–92, 2006.
4. T. D. Iordanov, J. L. Davis, A. M. E. Masunov, A. Levenson, O. V. Przhonska, and A. D. Kachkovski, "Symmetry breaking in cationic polymethine dyes, part 1: ground state potential energy surfaces and solvent effects on electronic spectra of streptocyanines," International Journal of Quantum Chemistry, vol. 109, no. 15, pp. 3592–3601, 2009.
5. O. V. Przhonska, S. Webster, L. A. Padilha, et al., Advanced Fluorescence Reporters in Chemistry and Biology I: Fundamentals and Molecular Design, Springer Series in Fluorescence, Springer, Berlin, Germany, 2010.
6. I. G. Davidenko, Yu. L. Slominskiy, A. D. Kachkovskiy, and A. I. Tolmachev, "Polymethine dyes—the derivatives of 7,8-dihydrobenzo [cd]-furo [2,3-f] indole," Ukrainian Chemical Journal, vol. 74, no. 4, pp. 105–113, 2008 (Russian).
7. A. V. Kulinich and A. A. Ischenko, "Merotsianinovyie krasiteli: sintez, stroenie, svoystva, primenenie,"Uspehi himii, vol. 78, no. 2, pp. 151–175, 2009 (Russian).
8. A. H. Tarnovskiy, T. K. Razumova, E. P. Schelkina, and T. V. Veselova, "Photophysical, photochemical, and lasing characteristics of symmetric and asymmetric di- and trikarbotsianin dyes," Optics and Spectroscopy, vol. 74, pp. 107–115, 1993 (Russian).
9. M. J. S. Dewar, E. G. Zoebisch, E. F. Healy, and J. J. P. Stewart, "AM1: a new general purpose quantum mechanical molecular model," Journal of the American Chemical Society, vol. 107, no. 13, pp. 3902–3909, 1985.
10. P. A. Kondratenko, S. Y. Lopatkin, Y. M. Lopatkin, and T. N. Sakun, "Fotoelektricheskie svoystva polimernyih sloyov s krasitelyami," Visnik Sums›kogo Derzhavnogo Universitetu, vol. 1, pp. 145–153, 2007 (Russian).
11. P. O. Kondratenko, Yu. M. Lopatkin, and T. M. Sakun, "Quasi-equilibrium processes in the high excited molecules of resazurin," Journal of Nano- and Electron Physics, vol. 4, no. 2, Article ID 02017, 7 pages, 2012 (Ukraine).
12. P. O. Kondratenko, Yu. M. Lopatkin, and T. M. Sakun, "Relaxation processes in the high-excited molecules of resazurin," Physics and Chemistry of Solid State, vol. 8, no. 1, pp. 100–108, 2007 (Ukrainian).
13. R. J. Cave, K. Burke, and E. W. Castner Jr., "Theoretical investigation of the ground and excited states of coumarin 151 and coumarin 120," Journal of Physical Chemistry A, vol. 106, no. 40, pp. 9294–9305, 2002.

14. S. Kumar, S. K. Jain, and R. C. Rastogi, "An experimental and theoretical study of excited-state dipole moments of some flavones using an efficient solvatochromic method based on the solvent polarity parameter, ETN," Spectrochimica Acta Part A: Molecular and Biomolecular Spectroscopy, vol. 57, no. 2, pp. 291–298, 2001.
15. D. Peceli, H. Hu, D. A. Fishman et al., "Enhanced intersystem crossing rate in polymethine-like molecules: sulfur-containing squaraines versus oxygen-containing analogues," Journal of Physical Chemistry A, vol. 117, no. 11, pp. 2333–2346, 2013.
16. V. I. Zemskiy, Y. U. L. Kolesnikov, and I. K. Meshkovskiy, Physics and Technology of Pulsed Dye Lasers, Petersburg State University of Information Technologies, Saint Petersburg, Russia, 2005 (Russian).
17. S. Webster, L. A. Padilha, H. Hu et al., "Structure and linear spectroscopic properties of near IR polymethine dyes," Journal of Luminescence, vol. 128, no. 12, pp. 1927–1936, 2008.
18. G. G. Dyadyusha, O. V. Przhonskaya, Y. A. Tikhonov, and M. T. Shpak, "The fluorescence intensity out the second excited state of molecular solutions of organic dyes," Journal of Experimental and Theoretical Physics, vol. 1, no. 14, pp. 330–333, 2008 (Russian)

12

CHAPTER

ON THE GROUP THEORETICAL APPROACH TO THE QUANTUM THEORY OF AN INTERACTING SPIN-0 PARTICLE

Giuseppe Nisticò

INTRODUCTION

A very satisfactory approach to the Quantum Theory of a free particle can be accomplished by following group theoretical methods, developed in particular by E.P. Wigner and G.W. Mackey. The key starting point is that Galilei's group G, within a non-relativistic theory, is a group of symmetry transformations for a free particle; then Wigner's theorem [1],[2] on the representation of symmetries and Mackey's imprimitivity theorem [3],[4],[5] apply. In so doing the specific Quantum Theory of a free particle is explicitly obtained [3],[6],[7],[8],[9] without invoking canonical quantization, but, instead, by means of a mathematical deduction based on symmetry principles. We recall the main mathematical tools that allow for such an approach in section 2, while the necessary formal implications of the concept of Galileian transformations in the Quantum Theory of a particle are outlined in section 3.

But the Galileian transformations are not symmetries for an interacting particle, so that an obstacle is found in extending the approach to this case. To overcome this difficulty, the approaches that in the literature extend the cited group theoretical method to the interacting case (e.g., [7],[8],[9]) adopt, among others, the following assumption.

(P) Each Galileian transformation $g \in G$ is assigned a unitary operator U_g which realizes the corresponding quantum transformations of states and observables according to $\rho \xrightarrow{g} U_g \rho U_g^{-1}$ and $A \xrightarrow{g} U_g A U_g^{-1}$; the correspondence $g \to U_g$ is a projective representation.

Citation: Giuseppe Nistic`o, On the group theoretical approach to the Quantum Theory of an interacting spin-0 particle, doi:10.1088/1742-6596/670/1/012039

Now, in section 4 we show that (P) forces the hamiltonian operator into the form $-\frac{1}{2\mu}\left(\frac{\partial^2}{\partial x_1^2}+\frac{\partial^2}{\partial x_2^2}+\frac{\partial^2}{\partial x_3^2}\right)+\Phi(x_1,x_2,x_3)$. Therefore a Quantum Theory where (P) is assumed cannot describe all empirically known interactions; in particular it excludes electromagnetic interactions. An empirically more adequate approach must abandon assumption (P).

Then in section 5 we undertake the task of establishing the Quantum Theory of an interacting particle through a group theoretical approach, but without assumption (P). The difficulties raised by the loss of symmetry are addressed through a theoretical development based on the passive interpretation of the Galileian transformations and on the concept of σ-conversion. We find that the form of the dynamical equation is constrained by the covariance properties preserved by the σ-conversion admitted by the interaction; in particular, we exactly identify which covariance properties characterize electromagnetic interaction among all possible ones.

GROUP THEORETICAL KEY RESULTS

Here we recall the main mathematical tools of the cited group theoretical approach.

Basic achievements

Given the Hilbert space H of the Quantum Theory of a physical system, we shall make use of the following operator's structures.

- The set Ω(H) of all self-adjoint operators of H, which represent quantum observables.
- The complete, ortho-complemented lattice Π(H) of all projections operators of H, i.e. quantum observables with possible outcomes in {0, 1}.
- The set Π1(H) of all rank one orthogonal projections of H.
- The set S(H) of all density operators of H, which represent quantum states.
- The set U(H) of all unitary operators of the Hilbert space H.

The main mathematical tools will be used in this work are the imprimitivity theorem of Mackey and Wigner's representation theorem.

Mackey's imprimitivity theorem is a representation theorem for imprimitivity systems relative to projective representations [4]. The following definition recalls the notion of projective representation.

Definition 2.1. Let G be a separable, locally compact group with identity element e. A correspondence U: G → U(H), g → U_g, with U_e = 1I, is a projective representation of G if the following conditions hold.

i) A complex function σ : G × G → C such that $|\sigma(g_1, g_2)| = 1$ for all $g_1, g_2 \in G$, called multiplier, exists such that $U_{g1g2} = \sigma(g_1, g_2)U_{g1}U_{g2}$;

ii) for all φ, ψ ∈ H, the mapping $g \to \langle U_g\phi \mid \psi\rangle$ is a Borel function in g.

A projective representation with multiplier σ is called σ-representation.

A projective representation is said to be continuous if for any fixed ψ ∈ H the mapping

$g \rightarrow U_g\psi$ from G into H is continuous with respect to g.

Let E be the Euclidean group, i.e. the semi-direct product $\mathcal{E} = \mathbf{R}^3 Ⓢ SO(3)$ between the group of spatial translations R^3 and the group of spatial proper rotations SO(3); each transformation g ∈ E bi-univocally corresponds to the pair (a, R) ∈ $R^3 \times SO(3)$ such that $Rx + a \equiv g(x)$ is the result of the transformation of the spatial point $x = (x_1, x_2, x_3)$ by g. The general imprimitivity theorem is an advanced mathematical result, but its formulation becomes very simple when adapted to the case of the Euclidean group E, that is the case we are interested to. Then we introduce the concept of imprimitivity system and the theorem for this simple case.

Definition 2.2. Let H be the Hilbert space of a σ-representation $g \rightarrow U_g$ of the Euclidean group E. A projection valued (PV) measure E: B (R^3) → Π (H), Δ → E (Δ) is an imprimitivity system for the σ-representation $g \rightarrow U_g$ if the relation

$$U_g E(\Delta) U_g^{-1} = E(g(\Delta)) \equiv E(R(\Delta) + \mathbf{a})$$

holds for all (a, R) ∈ E.

Mackey's theorem of imprimitivity. If a PV measure E : B(R^3) → Π(H) is an imprimitivity system for a continuous σ-representation $g \rightarrow U_g$ of the Euclidean group E, then a σ- representation L : SO(3) → U(H_0) exists such that, modulo a unitary isomorphism,

$$(M.1)\ \mathcal{H} = L_2(\mathbf{R}^3, \mathcal{H}_0),$$

$$(M.2)\ (E(\Delta)\psi)(\mathbf{x}) = \chi_\Delta(x)\psi(\mathbf{x}),\ \textit{where } \chi_\Delta \textit{ is the characteristic fu}$$

$$(M.3)\ (U_g\psi)(\mathbf{x}) = L_R\psi(g^{-1}(\mathbf{x})) \equiv L_R\psi(R^{-1}\mathbf{x} - R^{-1}\mathbf{a}),\ \textit{for every } g$$

In the literature different equivalent formulations of Wigner's theorem have been proved [2],[11]. Here we formulate those "Wigner's' theorems" we need for our work.

Wigner's theorem 1. If $S^{(1)}$: S (H) → S (H) and $S^{(2)}$: Ω (H) → Ω (H) are bijective mappings such that

$$Tr(S^{(1)}[\rho]S^{(2)}[A]) = Tr[\rho A] \tag{1}$$

holds for all (ρ, A) ∈ S(H) × Ω(H) for which T r(ρA) exists, then either a unitary operator or an anti-unitary operator U of H exists such that $S^{(1)}[\rho] = U\rho U^*$ and $S^{(2)}[A] = UAU*$ for all (ρ, A) ∈ S(H) × Ω(H), unique up a phase factor

Wigner's theorem 2. If S: Π(H) → Π(H) is an automorphism of Π(H), i.e. if it is a bijective mapping such that

$$E_1 \leq E_2 \Leftrightarrow S[E_1] \leq S[E_2] \quad \text{and} \quad S[E^\perp] = (S[E])^\perp, \quad \forall E_1, E_2, E \in \Pi(\mathcal{H}),$$

then either a unitary operator or an anti-unitary operator U of H exists such that S(E) = UEU^* for all E ∈ Π(H), unique up a phase factor.

particle. However, in the literature several proposals can be found [5],[7],[9] where the group theoretical methods are extended to the interacting case. The aforesaid obstacles are overcome by assuming, more or less implicitly, that the following statement holds (e.g., see [5], page 201 and [7], page 236).

(P) Each Galileian transformation g is represented in the formalism of the Quantum Theory by a unitary or anti-unitary operator U_g in such a way that

i) $R_g^{(2)}[A] = U_g A U_g^*$ is the result of the active transformation of the observable A by g;

ii) $R_{g_1 g_2}^{(2)} = R_{g_1}^{(2)} \circ R_{g_2}^{(2)}$, *for all* $g_1, g_2 \in \mathcal{G}$.

$$H = \frac{1}{2\mu}\sum_{\alpha=1}^{3}\left(-i\frac{\partial}{\partial x_\alpha} - a_\alpha(\mathbf{Q})\right)^2 + \Phi(\mathbf{x}). \tag{8}$$

Now we prove, instead, the following statement

(R) Assumption (P) forces the hamiltonian of the Quantum Theory of a spin-0 particle undergoing an interaction homogeneous in time into the form

$$H = \frac{1}{2\mu}\sum_{\alpha=1}^{3}\left(-i\frac{\partial}{\partial x_\alpha}\right)^2 + \Phi(\mathbf{x}).$$

To prove (R), first we imply from (P) that $g \to U_g$ is a projective representation of the Galileian group G. Then, according to section 2.2, the sub-groups T_α, R_α, B_α can be represented by the one parameter unitary sub-groups $\{e^{-iP_\alpha a}\}_{a\in\mathbf{R}}, \{e^{-iJ_\alpha \hat{\theta}}\}_{\theta\in\mathbf{R}}, \{e^{iG_\alpha u}\}_{u\in\mathbf{R}}$, in such a way that the hermitean generators P_α, J_α, G_α satisfy (2). Once defined the self-adjoint operators $F_\alpha = \frac{G_\alpha}{\mu}$, it can be proved that relations (2) imply that

$$U_g \mathbf{F} U_g^{-1} = g^{-1}(\mathbf{F}). \tag{9}$$

Since the $F\alpha$'s commute with each other, according to spectral theory, there is a unique PV measure P: $B(R^3) \to \Pi(H)$ such that $F_\alpha = \int \lambda dE_\lambda^{(\alpha)}$, where $E_\lambda^{(1)} = P((-\infty,\lambda] \times \mathbf{R}^2)$, $E_\lambda^{(2)} = P(\mathbf{R} \times (-\infty,\lambda] \times \mathbf{R})$, $E_\lambda^{(3)} = P(\mathbf{R}^2 \times (-\infty,\lambda])$. Then (9) easily implies that $\Delta \to P(\Delta)$ is an imprimitivity system for the restriction to E of $g \to Ug$; therefore Mackey's theorem applies. In so doing, the simplest choice for H_0, i.e. $H_0 = C$, leads to identify $\mathcal{H}, F_\alpha, P_\alpha, \hat{U}_g$ as

$$\mathcal{H} = L_2(\mathbf{R}^3), \quad (F_\alpha\psi)(\mathbf{x}) = x_\alpha\psi(\mathbf{x}), \quad P_\alpha = -i\frac{\partial}{\partial x_\alpha}, \quad (U_g\psi)(\mathbf{x}) = \psi\left(g^{-1}(\mathbf{x})\right). \tag{10}$$

Both F and P in (10) are complete systems of operators in $L_2(R^3)$, and (F, P) is an irreducible system of operators [5]. Then we can easily prove the following proposition.

Proposition 4.1. If (P) holds, then in the simplest Quantum Theory of a localizable interacting particle the equality F = Q holds.

Proof. If $g \in T_1$ and (P) holds, then $U_g = e^{-iP_\beta a} = \mathbb{I} - iP_\beta a + o_1(a)$, where $o_1(a)$ is an infinitesimal operator of order greater than 1 with respect to a; relation (5.ii) implies $[Q_\alpha, P_\beta] = i\delta_{\alpha\beta}\mathbb{I}$. So, by using (2.vi) we obtain $[F_\alpha - Q_\alpha, P_\beta] = 0$; but (5.ii) for $U_g = e^{iG_\beta u} = \mathbb{I} + iG_\beta u + o_2(u)$ implies also $[F_\alpha - Q_\alpha, F_\beta] = 0$, i.e. $F_\alpha - Q_\alpha = c_\alpha 1I \equiv$constant for the irreducibility of (F, P); on the other hand, (P.i) together with (2.iv) and (5.ii) for $U_g = e^{-iJ_\alpha\theta}$ imply $[J_\alpha, F_\beta - Q_\beta] = i\hat{\epsilon}_{\alpha,\beta,\gamma}(F_\gamma - Q_\gamma) = [J_\alpha, c_\beta \mathbb{I}] = \mathbf{0}$; thus, $F_\alpha - Q_\alpha = \mathbf{0}$.

Prop. 4.1 together with (5.i) is sufficient to determine the form of the hamiltonian operator H consistent with (P). First, we determine $[G_\alpha, \dot{Q}_\beta]$. Let us start with

$$e^{iG_\alpha u}\dot{Q}_\beta e^{-iG_\alpha u} = \dot{Q}_\beta + i[G_\alpha, \dot{Q}_\beta]u + o(u). \tag{11}$$

By making use of $\dot{Q}_\beta = i[H, Q_\beta] = \lim_{t\to 0}\frac{Q_\beta^{(t)} - Q_\beta}{t}$ and of $e^{iG_\alpha u}Q_\beta^{(t)}e^{-iG_\alpha u} = Q_\beta^{(t)} - \delta_{\alpha\beta}ut\mathbb{I}$, implied by (5.i), we also find

$$e^{iG_\alpha u}\dot{Q}_\beta e^{-iG_\alpha u} = \lim_{t\to 0} e^{iG_\alpha u}\frac{Q_\beta^{(t)} - Q_\beta}{t}e^{-iG_\alpha u} = \lim_{t\to 0}\frac{Q_\beta^{(t)} - \delta_{\alpha\beta}ut\mathbb{I} - Q_\beta}{t} = \dot{Q}_\beta - \delta_{\alpha\beta}u\mathbb{I}. \tag{12}$$

The comparison between (11) and (12) yields

$$[G_\alpha, \dot{Q}_\beta] = [Q_\alpha, \mu\dot{Q}_\beta] = i\delta_{\alpha\beta}\mathbb{I}, \text{ which implies } [F_\alpha, \mu\dot{Q}_\beta - P_\beta] = \mathbf{0}. \tag{13}$$

By using $U_\sigma = e^{-iP_\alpha a}$ and $U_\sigma = e^{-iJ_\alpha\theta}$ instead of $e^{iG_\alpha u}$ in this last argument, we obtain $[P_\alpha, \mu\dot{Q}_\beta - P_\beta] = \mathbf{0}$ and $[J_\alpha, \mu\dot{Q}_\beta] = i\hat{\epsilon}_{\alpha,\beta,\gamma}\mu\dot{Q}_\gamma$. These two relations, together with (13), imply $\dot{Q}_\beta = (1/\mu)P_\beta$. To determine H, by making use of (2.vi) we obtain

$$i[H, Q_\beta] = \dot{Q}_\beta = \frac{1}{\mu}P_\beta = i\left[(1/2\mu)(P_1^2 + P_2^2 + P_3^2), \frac{G_\alpha}{\mu}\right] \equiv i[(1/2\mu)(P_1^2 + P_2^2 + P_3^2), Q_\alpha]. \tag{14}$$

Then the completeness of Q implies that $H - (1/2\mu)(P_1^2 + P_2^2 + P_3^2)$ is a function of Q. Thus

$$H = -\frac{1}{2\mu}\left(\frac{\partial^2}{\partial x_1^2} + \frac{\partial^2}{\partial x_2^2} + \frac{\partial^2}{\partial x_3^2}\right) + \Phi(\mathbf{Q}). \tag{15}$$

Thus, assumption (P) forbids the description of electro-magnetic interactions.

QUANTUM THEORY OF AN INTERACTING PARTICLE

According to the conclusion of the last section, to develop a Quantum Theory of a particle, able to describe also electromagnetic interactions, assumption (P) must be abandoned besides the symmetry's implications (6) and (7). In this section we undertake such a development

First, in sect. 5.1, we replace the active transformation $A \to R_g^{(2)}[A]$ of observables by the passive transformation $A \to S_g[A]$. This allows us to prove, without making use of (7), that a unitary or an anti-unitary operator U_g exists such that $S_g[A] = U_g A U^*_g$, for every $g \in G$.

However, our correspondence $U : \mathcal{G} \to \mathcal{U}(\mathcal{H}),\ S_g[A] = U_g A U_g^{-1}$ is yet not a projective representation, so that Mackey's theorem cannot apply. In section 5.3 we argue how $g \to U_g$ can be changed into a projective representation by means of a σ-conversion. We show that the operators representing position are identifiable with the multiplication operators in the case that the interaction admits a σ-conversion that leaves unaltered the covariance properties of the position with respect to G. Therefore our approach explains such an identification by the existence of a determined covariance property related to the interaction.

In section 5.4 we determine the dynamical law for this class of interactions, whose hamiltonian operator turns out to be more general than the known hamiltonian of a particle in an electromagnetic field. However, in section 5.5 we completely characterize the electromagnetic interaction, among all possible ones, in terms of covariance properties preserved by the σ- conversion.

Passive interpretation of quantum transformations

To recall the meaning of the passive interpretation we make use of the Quantum Theory of a free particle, where $H = L_2(R^3)$ and the position operators Q_α are the multiplication operators defined by $(Q_\alpha \psi)(x) = x_\alpha \psi(x)$. In fact, these operators Q_α represent the position observables of the particle just as referred to a pre-specified spatial reference frame Σ0; if Σ_g is the frame related to such a pre-fixed Σ0 just by g, then the position operators with respect to Σg must be $g^{-1}(Q)$, in general different from Q. Such a dependence on the reference frame is not an exclusive feature of the position observables; for each g ∈ G, we can introduce a mapping

$$S_g : \Omega(\mathcal{H}) \to \Omega(\mathcal{H}), \quad A \to S_g[A] \tag{16}$$

with the following interpretation

PI) If the operator A represents a given observable with respect to the pre-fixed reference frame Σ_0, the operator $S_g[A]$ represents that observable with respect to Σ_g.

Statement (PI) expresses the passive interpretation of the transformation g. For instance, since remark 3.1 extends to the mapping S_g, then the operators $Q^{(t)}$ transform according to

$$S_g[\mathbf{Q}^{(t)}] = g_t^{-1}(\mathbf{Q}^{(t)}), \quad \text{and, as a particular case, } S_g[\mathbf{Q}] = g^{-1}(\mathbf{Q}). \tag{17}$$

The following statements are conceptually implied by the meaning of the passive quantum transformations.

(S.1) For every g ∈ G, the mapping S_g is bijective.

(S.2) Let f be any fixed real Borel function such that if A is a self-adjoint operator, then B = f(A) is a self-adjoint operator too. Since in Quantum Theory a measurement of the quantum observable f(A) can be performed by measuring A and then transforming the obtained outcome a by the purely mathematical transformation f into the outcome b = f(a) of f(A), the following equality has to hold:

$$f(S_g[A]) = S_g[f(A)]. \tag{18}$$

Conditions (S.1) and (S.2) are sufficient to show further properties of the mappings S_g, according to the following propositions.

Proposition 5.1. Let S: Ω (H) → Ω(H) be a bijective mapping such that S[f(A)] = f(S[A]), for every function as in (S2). Then the following statements hold.

i) If E ∈ Π(H) then S[E] ∈ Π[H], i.e., the mapping S is an extension of a bijection of Π(H]

ii) If A, B ∈ Ω(H) and A + B ∈ Ω(H), then [A, B] = 0 implies S[A + B] = S[A] + S[B]. This partial additivity immediately implies S[A] = 0 if and only if A = 0.

iii) For all E, F ∈ Π(H), EF = 0 implies S[E + F] = S[E] + S[F] ∈ Π(H); as a consequence, E ≤ F if and only if S[E] ≤ S[F].

iv) S[P] ∈ Π_1(H) if and only if P ∈ Π_1(H).

Proof.

(i) If E ∈ Π(H) and f(λ) = λ^2 then f(E) = E holds; so S[f(E)] = f(S[E]) implies $(S_g[E])^2$ ≡ f(S[E]) = S[E^2] ≡ S[E], i.e. S^2 [E] = S[E].

(ii) If [A, B] = 0 then C ∈ Ω(H) and two functions f_a, f_b exist so that A = f_a(C) and B = fb(C); once defined the function f = f_a + fb, we have S[A + B] ≡ S[f(C)] = f(S[C]) = f_a(S[C]) + f_b(S[C]) = S[f_a(C)] + S[f_b(C)] ≡ S[A] + S[C].

(iii) If EF = 0, then [E, F] = 0 and (E + F) ∈ Π(H) hold. Statements (i) and (ii) imply S[E + F] = S[E] + S[F] ∈ Π(H].

(iv) If P ∈ Π_1(H) then S[P] ∈ Π(H) by (i). If Q ∈ Π_1(H) and Q ≤ S([P] then P_0 ≡ S^{-1} [Q] ≤ P by (iii); but P is rank 1, therefore P_0 = P and Q = S[P].

Corollary 5.1. From prop. 5.1 immediately follows that the restriction of S to Π(H) is a bijection that also satisfies $S[0] = 0$, $S[I\!I] = I\!I$, $E \leq F$ iff $S[E] \leq S[F]$, $S[E^{\perp}] = (S[E])^{\perp}$

In virtue of corollary 5.1, by Wigner's Theorem 2 the following proposition is easily proved

Proposition 5.2. If a mapping S satisfies the hypothesis of Prop. 5.1, then a unitary or an anti-unitary operator exists such that S[A] = UAU* for every A ∈ Ω(H); if another unitary or anti-unitary operator V satisfies S[A] = V AV * for every A ∈ Ω, then V = $e^{i\theta}$U with θ ∈ R.

Further implications

Propositions 5.1 and 5.2 imply that for every transformation g ∈ G, the corresponding passive quantum transformation is given by $A \to S_g[A] = U_g A U_g^*$. If the correspondence g → S_g satisfied $S_{g1g2} = S_{g1} \circ S_{g2}$, so that g → U_g would be a projective representation, then it could be easily proved, according to [3],[5],[10],[12], that every Ug must be unitary.

But in presence of interaction g → U_g in general is not a projective representation. Now we prove that anti-unitary U_g can be excluded under the only hypothesis that the correspondence g → S_g is continuous according to the following continuity notion given by Bargmann [12].

Definition 5.1. Given P1, P2 ∈ Π_1 (H), the distance d (P_1, P_2) is defined as the minimal distance $\|\psi_1 - \psi_2\|$ between vectors ψ_1, ψ_2 such that $P_1 = |\psi_1\rangle\langle\psi_1|$ *and* $P_2 = |\psi_2\rangle\langle\psi_2|$, *i.e.*, $d(P_1, P_2) = [2(1 - |\langle\psi_1 \mid \psi_2\rangle|]^{1/2}$.

A correspondence g → T_g from G into the set of all automorphims of Π(H) is continuous if for any fixed P ∈ Π_1(H) the mapping from G into Π_1(H), g → T_g[P] is continuous in g with respect the distance d above defined on Π_1(H).

Bargmann proved that if a correspondence g → T_g from G into the set of all automorphisms of Π(H) is continuous then the arbitrary phase factor of the operator Ug for which $T_g[A] = U_g A U_g^*$ can be chosen so that $U_g\psi$ is continuous with respect to g in the topology of H [12].

Proposition 5.3. If the correspondence g ! Sg in (16) is continuous according to def. 5.1, then every operator U_g such that Sg[A] = $U_g A U^*_g$ for all A 2 (H) is unitary.

Proof. Since $S_e[A] = A = \mathbb{I}A\mathbb{I}^*$, we can choose $U_e = \mathbb{I}$ which is unitary. Hence, because of the continuity of g → U_g, a maximal neighborhood K_e of e must exist so that U_g is unitary for all g ∈ K_e; otherwise a sequence gn → e would exist with Ugn anti-unitary and so $\langle\psi \mid \varphi\rangle = \langle U_{g_n}\varphi \mid U_{g_n}\psi\rangle$, and then $\langle\psi \mid \varphi\rangle = \lim_{n\to\infty}\langle U_{g_n}\varphi \mid U_{g_n}\psi\rangle = \langle U_e\varphi \mid U_e\psi\rangle = \langle\varphi \mid \psi\rangle$ which cannot hold for all ψ, ϕ unless H is real. Now, such a neighborhood K_e has no boundary, and since G is a connected group, K_e = G; indeed, if $g_0 \in \partial K_e$, two sequences g_n → g0 and $h_n \to g_0$ would exist with U_{gn} unitary and Uhn anti-unitary; therefore, the continuity of U_g would imply that U_{g0} should simultaneously be unitary and anti-unitary.

σ-conversions

By switching to the passive interpretation, under a continuity condition we were able to establish the existence of a continuous correspondence U : G → U(H), g → U_g, such that S_g[A] = $U_g A U_g^{-1}$. The argument of section 4, which shows that assumption (P) is empirically inadequate, can be easily adapted to show that such a correspondence is not a σ-representation, in general, even if the passive interpretation is adopted. Without such a representation property we cannot apply Mackey's theorem to proceed with our approach. Now we address this obstacle.

The correspondence g → U_g, can be converted into a σ-representation if we multiply each operator U_g by a suitable unitary operator V_g of H; namely, V_g is a unitary operator such that the correspondence $g \to \hat{U}_g = V_g U_g$ turns out to be a σ-representation. The transition from $g \to U_g$ to $g \to \hat{U}_g = V_g U_g$ will be called σ-conversion; the mapping V: G → U (H), g → V_g that realizes the σ-conversion will be called σ-conversion mapping. If g → Vg is a σ-conversion mapping for g → U_g and θ : G → R is a real function, then also g → $e^{i\theta}(g)V_g$ is a σ-conversion mapping, provided that $e^{i\theta}(e) = 1$. In any case, $V_e = \mathbb{I}$ must hold.

If $g \to V_g$ is a σ-conversion mapping for U_g then, according to sect. 2.2, the σ-representation $g \to \hat{U}_g = V_g U_g$ has nine hermitean generators $\hat{P}_\alpha, \hat{J}_\alpha, \hat{G}_\alpha$ for which (2) hold. Moreover, the common spectral measure of the triple $F = \hat{G}/\mu$ is an imprimitivity system for the restriction of $g \to \hat{U}_g$ to $\mathcal{E}$. Then Mackey's theorem allows us to identify H with $L_2(R^3, H_0)$ and explicitly indicates the concrete form of F and of $\hat{P}_\alpha, \hat{J}_\alpha, \hat{G}_\alpha$. In the following, we restrict ourselves to the case $H_0 = C$; is so doing we can take $\mathcal{H} = L_2(\mathbf{R}^3)$, $(F_\alpha \psi)(\mathbf{x}) = x_\alpha \psi(\mathbf{x})$, $\hat{P}_\alpha = -i\frac{\partial}{\partial x_\alpha}$, $\hat{J}_\alpha = F_\beta \hat{P}_\gamma - F_\gamma \hat{P}_\beta$, where (α, β, γ) is a cyclic permutation of (1, 2, 3), and $\hat{G}_\alpha = \mu F_\alpha$.

But $\hat{U}_g$ is not the unitary operator which realizes the passive transformation: $S_g[A]$ 6= $\hat{U}_g A \hat{U}_g^{-1}$. Moreover, F is not the triple Q representing the position. So, our explicit realization of the mathematical formalism of the theory is, in general, devoid of physical significance. But the approach can go on if we restrict our investigation to those interactions which admit σ- conversions $U_g \to \hat{U}_g = V_g U_g$ which are Q-covariant, i.e. σ-conversions that leave unaltered the covariance properties of the position operators Q, i.e. such that

$$\hat{U}_g \mathbf{Q} \hat{U}_g^{-1} = g^{-1}(\mathbf{Q}) \ \forall g \in \mathcal{G}. \tag{19}$$

The proof of Prop. 4.1 can be easily adapted to show that if the interaction admits a Q covariant σ-conversion then F = Q in the present case too. Hence, the operator which represents the position in the mathematical formalism of the theory is the multiplication operator: $Q_\alpha \psi(x) = x_\alpha \psi(x)$. So, in the present approach the identification of the multiplication operator with the position operator is implied by the possibility that the interaction admits a σ-conversion which preserves the covariance properties of the position operators. The existence of interaction which are not Q-covariant is not excluded, of course.

However, the operators $\hat{U}_g$ continue to be not the representative of the transformations of G, i.e., $S_g[A] = \hat{U}_g A \hat{U}_g^{-1}$ does not hold. The following proposition specify how $\hat{U}_g$ relates to the unitary operator U_g that realizes the transformation g according to the passive interpretation

Proposition 5.4. The σ-conversion mapping of a Q-covariant σ-conversion has the form $g \to V_g = e^{i\theta(g,Q)}$, for some function θ.

Proof. The relation $V_g U_g \mathbf{Q} U_g^{-1} V_g^{-1} = g^{-1}(\mathbf{Q})$ implied by (19) and (17) imply $V_g(g^{-1}(\mathbf{Q}))V_g^{-1} = g^{-1}(\mathbf{Q})$, i.e. $[V_g, g^{-1}(\mathbf{Q})] = 0$. Then $[V_g, f(g^{-1}(\mathbf{Q}))] = 0$ for every sufficiently regular function f; by taking f = g we have $[V_g, Q] = 0$. The thesis follows from the completeness of Q and the unitary character of V_g.

Dynamical equation for Q-covariant interactions

Let us consider a velocity boost g ∈ G such that $\hat{U}_g = e^{i\hat{G}_\alpha u}$, for an interaction with Q-covariant σ-conversion. Since $\hat{G}_\alpha = \mu F_\alpha = \mu Q_\alpha$, we can write $\hat{U}_g = e^{i\mu Q_\alpha u}$; therefore

$$\hat{U}_g\dot{Q}_\beta\hat{U}_g^{-1} = \dot{Q}_\beta + i\mu[Q_\alpha, \dot{Q}_\beta]u + o_1(u). \tag{20}$$

On the other hand

$$\hat{U}_g\dot{Q}_\beta\hat{U}_g^{-1} = \lim_{t\to 0} V_g U_g \frac{(Q_\beta^{(t)} - Q_\beta)}{t} U_g^{-1} V_g^{-1}. \tag{21}$$

By making use of $U_g Q_\beta^{(t)} U_g^{-1} = Q_\beta^{(t)} - \delta_{\alpha\beta} ut \mathbb{I}$, implied by (17), in (21) and then comparing with (20) we obtain

$$\hat{U}_g\dot{Q}_\beta\hat{U}_g^{-1} = V_g\dot{Q}_\beta V_g^{-1} - \delta_{\alpha\beta}u\mathbb{I} = \dot{Q}_\beta + i\mu[Q_\alpha, \dot{Q}_\beta]u + o_1(u). \tag{22}$$

But Prop. 5.4 implies that $V_g = e^{i\varsigma_\alpha(u,\mathbf{Q})}$, for some fuction ς_α; replacing in (22) we obtain

$$\dot{Q}_\beta + i[\varsigma_\alpha(u,\mathbf{Q}), \dot{Q}_\beta] + o_2(u) - \delta_{\alpha\beta}u\mathbb{I} = \dot{Q}_\beta + i\mu[Q_\alpha, \dot{Q}_\beta]u + o_1(u). \tag{23}$$

Since $e^{i\varsigma_\alpha(0,\mathbf{Q})} = \mathbb{I}$, the expansion of ςα with respect to u yields $\varsigma_\alpha(u,\mathbf{Q}) = \frac{\partial\varsigma_\alpha}{\partial u}(0,\mathbf{Q})u + o_3(u)$; by replacing this last relation in (23) we obtain $\mu[Q_\alpha, \dot{Q}_\beta] = [\eta_\alpha(\mathbf{Q}), \dot{Q}_\beta] + i\delta_{\alpha\beta}\mathbb{I}$, where $\eta_\alpha(\mathbf{Q}) = \frac{\partial\varsigma_\alpha}{\partial u}(0,\mathbf{Q})$. By replacing Q $\dot{Q}_\beta = i[H, Q_\beta]$ in this equation we can apply Jacobi's identity, and in so doing we obtain $[Q_\beta, \mu\dot{Q}_\alpha] = [Q_\beta, \dot{\eta}_\alpha(\mathbf{Q})] + i\delta_{\alpha\beta}\mathbb{I}$, i.e.

$$[Q_\beta, \dot{\eta}_\alpha(\mathbf{Q}) - \mu\dot{Q}_\alpha] = -i\delta_{\alpha\beta}\mathbb{I} = [Q_\beta, -\hat{P}_\alpha]. \tag{24}$$

By the completeness of Q, from (24) we imply that an operator f_α(Q), function of Q, must exist such that the equation $\dot{\eta}(\mathbf{Q}) - \mu\dot{Q}_\alpha + \hat{P}_\alpha = f_\alpha(\mathbf{Q})$ holds; then we can rewrite (24) as

$$i[H, \mu Q_\alpha - \eta_\alpha(\mathbf{Q})] = \hat{P}_\alpha - f_\alpha(\mathbf{Q}). \tag{25}$$

This is a general dynamical equation for a localizable particle whose interaction admits Qcovariant σ-conversions.

Characterization of the electromagnetic interaction

Once derived the general dynamical law (25) for a localizable particle with Q-covariant and homogeneous in time interaction, it is worth to re-discover the equation currently assumed in quantum physics as a particular case of the general equation (25). The nowadays adopted Schroedinger equation for a spin-0 particle has the form

$$i\frac{d}{dt}\psi_t = \left\{\frac{1}{2m}\sum_{\alpha=1}^{3}[\hat{P}_\alpha - a_\alpha(\mathbf{Q})]^2 + \Phi(\mathbf{Q})\right\}\psi_t.$$

i.e. the Hamiltonian operator is $H = (1/2\mu)\sum_{\alpha=1}^{3}\{\hat{P}_\alpha - a_\alpha(\mathbf{x})\}^2 + \Phi(\mathbf{Q})$. The consequent Quantum Theory corresponds to the case that the function η_α in the general law (25) is a constant function. Let us show this result

Proposition 5.5. The hamiltonian operator H of an interacting particle which admits Qcovariant σ-conversion has the form $H = (1/2\mu)\sum_{\alpha=1}^{3}\{\hat{P}_\alpha - a_\alpha(\mathbf{x})\}^2 + \Phi(\mathbf{Q})$ if and only if the functions η_α in (25) are constant functions. In this case $a_\alpha = f_\alpha$.

Proof. If $\eta\alpha$ is a constant function, then (25) transforms into $i[H, \mu Q_\alpha] = \hat{P}_\alpha - f_\alpha(\mathbf{Q})$ which holds if $H_0 = \frac{1}{2\mu}\sum_{\alpha=1}^{3}[\hat{P}_\alpha - f_\alpha(\mathbf{Q})]^2$ replaces H. Hence the operator H − H0 must be a function Φ of Q because of the completeness of Q. Then $\eta_\alpha(Q) = c\alpha 1I$ implies $H = \frac{1}{2\mu}\sum_{\alpha=1}^{3}[\hat{P}_\alpha - f_\alpha(\mathbf{Q})]^2 + \Phi(\mathbf{Q})$.

Now we prove the converse. Let us suppose that $H = \frac{1}{2\mu}\sum_{\alpha=1}^{3}\{\hat{P}_\alpha - a_\alpha(\mathbf{Q})\}^2 + \Phi(\mathbf{Q})$; by replacing this H in (25) to obtain

$$
\begin{aligned}
i[H, \mu Q_\alpha - \eta_\alpha(\mathbf{Q})] = \hat{P}_\alpha - f_\alpha(\mathbf{Q}) = \\
= \quad & \frac{i}{2\mu}\sum_\beta [P_\beta^2, \mu Q_\alpha] - \frac{i}{2\mu}\sum_\beta [a_\beta \hat{P}_\beta, \mu Q_\alpha] - \frac{i}{2\mu}\sum_\beta [\hat{P}_\beta a_\beta, \mu Q_\alpha] + \frac{i}{2\mu}\sum_\beta [a_\beta^2, \mu Q_\alpha] + \\
& -\frac{i}{2\mu}\sum_\beta [P_\beta^2, \eta_\alpha] + \frac{i}{2\mu}\sum_\beta [a_\beta \hat{P}_\beta, \eta_\alpha] + \frac{i}{2\mu}\sum_\beta [\hat{P}_\beta a_\beta, \eta_\alpha] - \frac{i}{2\mu}\sum_\beta [a_\beta^2, \eta_\alpha] + \\
& + i[\Phi(\mathbf{Q}), \mu Q_\alpha - \eta_\alpha].
\end{aligned}
$$

The last term and the fourth terms of the second and third lines are zero. Then we have

$$
i[H, \mu Q_\alpha - \eta_\alpha(\mathbf{Q})] = \hat{P}_\alpha - f_\alpha(\mathbf{Q}) \tag{26}
$$

$$
\begin{aligned}
= \quad & \hat{P}_\alpha - \frac{i}{2}\sum_\beta (a_\beta \hat{P}_\beta Q_\alpha - Q_\alpha a_\beta \hat{P}_\beta + \hat{P}_\beta a_\beta Q_\alpha - Q_\alpha \hat{P}_\beta a_\beta) + \\
& -\frac{i}{2\mu}\sum_\beta [P_\beta^2, \eta_\alpha] + \frac{i}{2\mu}\sum_\beta (a_\beta \hat{P}_\beta \eta_\alpha - \eta_\alpha a_\beta \hat{P}_\beta + \hat{P}_\beta a_\beta \eta_\alpha - \eta_\alpha \hat{P}_\beta a_\beta) \\
= \quad & \hat{P}_\alpha - \frac{i}{2}\sum_\beta (a_\beta [\hat{P}_\beta, Q_\alpha] + [\hat{P}_\beta, Q_\alpha] a_\beta) - \frac{i}{2\mu}\sum_\beta [P_\beta^2, \eta_\alpha] + \frac{i}{2\mu}\sum_\beta (a_\beta [\hat{P}_\beta, \eta_\alpha] + [\hat{P}_\beta, \eta_\alpha] a_\beta) \\
= \quad & \hat{P}_\alpha - \frac{i}{2}(-2ia_\alpha) - \frac{i}{2\mu}\sum_\beta [P_\beta^2, \eta_\alpha] + \frac{i}{2\mu}\sum_\beta \left(-2i\, a_\beta \frac{\partial \eta_\alpha}{\partial q_\beta}\right) \\
= \quad & \hat{P}_\alpha - a_\alpha + \frac{1}{\mu}\sum_\beta a_\beta \frac{\partial \eta_\alpha}{\partial q_\beta} - \frac{i}{2\mu}\sum [P_\beta^2, \eta_\alpha].
\end{aligned}
$$

From the second and last members of this equations' chain we obtain $f_\alpha(\mathbf{Q}) = a_\alpha - \frac{1}{\mu}\sum_\beta a_\beta \frac{\partial \eta_\alpha}{\partial q_\beta} + \frac{i}{2\mu}\sum [P_\beta^2, \eta_\alpha]$, which implies that $\sum_\beta [P_\beta^2, \eta_\alpha]$ is a function of Q. Therefore we have

$$
\begin{aligned}
\sum_\beta [P_\beta^2, \eta_\alpha] = \phi_\alpha(\mathbf{Q}) \quad &= \quad \sum_\beta (\hat{P}_\beta [\hat{P}_\beta, \eta_\alpha] + [\hat{P}_\beta, \eta_\alpha] \hat{P}_\beta) = (-i)\sum_\beta \left(\hat{P}_\beta \frac{\partial \eta_\alpha}{\partial q_\beta} + \frac{\partial \eta_\alpha}{\partial q_\beta} \hat{P}_\beta\right) \\
&= \quad (-i)\sum_\beta \left(\left[\hat{P}_\beta, \frac{\partial \eta_\alpha}{\partial q_\beta}\right] + 2\frac{\partial \eta_\alpha}{\partial q_\beta} \hat{P}_\beta\right) = (-i)\sum_\beta \left((-i)\frac{\partial^2 \eta_\alpha}{\partial q_\beta^2} + 2\frac{\partial \eta_\alpha}{\partial q_\beta} \hat{P}_\beta\right).
\end{aligned}
$$

Then $\sum_\beta \frac{\partial \eta_\alpha}{\partial q_\beta}\hat{P}_\beta$ isafunctionofQ,andthisimplies $\sum_\beta \left[Q_\gamma, \frac{\partial \eta_\alpha}{\partial q_\beta}\hat{P}_\beta\right] = 0 = \frac{\partial \eta_\alpha}{\partial q_\gamma}[Q_\gamma, \hat{P}_\gamma] = i\frac{\partial \eta_\alpha}{\partial q_\gamma}$ for every γ; therefore $\frac{\partial \eta_\alpha}{\partial q_\gamma} = 0$; thus ηα is a constant function. By using this result in the equality between the second and the last members of (26) we obtain $a_\alpha = f_\alpha$.

Now we show that the case η_α = constant is completely characterized in terms of covariance properties tied to the interaction.

Proposition 5.6. The Q-covariant σ-conversions for which $\eta_\alpha(Q)$ = constant are those and only those which leave unaltered the covariant properties of $Q^{(t)}$ with respect to the Galileian boosts g ∈ G, at the first order in the boost's velocity, i.e. the σ-conversions such that

$$\hat{U}_g\mathbf{Q}^{(t)}\hat{U}_g^{-1} = g_t^{-1}(\mathbf{Q}^{(t)}) + o(u), \quad \text{where } \hat{U}_g = e^{i\hat{G}_\alpha u} \tag{27}$$

at the first order in u

Proof. Let $\hat{U}_g = e^{i\hat{G}_\alpha u} = V_g U_g$ be the σ-converted unitary operator associated with the Galileian boost g, where $Vg = e^{i\varsigma\alpha(u,Q)}$ according to Prop.2. By starting from (27) and by expanding $e^{\pm i\varsigma\alpha(u,Q)}$ with respect to u we obtain

$$e^{i\hat{G}_\alpha u}Q_\beta^{(t)}e^{-i\hat{G}_\alpha u} = V_gU_gQ_\beta^{(t)}U_g^{-1}V_g^{-1} = Q_\beta^{(t)} + i[\eta_\alpha(\mathbf{Q}), Q_\beta^{(t)}]u - \delta_{\alpha\beta}ut\mathbb{I} + o_1(u). \tag{28}$$

The covariance properties of $Q^{(t)}$ with respect to our Galileian boost g are expressed b $S_g[Q_\beta^{(t)}] = Q_\beta^{(t)} - \delta_{\alpha\beta}ut\mathbb{I}$. Hence the equation

$$e^{i\hat{G}_\alpha u}Q_\beta^{(t)}e^{-i\hat{G}_\alpha u} = Q_\beta^{(t)} - \delta_{\alpha\beta}ut\mathbb{I} + o_2(u) \tag{29}$$

is the condition in order that the σ-conversion leave unaltered the covariance properties of $Q^{(t)}$ with respect to the Galileian boosts, at the first order in u; the comparison between (28) and (29) implies that such a condition holds if and only if

$$[\eta_\alpha(\mathbf{Q}), Q_\beta^{(t)}] = 0. \tag{30}$$

If η_α=constant then (30) holds, of course. Therefore, in order to prove the proposition, it is sufficient to prove the inverse implication. Hence we suppose that (30) holds. By expanding $e^{\pm i\hat{G}_\alpha u}$ with respect to u we find $e^{i\hat{G}_\alpha u}Q_\beta^{(t)}e^{-i\hat{G}_\alpha u} = Q_\beta^{(t)} + i[\hat{G}_\alpha, Q_\beta^{(t)}]u + o_3(u)$, so that (29) holds if and only if $i[\hat{G}_\alpha, Q_\beta^{(t)}] = -\delta_{\alpha\beta}t\mathbb{I}$, i.e. if and only if $[Q_\alpha, (\mu Q_\beta^{(t)})/t] = i\delta_{\alpha\beta} \equiv [Q_\alpha, \hat{P}_\beta]$; then the completeness of Q implies that a function ϕ_β must exist such that $(\mu Q_\beta^{(t)})/t - \hat{P}_\beta = \varphi_\beta(\mathbf{Q})$ and hence the following equation holds

$$Q_\beta^{(t)} = \frac{t}{\mu}(\varphi_\beta(\mathbf{Q}) + \hat{P}_\beta). \tag{31}$$

Now, the condition $[\eta_\alpha(Q), Q_\alpha] = 0$ is obviously satisfied for all α; but it and the necessary and sufficient condition (30), where (31) is used, imply also $[\eta_\alpha(\mathbf{Q}), \hat{P}_\beta] = 0$; therefore ηα(Q) is a constant operator $c_\alpha\mathbb{I}$

REFERENCES

1. Wigner E P 1959 Group Theory and its Applications to the Quantum Theory of Atomic Spectra (Boston: Academic Press)
2. Cassinelli G, De Vito E, Lahti P and Levrero A 1997 Rev. Math. Phys. 8 921
3. Mackey G W 1968 Induced Representations of Group and Quantum Mechanics (New York: W.A Benjamin, Inc)
4. Mackey G W 1958 Acta Mathematica 99 265
5. Mackey G W 1989 Unitary Representations in Physics, Probability and Number Theory (Reading: Addison Wesley Pub Co)
6. Varadarjan V S 1968 Geometry of Quantum Theory (Berlin: Springer)
7. Jauch J M 1968 Foundations of Quantum Mechanics (Reading: Addison Wesley Pub Co)
8. Mackey G W 1993 Int. J. Th. Phys 32 1643
9. Ballentine L E 2006 Quantum Mechanics - A Modern Development (Singapore: World Scientific)
10. Simon B 1976 in Studies in Mathematical Physics: Essays in Honor of Valentine Bargmann ed Lieb E H, Simon B and Wightman A S (Princeton: Princeton Un. Press) pp. 327-350
11. Moln´ar L 2007 Lecture Notes in Matemathics 1895 (Berlin: Springer)
12. Bargmann V 1954 Ann. of Math. 59 1

13
CHAPTER

A GENERAL GROUP THEORETICAL METHOD TO UNFOLD BAND STRUCTURES AND ITS APPLICATION

Huaqing Huang, Fawei Zheng, Ping Zhang, Jian Wu, Bing-Lin Gu and Wenhui Duan

Keywords: Band unfolding, electronic structure, group theoretical method

INTRODUCTION

The electronic energy band structure (EBS) is a basic concept in textbooks of condensed matter physics [1] and can be used to investigate various physical properties of crystal materials. Furthermore, the EBS can be directly compared with the results of angle resolved photoemission spectroscopy (ARPES) measurements, which give the spectral function within the quasiparticle picture. Because crystal materials have translational symmetry, we can introduce a wave vector $\vec{k}$ based on Bloch's theorem and get the energy $E(\vec{k})$ as a function of wave vector $\vec{k}$ (i.e., the EBS). With the development of computational techniques and the progress in condensed matter theory, one can easily calculate the EBS of perfect crystal materials using either the empirical tight binding (TB) model [2] or first-principles methods within the framework of density functional theory (DFT) [3, 4].

However, when the translational symmetry of the physical systems is destroyed by defects (such as substitutional doping, impurity, vacancy, dislocation, etc.) [5–8], disorder [9–11], interfacial reconstructoin [12–14], and even different spin configurations [15], one cannot define $\vec{k}$ as well as EBSs in the first Brillouin zone (FBZ) of the primitive lattice. It is common practice to use the supercell approximation with periodic

Citation: Huang, Fawei Zheng, Ping Zhang, Jian Wu,Bing-Lin Gu and Wenhui Duan, A general group theoretical method to unfold band structures and its application, doi:10.1088/1367-2630/16/3/033034

boundary conditions in the computational exploration of those aperiodic systems. As the supercell becomes larger, the corresponding FBZ shrinks and the calculated EBS attains gradually denser energy levels. Consequently, it is hard to extract useful information from such heavily folded EBSs and compare it with ARPES results. In recent years, much effort has been devoted to resolving this issue. Boykin et al developed basic ideas of unfolding under the tight binding approximation [16–18]. Ku et al [19] developed an effective algorithm to unfold the EBS of a supercell to the FBZ of the primitive lattice via localized Wannier functions. Popescu et al [20] also presented a method to extract an effective band structure from supercell calculations on random alloys. Allen et al [21] provided a convenient notation and useful theoretical formulas of band unfolding, which were applied in electron, phonon and slab systems.

Generally, the supercell approximation introduces an artificial translational symmetry of the supercell lattice. Since the translational symmetry operations of primitive and supercell lattices form two Abian groups, we can deal with the unfolding process from the aspect of group theory. In this work, by investigating the relationship between these two translational groups, we propose an approach to unfold the EBS of a supercell calculation to a primitive Brillouin zone. This approach describes an unfolding procedure using different basis sets uniformly, and thus is easy to implement in all TB models and ab initio code packages, where different basis sets such as plane waves, atomic orbitals or Wannier functions are employed. We then demonstrate the validity of this method in a TB calculation of a 30 × 30 graphene supercell. As a more realistic example, we further investigate the effect of a magnetic order of monolayer FeSe on $SrTiO_3$substrate, and find that checkerboard antiferromagnetic (AFM) instead of collinear AFM order might be the ground state spin configuration, which yields the EBS similar to recent experimental observations [22–24].

METHODOLOGY

Translation GroupG_p, G_s, and G_f

Periodic materials, like crystals, have translational symmetry. The translation operation $T_{\vec{r}}$ leaves the lattice invariant. All the translation operations for a primitive lattice form an infinite Abelian group, named as G_p. Besides the primitive lattice, a supercell lattice is also frequently used in electronic structure calculations, which is suitable to simulate complicated systems such as defects and alloys. The translation operations$T_{\vec{R}}$, which leave the supercell lattice invariant, form another group, G_s. The translation vectors for group G_p and $G_s$can be written as

$$\vec{r} = \sum_{i=1}^{D} n_i \vec{a}_i, \tag{1}$$

$$\vec{R} = \sum_{i=1}^{D} N_i \vec{A}_i, \tag{2}$$

where n_i and N_i are integers, and D stands for the dimension of the system. The supercell basis vectors $\vec{A}_i$ and primitive basis vectors $\vec{a}_i$ are related by an integer matrix N, specifically, $\vec{A}_i = \Sigma_j^D N_{ij}\vec{a}_j$.

Actually, all the elements of group G_s are also elements of group G_p, and thus G_s is a subgroup of G_p. Because both G_s and G_p are Abelian groups, there should exist a group isomorphic to the factor group (i.e., $G_p/G_s \cong G_f$), which is denoted by G_f for simplicity. G_p is the direct product of G_s and G_f,

$$G_p = G_s \otimes G_f. \tag{3}$$

The translation vector of group G_f has the same form of $\vec{r}$:

$$\vec{\mathcal{R}} = \sum_i^D n_i \vec{a}_i. \tag{4}$$

Different from infinite G_p and G_s, G_f is a finite group with the translation operations obeying

$$T_{\vec{\mathcal{R}}} = E\left(\text{for } \vec{\mathcal{R}} = \vec{A}_i\right), \tag{5}$$

where E is the unit element of G_f.

Group representation and energy bands unfolding

For Abelian groups, each group element is a class by itself, and then the dimension of irreducible unitary matrix representation is 1 × 1 (just a complex number with norm 1). Thus, the irreducible representations for G_p, G_s, and G_f can be written as

$$D_{\vec{k}}\left(T_{\vec{r}}\right) = e^{i\vec{k}\cdot\vec{r}} \quad \left(\vec{k} \in \omega\right), \tag{6}$$

$$D_{\vec{K}}\left(T_{\vec{R}}\right) = e^{i\vec{K}\cdot\vec{R}} \quad \left(\vec{K} \in \Omega\right), \tag{7}$$

$$D_{\vec{\mathcal{K}}}\left(T_{\vec{\mathcal{R}}}\right) = e^{i\vec{\mathcal{K}}\cdot\vec{\mathcal{R}}} \quad \left(\vec{\mathcal{K}} \in \Lambda\right), \tag{8}$$

where $\vec{k}$ is in the FBZ of primitive lattice (ω), and $\vec{K}$ is in the FBZ of supercell lattice (Ω). From equation (5), the vector $\vec{\mathcal{K}}$ can be written as $\vec{\mathcal{K}} = \Sigma_i^D N_i\vec{B}_i$, where $\vec{B}_i$ are reciprocal lattice vectors of the supercell lattice. Herein N_i is an integer and should be chosen to ensure that $\vec{\mathcal{K}}$ is in the primitive cell FBZ ω. The total number of choices for $\{N_1, \ldots, N_D\}$ is equal to the volume ratio $|\omega|/|\Omega|$, which is also the order of group G_f. We denote the set of all the possible $\vec{\mathcal{K}}$ by Λ.

According to Bloch's theorem, the basis function of the irreducible representation of G_p can be written as

$$\psi_{\vec{k}}(\vec{x}) = e^{i\vec{k}\cdot\vec{x}} u_{\vec{k}}(\vec{x}), \tag{9}$$

where $u_{\vec{k}}(\vec{x})$ is a periodic function in the primitive lattice (i.e., $T_{\vec{r}} u_{\vec{k}}(\vec{x}) = u_{\vec{k}}(\vec{x})$). As a result of the fact that G_s is a subgroup of G_p, the function $\psi_{\vec{k}}(\vec{x})$ is also the basis function of group G_s,

$$\psi_{\vec{k}}(\vec{x}) = e^{i(\vec{K}+\vec{\mathcal{K}})\cdot\vec{x}} u_{\vec{K}+\vec{\mathcal{K}}}(\vec{x}) = e^{i\vec{K}\cdot\vec{x}}\left[e^{i\vec{\mathcal{K}}\cdot\vec{x}} u_{\vec{K}+\vec{\mathcal{K}}}(\vec{x})\right] = e^{i\vec{K}\cdot\vec{x}} u'_{\vec{K}}(\vec{x}). \tag{10}$$

Herein the function $u'_{\vec{K}}(\vec{x})$ is periodic in the supercell lattice $\left[T_{\vec{R}} u'_{\vec{K}}(\vec{x}) = u'_{\vec{K}}(\vec{x})\right]$, and it is also the basis function of group G_f with wave vector $\vec{\mathcal{K}}$.

In supercell electronic structure calculations, we get a series of functions $u'_{\vec{K}}(\vec{x})$ and eigenenergies for each $\vec{K}$. As discussed previously, there is a hidden parameter $\vec{\mathcal{K}}$ in $u'_{\vec{K}}(\vec{x})$. Once we get the parameter $\vec{\mathcal{K}}$, the Bloch function and the corresponding energy for vector $\vec{k} = \vec{K} + \vec{\mathcal{K}}$ in ω are known. In other words, the energy bands are unfolded. Since any primitive wavevector $\vec{k}$ in the primitive FBZ belongs to a unique $\vec{K}$ in the supercell FBZ, an easy search in supercell reciprocal lattices can yield the proper $\vec{\mathcal{K}}$, so that $\vec{k} - \vec{\mathcal{K}}$ is in the supercell FBZ. More detailed analysis about wavevectors can be found in [25, 26].

In the following, we will present a method to identify the wave vector $\vec{\mathcal{K}}$ for $u'_{\vec{K}}(\vec{x})$. We choose a set of normalized orthogonal periodic functions $v_i(\vec{x})$ (i = 1, 2, 3 ...), which are complete in the primitive cell (for the nonorthogonal case, it is straightforward to perform the standard orthonormalization procedure priorly). The basis function of group G_f can be written as $|i, \vec{\mathcal{K}}'\rangle = e^{i\vec{\mathcal{K}}'\cdot\vec{x}} v_i(\vec{x})$. Then we construct a group of projection operators $P(\vec{\mathcal{K}}')$ as

$$P(\vec{\mathcal{K}}') = \sum_i |i, \vec{\mathcal{K}}'\rangle\langle i, \vec{\mathcal{K}}'| \quad (\vec{\mathcal{K}}' \in \Lambda). \tag{11}$$

The expectation value of $P(\vec{\mathcal{K}}')$ for function $u'_{\vec{K}}(\vec{x}) = e^{i\vec{\mathcal{K}}\cdot\vec{x}} u_{\vec{K}+\vec{\mathcal{K}}}(\vec{x})$ is $\delta_{\vec{\mathcal{K}}\vec{\mathcal{K}}'}$. From this expectation value, we can identify the vector $\vec{\mathcal{K}}$, then further unfold the energy bands. The unfolded energy bands should be identical to the energy bands calculated using the primitive cell for a perfect lattice. However, when the translational symmetry is broken (due to impurity, structure reconstruction, magnetic reconfiguration, etc.), the above unfolding procedure produces new energy bands, which are similar to (but need not be identical to) energy bands calculated using the primitive cell. The expectation value of $P(\vec{\mathcal{K}}')$ for $u'_{\vec{K}}(\vec{x})$ is between 0 and 1 instead of 0 or 1, representing the weight of $u'_{\vec{K}}(\vec{x})$ at point $\vec{K} + \vec{\mathcal{K}}'$. Thus, the energy bands become fuzzy and may break into separate parts. In this case, the unfolded energy bands contain information of the defects or reconstruction and can be directly compared with ARPES or other measurements. Since the main difference of computing real and complex bands lies in the choice of basis, our method can be used for complex bands too, if some modifications are included [27, 28].

In practice, the function $v_i(\vec{x})$ can be taken as plane waves, atomic orbitals, Wannier functions, or any other basis sets in quantum chemistry. Thus our method should be easy to implement in both TB models [16–18] and existing ab initio code packages employing different basis sets [29–32]. Note that when we choose $e^{i\vec{K}\cdot\vec{x}}v_i(\vec{x})$ to be the eigenvectors of the primitive cell calculations, and they are complete and can be represented by a plane wave expansion, then we can get similar weight expression to [20] (see the appendix). In fact, the basis sets in our method should be complete periodic functions but are not required to be eigenstates. It turns out that the method is even simpler when the local basis sets are used. For example, when Wannier functions of the primitive cell are used as the $v_i(\vec{x})$, the method turns out similar to Ku et al's approach [19]. Other localized basis sets can also be adapted, which results in a modified formula of weight factor [33]. Hence, our method provides a more general strategy to unfold EBSs by constructing a group of universal projector operators that are not restricted to specific basis sets.

NUMERICAL IMPLEMENTATION AND APPLICATIONS

In this section, we discuss the application of the previously described method. First, we illustrate the validity of the approach in 30 × 30 graphene supercell calculations within the TB approximation. We then apply the method to investigate the magnetic order of monolayer FeSe on a TiO_2-terminated $SrTiO_3$ (001) surface, which shows signatures of high-temperature superconductivity with $T_c > 50$ K [34], within the framework of DFT.

Unfolding bands of perfect graphene in the TB calculations

Graphene is a one-atom-thick, two-dimensional allotrope of carbon with a unique Dirac cone in its EBS, which can provide a simple but nontrivial test for our unfolding method. We consider the 30 × 30 supercell for graphene in the TB calculation. Herein, only one p_z orbital for each carbon atom is used, and the TB parameters are taken as transfer integral $t = -3.03$ eV and overlap integral s = 0.129 [35]. Figure 1(b) shows the EBS from the 30 × 30 supercell calculation. With the supercell EBS and corresponding wave functions, we unfold the energy bands to the FBZ of the primitive, and the results are shown in figure 1(a) by red circles. Evidently, without symmetry breaking, the EBS unfolding of the supercell calculations indeed reproduces exactly the same EBS as that from the primitive cell calculation, indicating the validity of our unfolding approach.

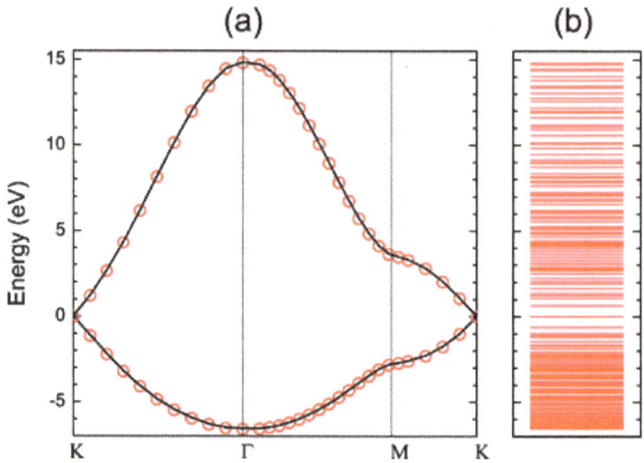

Figure 1: The TB energy bands of graphene from the primitive cell calculation (black line in (a)) and supercell calculations with band unfolding (red circles in (a)). The unfolded energy bands are obtained from the single Γ point calculation of 30 × 30 supercell (b).

Unfolding bands of monolayer FeSe on $SrTiO_3$ substrate for *ab initio* calculations

Now we turn to a realistic but more complicated system. Recently, high-temperature superconductivity ($T_c \sim 65$ K) in monolayer FeSe grown on $SrTiO_3$ substrate (FeSe/STO) by molecular beam epitaxy was reported [34]. This simplest iron-based superconductor shows quite different Fermi surface topology from other iron-based superconductors [22–24], presumably implying a different mechanism. Moreover, there are several intriguing issues that merit further studies. One is that the magnetic order of that system is still controversial. A recent ARPES experiment suggested that the ground state of FeSe/STO is the spin density wave state [24], similar to its bulk counterpart [36]. By using first-principles calculations, Liu et al [15] illustrated that the spin configuration of ground state FeSe/STO is the collinear AFM state. The first-principles calculations by Bazhirov and Cohen [37] for the FeSe monolayer film without STO substrate, however, showed that the experimentally observed Fermi surface is best described by the checkerboard AFM spin configuration. Zheng et al [38] further demonstrated the effects of charge doping and electric field on the Fermi surface of a monolayer checkerboard AFM FeSe on $SrTiO_3$.

Actually, FeSe/STO systems in different magnetic states have different supercells and thus different FBZs. This makes it difficult to compare the calculated EBSs and Fermi surfaces in different FBZs with those from ARPES measurements which are in the FBZ of the primitive cell [39]. Herein, by combining ab initiocalculation with our group-theory-based unfolding method, we studied FeSe/STO in the checkerboard (Néel) and collinear (striped) AFM spin configurations.

The calculations were carried out within the framework of DFT as implemented in the QUANTUM ESPRESSO code package [30]. A kinetic energy cutoff of 120 Ry was used for the plane wave basis to achieve a balance between calculation efficiency and accuracy, and we adopted the generalized gradient approximation with the Perdew–Burke–Ernzerhof exchange correlation functional [40]. The norm-conserving pseudopotentials [41] were employed to describe the electron–ion interactions. The convergence thresholds of energy and force were set to 5.0×10^{-5} Ry and 5.0×10^{-4} Ry/Bohr for structure optimization. We constructed a six-layer $SrTiO_3$ (001) slab with monolayer FeSe on the TiO_2 termination, where the four bottom atomic layers are fixed at their bulk positions. To model checkerboard and collinear AFM configurations, we used 1×1 and $\sqrt{2} \times \sqrt{2}$ supercells with a vacuum layer of about 10 Å. $9 \times 9 \times 1$ and $6 \times 6 \times 1$ Monkhorst-Pack k-point meshes [42] for Brillouin-Zone sampling were used for the 1×1 and $\sqrt{2} \times \sqrt{2}$ supercells, respectively.

The optimized structure of $\sqrt{2} \times \sqrt{2}$ FeSe/STO is shown in figure 2(a). The lower Se (Fe) atoms are on the top of the Ti(O) atoms with an average vertical distance of 3.26 Å (4.56 Å). The checkerboard and collinear AFM spin configurations of the Fe atoms are presented in figures 2(b) and (c), respectively. The calculated magnetizations of the checkerboard and collinear AFM states are, respectively, 2.77 and 3.02 μ_B per Fe atom, consistent with previous theoretical results [37, 38].

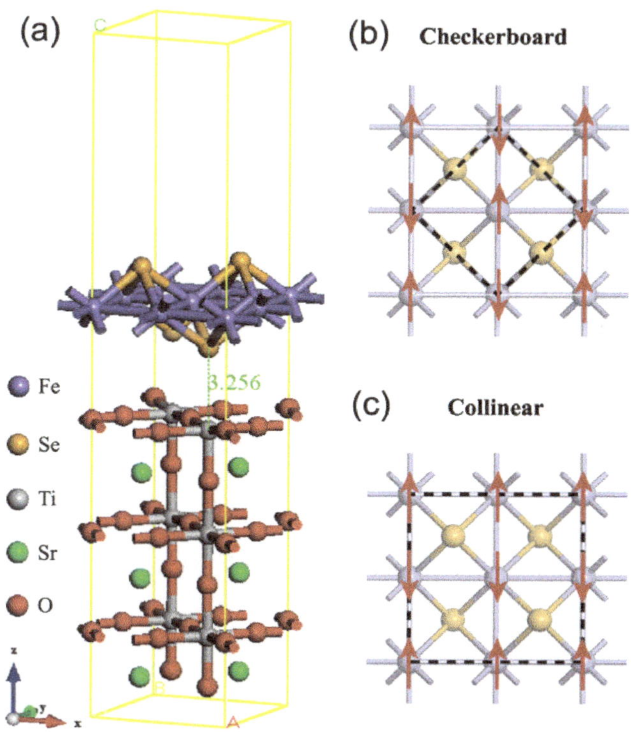

Figure 2: Atomic structure (a) and spin configurations of (b) checkerboard and (c) collinear AFM orders of Fe atoms for monolayer FeSe on TiO_2 terminated $SrTiO_3$ (001) surface. Spin-up and spin-down are marked by upward and downward red arrows, respectively. The dashed-line squares in (c) and (d) denote the 1×1 primitive cell and $\sqrt{2} \times \sqrt{2}$ supercell, respectively.

The calculated spin-up EBSs are shown in figure 3. Note that the spin-down EBS is similar to spin-up one and thus is not shown for simplicity. As shown in figure 3(a), FeSe/STO in the checkerboard AFM state has a Fermi surface pocket near the zone corners (i.e., M and M' points), but without any indication of pockets around the zone center (i.e., Γ point), which is consistent with recent experimental observations [22]. Then, based on the spin-up EBS of the collinear AFM state with $\sqrt{2} \times \sqrt{2}$ supercell (see figure 3(d)), we get the unfolded EBS (see figure 3(c)) in the primitive FBS, where the shade and radius of dots represent the weight of each eigenvalues. Comparing the EBS of the checkerboard AFM state with the unfolded EBS of the collinear AFM state, we can see that hole like bands appear in both the zone corner and center in figure 3(c). Moreover, the unfolded bands cross the Fermi level along $\Gamma M'$, indicating extra Fermi surface pockets located around Γ towards M'. Our analysis suggests that the checkerboard instead of the collinear AFM state can yield the EBS compatible with recent experimental measurements [22–24]. This reveals that the ground state spin configuration of monolayer FeSe on a TiO_2 terminated $SrTiO_3$ (001) surface may have checkerboard AFM order, which would be helpful to understand the superconductivity mechanism in low-dimensional Fe-based superconductors.

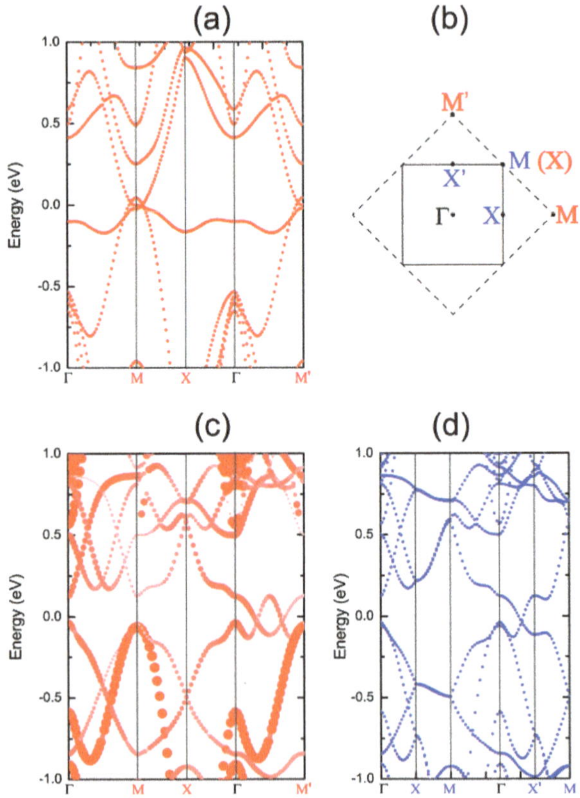

Figure 3: The spin-up EBS of monolayer $FeSe/SrTiO_3$ in different magnetic orders. (a) The spin-up EBS of the $FeSe/SrTiO_3$ system in the checkerboard AFM state along the high symmetry

direction of the primitive cell FBZ. (b) The FBZ of the primitive 1 × 1 ($\sqrt{2} \times \sqrt{2}$) lattice with high symmetry k points labeled by red (blue). (c) The unfolded and (d) folded spin-up EBSs of the system in the collinear AFM state. The shade and radius of dots in (c) represent the spectral weight of each eigenstate. The Fermi level is set to zero.

CONCLUSION

To summarize, we present a method to unfold the EBS from supercell calculations to the FBZ of the primitive lattice. Directly derived from the translation group symmetry, this method gives a uniform description of unfolding approaches based on various basis sets (such as plane waves, atomic orbitals and Wannier functions), which makes it easy to implement in both tight-binding models and existing ab initiocode packages using different basis sets. Based on this method, we can easily compare unfolded EBSs of different supercell calculations and connect the theoretical results with ARPES measurements. As an example, we apply this method to investigate magnetic order in monolayer FeSe on a $SrTiO_3$ (001) surface, finding that the checkerboard AFM state rather than the collinear AFM state has the EBS compatible with recent ARPES data. Our method can be further employed in the study of defects, disorder, interfacial reconstructions, and other systems that require the use of supercells.

ACKNOWLEDGMENTS

HH and FZ equally contributed to this work. We thank Yong Xu for valuable discussions. This work is supported by the Ministry of Science and Technology of China (grant nos. 2011CB921901 and 2011CB606405) and the National Natural Science Foundation of China (grant nos. 11004013 and 11074139).

APPENDIX. UNFOLDING FORMULA ON THE CHOICE OF PLANE WAVE AS BASIS FUNCTIONS

To prove the equivalence and generality of our method, we derive the unfolding formula on the plane wave basis and compare our results with previous work. The periodic part of the supercell Bloch function can be expressed as:

$$|u'_{\vec{K}}\rangle = \sum_{\vec{G}} C_{\vec{K}}\left(\vec{G}\right) e^{i\vec{G}\cdot\vec{x}}, \tag{A.1}$$

where $\vec{G} = \Sigma_i^D N_i \vec{B}_i$ are supercell reciprocal lattice vectors. We construct the projection operator $P\left(\vec{\mathcal{K}}'\right)$ using a set of plane wave functions with primitive cell period

$$|\vec{g}, \vec{\mathcal{K}}'\rangle = e^{i\left(\vec{\mathcal{K}}'+\vec{g}\right)\cdot\vec{x}}, \tag{A.2}$$

where $\vec{g} = \Sigma_i^D n_i \vec{b}_i$ are primitive cell reciprocal lattice vectors. The expectation value of $P\left(\vec{\mathcal{K}}'\right)$ is

$$\left\langle u'_{\vec{K}} \middle| P(\vec{\mathcal{K}}') \middle| u'_{\vec{K}} \right\rangle = \sum_{\vec{g}} \left\langle u'_{\vec{K}} \middle| \vec{g}, \vec{\mathcal{K}}' \right\rangle \left\langle \vec{g}, \vec{\mathcal{K}}' \middle| u'_{\vec{K}} \right\rangle = \sum_{\vec{g}} \left| \left\langle \vec{g}, \vec{\mathcal{K}}' \middle| u'_{\vec{K}} \right\rangle \right|^2. \tag{A.3}$$

We use (A.1) and (A.2) to get

$$\begin{aligned} \left\langle \vec{g}, \vec{\mathcal{K}}' \middle| u'_{\vec{K}} \right\rangle &= \sum_{\vec{G}} \int d\vec{x} C_{\vec{K}}(\vec{G}) e^{i\vec{G}\cdot\vec{x}} e^{-i(\vec{\mathcal{K}}'+\vec{g})\cdot\vec{x}} \\ &= \sum_{\vec{G}} C_{\vec{K}}(\vec{G}) \delta(\vec{G} - \vec{g} - \vec{\mathcal{K}}') \\ &= C_{\vec{K}}(\vec{g} + \vec{\mathcal{K}}'), \end{aligned} \tag{A.4}$$

thus

$$\left\langle u'_{\vec{K}} \middle| P(\vec{\mathcal{K}}') \middle| u'_{\vec{K}} \right\rangle = \sum_{\vec{g}} \left| C_{\vec{K}}(\vec{g} + \vec{\mathcal{K}}') \right|^2. \tag{A.5}$$

(A.5) is exactly the same as equation (15) of [20], implying that our method can derive the equivalent results to previous work explicitly if plane wave functions are adapted as a basis. What is more, our method does not need any information about the primitive eigenfunctions, which simplifies the derivation procedure.

REFERENCES

1. Ashcroft N W and Mermin N D 1976 Solid State Physics (London: Thomson Learning)
2. Slater J C and Koster G F 1954 Phys. Rev. 94 1498
3. Hohenberg P and Kohn W 1964 Phys. Rev. 136 B864
4. Kohn W and Sham L J 1965 Phys. Rev. 140 A1133
5. van Heumen E et al 2011 Phys. Rev. Lett. 106 027002
6. Berlijn T, Lin C-H, Garber W and Ku W 2012 Phys. Rev. Lett. 108 207003
7. Berlijn T, Hirschfeld P J and Ku W 2012 Phys. Rev. Lett. 109 147003
8. Konbu S, Nakamura K, Ikeda H and Arita R 2012 Solid State Commun. 152 728
9. Popescu V and Zunger A 2010 Phys. Rev. Lett. 104 236403
10. Berlijn T, Volja D and Ku W 2011 Phys. Rev. Lett. 106 077005
11. Haverkort M W, Elfimov L S and Sawatzky G A 2011 arXiv:1109.4036
12. Kim S, Ihm J, Choi H J and Son Y-W 2008 Phys. Rev. Lett. 100 176802
13. Qi Y, Rhim S H, Sun G F, Weinert M and Li L 2010 Phys. Rev. Lett. 105 085502
14. Cahangirov S, Audiffred M, Tang P, Iacomino A, Duan W, Merino G and Rubio A 2013 Phys. Rev. B 88 035432
15. Liu K, Lu Z-Y and Xiang T 2012 Phys. Rev. B 85 235123
16. Boykin T B and Klimeck G 2005 Phys. Rev. B 71 115215
17. Boykin T B, Kharche N and Klimeck G 2007 Phys. Rev. B 76 035310
18. Boykin T B, Kharche N, Klimeck G and Korkusinski M 2007 J. Phys.: Condens.

Matter 19 036203

19. Ku W, Berlijn T and Lee C-C 2010 Phys. Rev. Lett. 104 216401
20. Popescu V and Zunger A 2012 Phys. Rev. B 85 085201
21. Allen P B, Berlijn T, Casavant D A and Soler J M 2013 Phys. Rev. B 87 085322
22. Liu D et al 2012 Nature Commun. 3 931
23. He S et al 2013 Nature Mater. 12 605
24. Tan S et al 2013 Nature Mater. 12 634
25. Boykin T B, Kharche N and Klimeck G 2006 Eur. J. Phys. 27 5
26. Boykin T B, Kharche N and Klimeck G 2009 Physica E 41 490
27. Ajoy A and Karmalkar S J 2010 J. Phys.: Condens. Matter 22 435502
28. Ajoy A, Murali K V R M and Karmalkar S 2012 J. Phys.: Condens. Matter 24 055504
29. Kresse G and Furthmüller J 1996 Comput. Mater. Sci. 6 15
30. Giannozzi P et al 2009 J. Phys. Condens. Matter 21 395502
31. Soler J M, Artacho E, Gale J D, Garca A, Junquera J, Ordejón P and Sánchez-Portal D 2002 J. Phys.: Condens. Matter 14 2745
32. Andersen O K 1975 Phys. Rev. B 12 3060
33. Lee C-C, Yamada-Takamura Y and Ozaki T 2013 J. Phys.: Condens. Matter 25 345501
34. Wang Q-Y et al 2012 Chin. Phys. Lett. 29 037402
35. Saito R, Dresselhaus G and Dresselhaus M S 1998 Physical Properties of Carbon Nanotubes vol 4 (London: Imperial College Press)
36. Ma F, Ji W, Hu J, Lu Z-Y and Xiang T 2009 Phys. Rev. Lett. 102 177003
37. Bazhirov T and Cohen M L 2013 J. Phys.: Condens. Matter 25 105506
38. Zheng F, Wang Z, Kang W and Zhang P 2013 Sci. Rep. 3 2213
39. Lin C-H, Berlijn T, Wang L, Lee C-C, Yin W-G and Ku W 2011 Phys. Rev. Lett. 107 257001
40. Perdew J P, Burke K and Ernzerhof M 1996 Phys. Rev. Lett. 77 3865
41. Rappe A M, Rabe K M, Kaxiras E and Joannopoulos J D 1990 Phys. Rev. B 41 1227
42. Monkhorst H J and Pack J D 1976 Phys. Rev. B 13 5188

14
CHAPTER

SYMMETRY CLASSIFICATION OF ENERGY BANDS IN GRAPHENE AND SILICENE

Eugene Kogan

Keywords: Graphene, Silicene, Group Theory, Dirac Points

INTRODUCTION

Since graphene was first isolated experimentally [1], it is in the focus of attention of both theorists and experimenttalists. Obviously, understanding of the symmetries of the electrons dispersion law in graphene is of crucial importance. Actually, the symmetry classification of the energy bands in graphene (or "two-dimensional graphite") was presented nearly 60 years ago by Lomer in his seminal paper [2]. Later the subject was analyzed by Slonczewski and Weiss [3], Dresselhaus and Dresselhaus [4], Bassani and Parravicini [5]. Recent approaches to the problem are presented in the papers by Malard et al. [6], Manes [7] and in our publication [8].

The present work has two aspects: a pragmatic and a pedagogical one. The first aspect is connected with the recent synthesis of silicene, the counterpart of graphene for silicon, with buckled honeycomb geometry. This novel two-dimensional material has attracted recently considerable attention, both theoretically [9,10] and experimentally, due to its exotic electronic structure and promising applications in nanoelectronics as well as its compatibility with current silicon-based electronic technology. So we present the symmetrty analysis of the silicene electron bands.

Citation: E. Kogan, "Symmetry Classification of Energy Bands in Graphene and Silicene," Graphene, Vol. 2 No. 2, 2013, pp. 74-80. doi: 10.4236/graphene.2013.22011.

The pedagogical aspect is connected with the fact that different approaches to the symmetry classification, even if giving the same results, are based on different methods of applications of group theory. Thus in our previous paper [8] the labeling of the bands was based on com patibility relations and guesses. In the present work we show that in the framework of the tight-binding approximation the representations of the little group in the symmetry points can be rigorously found in the framework of the group theory algebra. Though the idea of using the tight-binding approximation is by no means new (it was used already in the work by Lomer), our mathematical approach is totally different, as one can easily see comparing the present work with [2], and, to our opinion, more convenient for applications. This statement is supported by the analysis of the symmetry of the energy band in silicene.

We also generalize the symmetry classification by taking into account the spin-orbit coupling both for graphene and for silicene. This, to the best of our knowledge, wasn't done before even for graphene. Though in graphene the spin-orbit coupling is very weak, the problem is interesting in principle. One can expect that in silicenr the coupling is stronger, and it will become even more so for graphene related materials from heavier elements, provided they can be synthesized.

To remind to a reader a few basic things, important for the symmetry classification of the bands in any crystal, consider a point sub-group R of the space group characterizing the symmetry of a crystal (we restrict ourselves with the consideration of symmorphic space groups). Any operation of the group R (save the unit transformation) takes a general wavevector k into a distinct one. However, for some special choices of k some of the operations of the group R will take k into itself rather than into a distinct wave vector. These particular operations are called the group of k; it is a subgroup of the group R. Points (lines) in the Brillouin zone for which the group of the wave vector contains elements other than the unit element are called symmetry points (lines). We may use a state (states) corresponding to such a special wavevector to generate a representation for the group of k [11,12]. In this paper we consider crystals with the hexagonal Brillouin zone. In this case the symmetry points are Γ—the center of the Brillouin zone, the points K which are corners of the Brillouin zone and the points M which are the centers of the edges of the Brillouin zone.

TIGHT-BINDING MODEL

We'll deal with the materials with a basis of two atoms per unit cell, and we'll search for the solution of Schroedinger equation as a linear combination of the functions

$$\psi^{j}_{\beta;k} = \sum_{R_j} \psi_\beta \left(\boldsymbol{r} - \boldsymbol{R}_j \right) e^{i\boldsymbol{k}\cdot\boldsymbol{R}_j}, \tag{1}$$

where ψ_β are atomic orbitals, $j = A, B$ labels the sublattices, and $\boldsymbol{R}_j$ is the radius vector of an atom in the sublattice j.

A point symmetry transformation of the functions $\psi^{j}_{\beta;k}$ is a direct product of two transformations: the transformation of the sub-lattice functions $\varphi_k^{A,B}$, where

$$\phi_k^j = \sum_{R_j} e^{ik \cdot R_j}, \tag{2}$$

and the transformation of the orbitals ψ_β. Thus the representations realized by the functions (1) will be the direct product of two representations. Generally, these representations will be reducible. To decompose a reducible representation into the irreducible ones it is convenient to use equation

$$a_\alpha = \frac{1}{g}\sum_G \chi(G)\chi_\alpha^*(G), \tag{3}$$

which shows how many times a given irreducible representation α is contained in a reducible one [13]. Additional information about the representations can be obtained if we use projection operator [14]

$$O_\alpha = \frac{n_\alpha}{g}\sum_G \chi_\alpha^*(G)P(G), \tag{4}$$

where n_α is the dimensionality of the irreducible representation α and $P(G)$ is the operator corresponding to a transformation G. The operator projects a given function to the linear space of the representation α. For a one dimensional representation the operator thus gives basis of the representation.

GROUP THEORY ANALYSIS IN THE TIGHT-BINDING MODEL WITHOUT THE SPIN-ORBIT COUPLING

Our tight-binding model space includes four atomic orbitals: $|s,p\rangle$. Notice that we assume only symmetry of the basis functions with respect to rotations and reflections; the question how these functions are connected with the atomic functions of the isolated carbon atom is irrelevant.

Graphene

The Hamiltonian of graphene being symmetric with respect to reflection in the graphene plane, the bands built from the $|z\rangle$ orbitals decouple from those built from the $|s,x,y\rangle$ orbitals. The former are odd with respect to reflection, the latter are even. In other words, the former form π bands, and the latter form σ bands.

The group of wave vector k at the point Γ is D_{6h}, at the point K is D_{3h}, at the lines Γ-K is C_{2v} [8,15]. The representations of the groups D_{3h} and D_{6h} can be obtained on the basis of identities

$$D_{3h} = D_3 \times C_s, \; D_{6h} = C_{6v} \times C_s. \tag{5}$$

the irreducible representations of the group D_3 are presented in the **Table 1**. Each representation of the group, say A_1, begets two representations of the group D_{3h}: A_1' and A_1'' ; prime means that the representation is even with respect to reflection σ_h, double prime means that it is odd.

The irreducible representations of the group C_{6v} are presented in the **Table 2**. Because the inversion transformation I can be presented as

$$I = C_2\sigma_h, \tag{6}$$

the representations of the group D_{6h} can be classified as symmetric (g) or antisymmetric (u) with respect to inversion. Thus each representation of the group C_{6v}, say A_1, begets two representations of the group D_{6h}: A_{1g} and A_{1u}.

Notice that the orbitals $|s\rangle$ (or $|z\rangle$) realize A_1 representation both of the group D_3 and of the group C_{6v}, hence the representations of the groups realized by the functions $\psi^{A,B}_{s,z;k}$ will be identical to those realized by the sub-lattice functions $\varphi^{A,B}_k$.

Let us start from the symmetry analysis at the point Γ. Because the transformations C_2, C_6, σ_v change sublattices, the characters corresponding to these transfor of the group C_{6v}.

Table 1: Characters table for irreducible representations of D_3 point groups.

D_3	E	$2C_3$	$3U_2$
A_1	1	1	1
A_2	1	1	−1
E	2	−1	0

Table 2: Characters table for irreducible representations of C_{6v} point group.

C_{6V}	E	C_2	$2C_3$	$2C_6$	$2\sigma_v$	$3\sigma_v'$
A_1	1	1	1	1	1	1
A_2	1	1	1	1	−1	−1
B_2	1	−1	1	−1	1	−1
B_1	1	−1	1	−1	−1	1
E_2	2	2	−1	−1	0	0
E_1	2	−2	−1	1	0	0

Mations are equal to zero. The transformations $E, C_3, \sigma_{v'}$ leave the sub-lattices as they were. Hence from **Table 2**, we see that the functions $\phi_0^{A,B}$ realizes reducible representation

$$R_\Gamma = A_1 + B_2 \tag{7}$$

Taking into account the symmetry of the states relative to reflection in the plane of graphene, we obtain that at the point Γ the functions ψ_z (here and further on, when this is not supposed to lead to a misunderstanding, we'll suppres the index k in $\psi_{B;k}$) realize A_{1u} and B_{2g} representations of the group D_{6h}, characterizing π band; the functions ψ_x realize A_{1g} and B_{2u} representations of the group D_{6h}, characterizing σ band.

Acting by projection operators O_A and O_B on a function ψ^j, we obtain that the irreducible representation A_1 is realized by symmetric combination of the A and B orbitals, and the irreducible representation B_2 by the antisymmetric combination. One can expect that the first case occurs in the hole band, and the second in the electron band.

The orbitals $|x, y\rangle$ realize representation E_1 of the group C_{6v} [13]. Hence, representation of the group realized by the functions $\psi_{x,y}^{A,B}$ can be decomposed as

$$E_1 \times R_\Gamma = E_1 + E_2. \tag{8}$$

Taking into account the symmetry of the states relative to reflection in the plane of graphene, we obtain that at the point Γ the functions $\psi_{x,y}$ realize E_{1u} and E_{2g} representations of the group D_{6h}, characterizing σ bands.

To find wavefunctions, realizing each of the irreducible representations, we apply the projection operators and obtain

$$O_{E_1}\psi_\pm^j \sim \psi_\pm^j + \psi_\pm^{\bar{j}},\ O_{E_2}\psi_\pm^j \sim \left(\psi_\pm^j - \psi_\pm^{\bar{j}}\right), \tag{9}$$

where $\bar{j} = B$ if $j = A$, and vice versa. Thus representation $E_1(E_2)$ is realized by symmetric (antisymmetric) combinations of $|x, y\rangle$ orbitals. One can expect that the first representation is realized at the hole band, and the second at the valence band.

Now let us perform the symmetry analysis at the point K. The representation of the group realized by the functions $\phi_K^{A,B}$ is determined by the transformation law of the exponentials $e^{iK\cdot R_j}$ under the symmetry operations. Rotation of the radius vector by the angle $2\pi/3$ anticlockwise, is equivalent to rotation of the vector K in the opposite direction, that is to substitution of the three equivalent corners of the Brillouin zone: $K \to K_2 \to K_3 \to K$, where $K = \left(2\pi/3a, 2\pi/3\sqrt{3}a\right)$, $K_2 = \left(0, -4\pi/3\sqrt{3}a\right)$ and $K_3 = \left(-2\pi/3a, 2\pi/3\sqrt{3}a\right)$. The rotation multiplies each basis vector by the factor $e^{2\pi i/3}$. Using Eqation (3), we obtain

$$a_E = (1/3)\left(2 - e^{2\pi i/3} - e^{-2\pi i/3}\right) = 1. \tag{10}$$

Hence, the functions $\phi_K^{A,B}$ realize irreducible representation E of the group D_3.

Taking into account the symmetry of the states relative to reflection in the plane of graphene, we obtain representation E'' of the group D_{3h}, realized by the ψ_z functions (merging π bands), and representation E', realized by the ψ_s functions, characterizing σ bands.

The orbitals $|x,y\rangle$ realize representation E of the group D_3 [13]. Hence, representation of the group realized by the quartet of functions $\psi_{x,y}^{A,B}$ can be decomposed as

$$E \times E = A_1 + A_2 + E. \tag{11}$$

taking into account the symmetry of the states relative to reflection in the plane of graphene, we obtain representations A_1', A_2' and E' of the group D_{3h}, realized by the $\psi_{x,y}$ functions, characterizing σ bands.

Acting by projection operators, we obtain that the representation A_1 is realized by the vector space with the basis vector $\psi_+^A + \psi_-^B$, and the representation A_2 is realized by the vector space with the basis vector $\psi_-^A - \psi_+^B$. The vector spaces realizing representations A_1 and A_2 being found, the representation E is obviously realized by the vector space spanned by the vectors $\psi_+^A - \psi_-^B, \psi_-^A + \psi_+^B$.

Because the irreducible representation E' is realized both by ψ_s and $\psi_{x,y}$ functions, these representations should be considered together. According to Wigner theorem [16] we still have two E' representations, each of them being realized by two functions from a quartet $\psi_s^A, \psi_s^B, \psi_+^A - \psi_-^B, \psi_+^A - \psi_-^B$. Each E' representation characterizes two σ bands, merging at the point K.

The symmetry of the electron bands at the points Γ and K being determined, the symmetry at the lines Γ-K follows unequivocally from the compatibility relations, presented in **Table 3** [8,13]. The table shows compatibility of the representations of the point group C_{2v}, realized at the symmetry line Γ-K, with those realized at the symmetry points Γ and K.

The results of this section are presented on **Figure 1**, reproduced from [8].

Silicene

The difference between silicene (or symmetrically equivalent to it buckled graphene) and graphene for our consideration is due solely to the decreased symmetry of the former. The group of the wavevector at the point K in silicene is D_3, at the point Γ – D_{3d} (this is also the point group of silicene). The representations of the group D_{3d} we can obtain on the basis of identity

$$D_{3d} = D_3 \times C_i \tag{12}$$

The direct product has twice as many representations as the group D_3, half of them being symmetric (denoted by the suffix g), and the other half antisymmetric (suffix u) with respect to inversion. The characters of the representations of the group D_{3d} are presented in the **Table 4**.

The symmetry analysis in silicene parallels that in graphene, so we'll be brief.

At the point K the functions $\phi_K^{1,B}$ realize E graphene, so we'll be brief.

Table 3: Compatibility relations.

C_2v			D_6h			D_3h
A_1	A_1g	B_2u	E_1u	E_2g	A_1'	E'
A_2	A_1u	B_2g	E_1g	E_2u	A''	E''
B_1	B_1u	A_2g	E_1u	E_2g	A_2'	E'
B_2	A_2u	B_1g	E_1g	E_2u	A''	E''

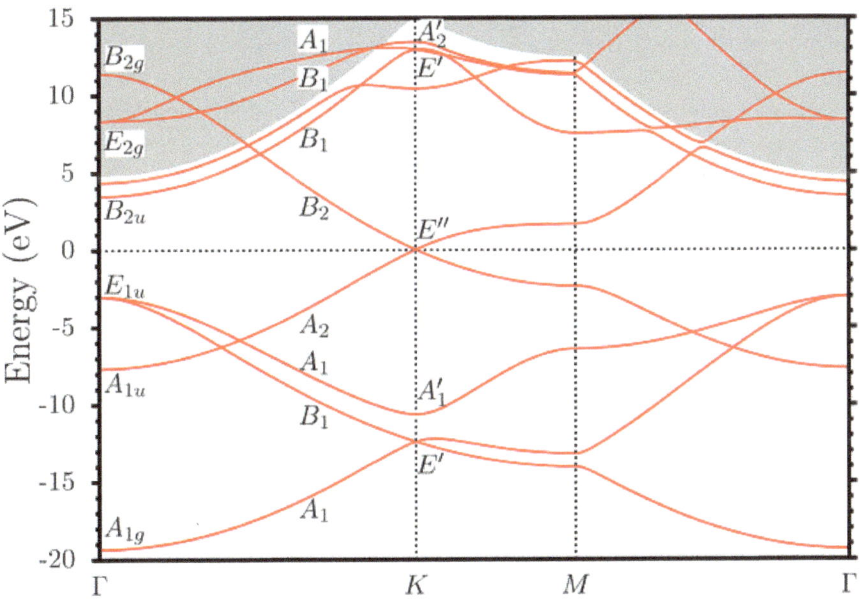

Figure 1: (Color online) Graphene band structure evaluated with use of the FP-LAPW method. The dashed line shows the Fermi energy [reproduced from [8]].

Table 4: Characters table for irreducible representations of D_{3d} point group.

Rep	E	$2C_3$	$3U_2$	I	$2S_6$	$3\sigma_d$
A_{1g}	1	1	1	1	1	1
A_{2g}	1	1	−1	1	1	−1
A_{1u}	1	1	1	−1	−1	−1
A_{2u}	1	1	−1	−1	−1	1
E_g	2	−1	0	2	−1	0
E_u	2	−1	0	−2	1	0

Orbitals $|s\rangle$ and $|z\rangle$ realize A_1 representation, and the orbitals $|x,y\rangle$ realize E representation of the group. Thus at the point K the functions ψ_s (and ψ_z) realize representation E of the group D_3. Reducible representation realized by the functions $\psi_{x,y}$ can be decomposed into the irreducible ones:

$$E\times E = A_1 + A_2 + E. \tag{13}$$

So when the symmetry is reduced by going from graphene to silicene, the representations A_1' and A_2' turn into A_1 and A_2. Representation E'' and two representations E' turn into three representations E. Loosing the reflection in plane symmetry, we can not claim now that one representation is realized exclusively by $|z\rangle$ orbitals. All the E representations mix $|s,p\rangle$ orbitals.

At the point Γ the the functions $\phi_0^{A,B}$ realizes reducible representation of the group D_{3d}:

$$R_\Gamma = A_{1g} + A_{2u}. \tag{14}$$

Orbitals $|s\rangle$ realize A_{1u}, orbitals $|z\rangle$ – A_{2u}, and orbitals $|x,y\rangle$ – E_u representations of the group. Thus the functions (1) realize reducible representation of the group D_{3d} which can be decomposed as

$$\left(A_{1u} + A_{2u} + E_u\right)\times\left(A_{1g} + A_{2u}\right)$$
$$= A_{1u} + A_{2g} + A_{2u} + A_{1g} + E_g + E_u. \tag{15}$$

So when the symmetry is reduced by going from graphene to silicene, the representations E_{2g} and E_{1u} turn into the representations E_g and E_u respectively.

The band structure of silicene is being different from that of graphene, the merging of the bands is no different. The statement becomes clear when comparing **Figure 2**, reproduced from [10], with **Figure 1**.

GROUP THEORY ANALYSIS IN THE TIGHT-BINDING MODEL WITH THE SPIN-ORBIT COUPLING

In the absence of spin-orbit coupling, electron spin can be taken into account in a trivial way: each band we considered was doubly spin degenerate.

When the spin-orbit coupling is taken into account, the symmetry, and the representations realized by the sub- lattice functions (2) remaim the same.

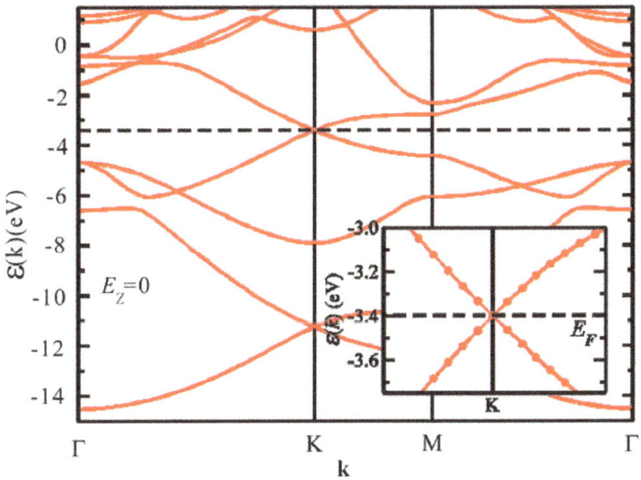

Figure 2: (Color online) DFT-PBE band structures for silicene. The dashed line shows the Fermi energy and the insets show the spectrum near the Fermi level in the vicinity of the K point [reproduced from [10] with permission].

However, instead of atomic orbitals we should consider atomic terms. So for the case of sp hybridization, β enumerates states from doublets $|s,p\rangle^{(1/2)}$ and quartet $|p\rangle^{(3/2)}$.

Due to the semi-integer value of the angular momentum J we have to consider double-valued representations realized by the atomic terms (and by the crystal wave functions). We remind that in this case it is convenient to introduce the concept of a new element of the group (denoted by Q); this is a rotation through an angle 2π about an arbitrary axis, and is not the unit element, but gives the latter when applied twice: $Q^2 = E$.

The characters of the rotation by angle ϕ applied to the term $|\rangle^{(J)}$ is

$$\chi^j(\phi) = \sin\left(J + \frac{1}{2}\right)\phi \Big/ \sin\frac{1}{2}\phi. \tag{16}$$

With respect to the inversion I the character is

$$\chi^j(I) = \pm(2J+1), \tag{17}$$

where the sign plus corresponds to the s states, and the sign minus to the p states. Finally, the charecters corresponding to reflection in a plane σ and rotary reflection through an angle ϕ are found writing these symmetry transformations as

$$\sigma = IC_2, \quad S(\phi) = IC(\phi + \pi). \tag{18}$$

Both in graphene and in silicene we'll restrict ourselves by the symmetry analysis at the point K.

Graphene

The sub-lattice functions $\phi_K^{A,B}$ realize E_g representation of point group D_{3h}. The electron terms realize two-valued representations of the group, which are presented in **Table 5** [17].

Doublet $|s\rangle^{(1/2)}$ realizes Γ_7 representation of the group; doublet $|p\rangle^{(1/2)}$ realizes Γ_8 representation, quartet

$|p\rangle^{(3/2)}$ realizes Γ_7 representation twice (We decided to use chemical notation for the single-valued representation, and BSW notation for double-valued representations [9]). The sub-lattice functions realize representation E_g; from Equation (3) we obtain

$$\Gamma_7 \times E_g = \Gamma_8 \times E_g = \Gamma_7 + \Gamma_8. \tag{19}$$

Thus at the point K four bands realize representation Γ_7 of the group D_3' each, and four bands realize representation Γ_8 each. In particular, we obtained the (well known) result that the four-fold degeneracy (including spin) of the bands merging at the point K is partially removed by the spin-orbit coupling, and only two-fold (Kramers) degeneracy is left.

Silicene

The two-valued representations of D_3' are presented in **Table 6** [17]. Each of the doublets $|s,p\rangle^{(1/2)}$ realizes Γ_4 representation of the group. Quartet $|p\rangle^{(3/2)}$ realizes this representation twice. The sub-lattice functions realize representation E of the group. From Equation (3) we obtain

$$\Gamma_4 \times E = \Gamma_4 + \Gamma_5 + \Gamma_6. \tag{20}$$

(For the same reasons as for ordinary representations, two complex conjugate two-valued representations $\Gamma_5 + \Gamma_6$ must be regarded as one physically irreducible representation of twice the dimension).

Thus at the point K four bands which realize representation Γ_4 of the group D_3' each, and four bands realize representation $\Gamma_5 + \Gamma_6$ each.

DIRAC POINTS

In this final part of the paper we would like to clarify the relation between the symmetry and the existence of dirac points.

According to the classical approach [18,19], the merging of the bands at a point k_0 is connected with the multi (higher than one)-dimensional representation of the space group G_0, realized in this point. Looking for a linear dispersion point in the vicinity of the merging point we may use the degenerate $k \cdot p$ perturbation theory. Let a two-dimensional irreducible representation is realized at a point k_0. Expanding the wavefunction with respect to the basis of the representation

$$\psi(k)=\sum_{i=1}^{2} c_i(k)\psi_i(k_0), \tag{21}$$

for the expansion coefficients we obtain equation in the form

Table 5: Characters for two-valued irreducible representations of group D'_{3h}.

D'_{3h}	E	Q	σ_h $\sigma_h Q$	C_3 $C_3 Q$	C_3^2 $C_3^2 Q$	$S3$ $S_3^2 Q$	S_3^2 $S_3 Q$	$3U_2$ $3U_2 Q$	$3\sigma_v$ $3\sigma_v Q$
Γ_7	2	−2	0	1	−1	$\sqrt{3}$	$-\sqrt{3}$	0	0
Γ_8	2	−2	0	1	−1	$-\sqrt{3}$	$-\sqrt{3}$	0	0
Γ_9	2	−2	0	−2	2	0	0	0	0

Table 6: Characters for two-valued irreducible representations of group D'.

Rep	E	Q	C_3 $C_3^2 Q$	C_3^2 $C_3 Q$	$3U_2$	$3U_2 Q$
Γ_5	1	−1	−1	1	i	$-i$
Γ_6	1	−1	−1	1	$-i$	i
Γ_4	2	−2	1	−1	0	0

$$\sum_{j=1}^{2} \frac{\boldsymbol{k} \cdot \boldsymbol{p}_{ij}}{m} c_j(k) = \varepsilon(k) c_i(k), \tag{22}$$

where $p_{ij} = \langle \psi_i(k_0) | p | \psi_j(k_0) \rangle$ (of course, we need the absence of inversion symmetry at the point, for the matrix elements to be different from zero). The dispersion law is given by the equation

$$\varepsilon(\boldsymbol{k}) = \sum_{\alpha} a_\alpha k_\alpha \pm \sqrt{\sum_{\alpha\beta} \gamma_{\alpha\beta} k_\alpha k_\beta} \tag{23}$$

where α, β are cartesian indexes x, y. Equation (24) should contain only combinations of wavevector components which are invariant with respect to all elements of the group G_0. In the case when the group G_0 does not have any vector invariants, and the only tensor invariant is the quantity $k_x^2 + k_y^2$, we obtain the dispersion law

$$\varepsilon(\boldsymbol{k}) = \pm v k, \tag{24}$$

which, like it was shown by Dirac himself in 1928, guaranties that Equation (25) is Dirac equation, in the sense the the matrices p_x and p_y satisfy anticommutation relations

$$p_x^2 = p_y^2 \propto I, \quad \{p_x, p_y\}_+ = 0, \tag{25}$$

where I is the unity matrix.

To be more specific, consider the groups of wavevector at the point Γ; in graphene it is D_{3h}, and in silicene it is D_3. In both cases, to find the dispersion law at the point Γ it is enough to study invariants of the group D_3. And we can easily checkup that both conditions, necessary for the existence of the Dirac point, are satisfied. This explains, in particular, why the band calculations show the existence of Dirac points in silicene [9,10], which has a lower symmetry than graphene.

In general, the role of the tight binding approximation in symmetry classification of the bands in graphene, like its role in symmetry classification of bands in other crystals, is only auxiliary. The approximation greatly helps in the classification and sheds additional light on the nature of the bands. but one must remember that there are more important things that this or that approximation and this is symmetry.

CONCLUSION

This paper presents an applications of group theory to very important cases of graphene and silicene.

ACKNOWLEDGEMENTS

Discussions with J. L. Manes, V. Falko and Hua Jiang were very illuminating for the author.

REFERENCES

1. K. S. Novoselov, A. K. Geim, S. V. Morozov, D. Jiang, Y. Zhang, S. V. Dubonos, I. V. Grigorieva and A. A. Firsov, "Electric Field Effect in Atomically Thin Carbon Films", Science, Vol. 306, No. 5696, 2004, pp. 666-669. doi:10.1126/science.1102896
2. W. M. Lomer, "The Valence Bands in Two-Dimensional Graphite," Proceedings of the Royal Society A, Vol. 227, No. 1170, 1955, pp. 330-349. doi:10.1098/rspa.1955.0014
3. J. C. Slonczewski and P. R. Weiss, "Band Structure of Graphite," Physical Review, Vol. 109, No. 2, 1958, pp. 272-279. doi:10.1103/PhysRev.109.272
4. G. Dresselhaus and M. S. Dresselhaus, "Spin-Orbit Interaction in Graphite," Physical Review, Vol. 140, No. 1A, 1965, pp. A401-A412. doi:10.1103/PhysRev.140.A401

5. F. Bassani and G. P. Parravicini, "Band Structure and Optical Properties of Graphite and of the Layer Compounds GaS and GaSe," Nuovo Cimento B, Vol. 50, No. 1, 1967, pp. 95-128. doi:10.1007/BF02710685

6. L. M. Malard, M. H. D. Guimaraes, D. L. Mafra, M. S. C. Mazzoni and A. Jorio, "Group-Theory Analysis of Electrons and Phonons in N-Layer Graphene Systems," Physical Review B, Vol. 79, No. 12, 2009, pp. 125426-125433. doi:10.1103/PhysRevB.79.125426

7. J. L. Manes, "Existence of Bulk Chiral Fermions and Crystal Symmetry," Physical Review B, Vol. 85, No. 15, 2012, pp. 155118-115125. doi:10.1103/PhysRevB.85.155118

8. E. Kogan and V. U. Nazarov, "Symmetry Classification of Energy Bands in Graphene," Physical Review B, Vol. 85, No. 11, 2012, pp. 115418-115423. doi:10.1103/PhysRevB.85.115418

9. C.-C. Liu, H. Jiang and Y. G. Yao, "Low-Energy Effective Hamiltonian Involving Spin-Orbit Coupling in Silicene and Two-Dimensional Germanium and Tin," Physical Review B, Vol. 84, No. 19, 2011, pp. 195430- 195440. doi:10.1103/PhysRevB.84.195430

10. N. D. Drummond, V. Zolyomi and V. I. Falko, "Electrically Tunable Band Gap in Silicone," Physical Review B, Vol. 85, 2012, pp. 075423-1-075423-7.

11. C. Kittel, "Quantum Theory of Solids," John Wiley & and Sons, Inc., New York, London, 1963.

12. W. A. Harrison, "Solid State Theory," McGraw Hill Book Company, New York, London, Toronto, 1970.

13. L. D. Landau and E. M. Lifshitz, "Quantum Mechanics," Pergamon Press, Oxford, 1991.

14. R. S. Knox and S. Gold, "Symmmetry in the Solid State", W. A. Benjamin, Inc., New York, Amsterdam, 1964.

15. C. Thomsen, S. Reich and J. Maultzsch, "Carbon Nanotubes: Basic Concepts and Physical Properties," Wiley Online Library, Wiley-VCH Verlag GmbH, 2004.

16. M. I. Petrashen, E. D. Trifonov and J. L. Trifonov, "Applications of Group Theory in Quantum Mechanics," Dover Publications, New York, 2009.

17. G. F. Koster, J. O. Dimmock, R. G. Wheeler and H. Statz, "Properties of the Thirty-Two Point Groups," MIT Press, Cambridge, 1964.

18. F. Hund, "Über den Zusammenhang Zwischen der Symmetrie eines Kristallgitters und den Zuständen seiner Elektronen," Zeitschrift für Physik, Vol. 99, No. 1-2, 1936, pp. 119-136.doi:10.1007/BF01847819

19. C. Herring, "Accidental Degeneracy in the Energy Bands of Crystals," Physical Review B, Vol. 52, No. 4, 1937, pp. 365-373. doi:10.1103/PhysRev.52.365

15
CHAPTER

EFFECT OF SYMMETRY BREAKING ON ELECTRONIC BAND STRUCTURE: GAP OPENING AT THE HIGH SYMMETRY POINTS

Guillaume Vasseur, Yannick Fagot-Revurat, Bertrand Kierren, Muriel Sicot and Daniel Malterre

Keywords: Symmetry, electronic structure, gap opening

INTRODUCTION

Symmetry is probably one of the most general and fundamental concepts in physics, and its central role was only recognized in the 20th century. The most emblematic example is particle physics, for which the space-time symmetry and the internal symmetries are essential to understand the interactions among elementary particles. The fundamental equations of physics, like Maxwell or Dirac equations, can be considered as a direct consequence of the symmetry principles. Group theory is then a powerful tool to exploit the symmetry concepts in all domains of physics. As an example, the representation theory shows that the essential degeneracy of an energy level is given by the dimension of the corresponding irreducible representation of the symmetry group.

Moreover, symmetry breaking emerged more recently and was shown to be a generic behavior in matter. In his famous article "More is different", Anderson, P.W. discussed the broken symmetry and the nature of the hierarchical structure of science [1]. He showed that for large systems, the physical states are usually less symmetrical than the laws. A trivial example is crystals, which clearly violate the homogeneity and isotropy of space and physical laws. Such a behavior is an example of the spontaneous symmetry breaking mechanism. More elaborate is the Higgs mechanism and the

Citation: Guillaume Vasseur, Yannick Fagot-Revurat, Bertrand Kierren, Muriel Sicot and Daniel Malterre, Effect of Symmetry Breaking on Electronic Band Structure: Gap Opening at the High Symmetry Points, doi:10.3390/sym5040344.

generation of mass for bosons in particle physics [2]. Such a mechanism is also useful in condensed matter physics, in particular, for phase transitions [3]. Another kind of symmetry breaking can be defined: intrinsic broken symmetry corresponding to the modification of the symmetry of the Hamiltonian. As an example, an external magnetic field yields the breaking of time reversal symmetry (Zeeman effect); it slightly affects the energy spectra of atoms, but usually changes the degeneracy. In the following, we will discuss this kind of symmetry breaking, yielding the modification of electronic properties at the surface of a crystal.

In this paper, we study the interplay between symmetry and the electronic properties of 2D systems with hexagonal periodic arrangements. Such honeycomb lattices can exhibit, in the most symmetrical case, singular electronic properties, which are associated with crossing bands at the *K*points of the Brillouin zone (BZ) with linear dispersions. When this band crossing occurs at the Fermi energy, the low energy excitations can be described by an effective 2D Dirac equation of massless fermions, like in graphene [4]. This singular behavior is directly related to the symmetry of the wave vector group at the *K* points and to the dimension of the corresponding irreducible representation [5]. A symmetry breaking with a change of the wave vector group can lead to non-crossing bands and, then, to gap opening and curvature of the band dispersion. This was observed, for example, by growing a graphene layer on a crystal surface [6]. The broken symmetry is then reflected by a generation of mass, since the Dirac equation governing the low energy excitations corresponds to finite mass associated with the gap width. The system we studied in this paper, the reconstructed Ag/Cu(111) surface, exhibits this kind of symmetry. Due to the misfit of lattice parameters of Ag and Cu, one monolayer of Ag epitaxially grown on Cu(111) induces a reconstruction consisting of an hexagonal lattice of dislocation loops [7]. In the (111) surface of noble metals, the surface states exhibit a nearly free electron-like behavior (Shockley states), and in Ag/Cu(111), its band structure is characterized by gaps at the boundary of the reconstructed Brillouin zone. Band gap opening reflects the symmetry of wave vector groups, and we show that the gap magnitudes can be used to obtain the surface potential or at least their first Fourier components. After a general discussion of the symmetry on the hexagonal 2D lattices and, in particular, the relation between the gap at the *K* point and the group of the wave vector, we present angle-resolved photoemission spectroscopy (ARPES) and scanning tunneling microscopy and spectroscopy (STM/STS) results obtained on Ag/Cu(111) and K/Ag/Cu(111). For these interfaces, minigaps appear at the *K* points of the surface Brillouin zone. We discuss, in detail, this system, especially the relation between symmetry and the surface potential.

EXPERIMENTAL DETAILS AND METHOD OF CALCULATIONS

Experimental Details

The angle-resolved photoemission experiments were performed at 80 K with a Scienta 200 high-resolution hemispherical analyzer using photon energy of h = 21.2 eV. The presented ARPES maps correspond to second derivative signals in order to emphasize the energy gaps. Data points in Figure 3 and Figure 4 are obtained from standard line fits of individual energy distribution curves (EDCs), using Lorentzian functions and Fermi edge functions. STM and STS experiments have been carried out using Low Temperature-Omicron STM. The dI/dV spectra and maps were recorded at 5 K in the open feedback loop mode using the lock-in technique with a bias modulation of 5 mV at 2300 Hz (stabilization parameters: tunneling current 1 nA, bias voltage 1.0 V). Before transferring to the STM cryostat at 5 K, the Ag monolayer is evaporated at 400 K on the clean Cu(111) crystal. This process leads to the sharp ~(9.5 × 9.5) hexagonal lattice of triangular dislocations. Evaporation of K atoms on the Ag/Cu(111) surface at room temperature was achieved by heating a standard K getter.

Calculation Method

Calculations of the electronic properties of the Shockley states are carried out by a direct diagonalization of the Schrödinger equation with a potential defined by the first Fourier components we experimentally deduced from the gap widths. As Shockley states exhibit nearly free electron-like behavior, a plane wave basis is appropriate, and we limit to the first 36 waves around the Γ point. Additional plane waves have negligible effects. Local Density of states (LDOS) maps and spectra are obtained resolving the Hamiltonian for more than 1500 wave vectors in the first Brillouin zone (BZ) and by taking into account a 15 meV Gaussian energy broadening.

RESULTS AND DISCUSSION

Symmetry Analysis

Let us consider a 2D hexagonal lattice. The unit cell is diamond-shaped but, the Wigner–Seitz cell is hexagonal. The Brillouin zone, which is only related to translational periodicity, is also hexagonal, but rotated by π/6π/6 with respect to the Wigner–Seitz cell, and the symmetry elements of direct and reciprocal space are the same. We will consider the two high symmetry *M* and *K* points. We have three inequivalent (not related by a reciprocal space vector) *M* points (*M*, M′M′, M”M″) and two inequivalent *K* points (*K*, K′K′), as shown in Figure 1a. We would like to point out the peculiar property of the *K* point with respect to time reversal symmetry. By contrast to the *M* point, time reversal symmetry couples the two non-equivalent *K* and K′K′ points. Let us choose the 2D highly symmetric group, C^1_{6v} or P6mm. This is the 2D-space group of graphene. Among the symmetry elements, one has two families of three-fold mirrors, σv and σh, as illustrated in Figure 1b, one (σ_v) crossing all *M* points, the other one (σ_h) all *K* points. For the electronic band structure, we have to consider the group of the wave vector at the high symmetry points of the BZ. The wave vector group is the group of transformations conserving the wave vector in the first Brillouin zone (*i.e.*, modulo a reciprocal space vector). At the Γ point, the symmetry group of the wave vector is C_{6v}; at the *M* point it is C_{2v}, and at the *K* point it is C_{3v}. These considerations are important for the essential band degeneracy, since the band energy levels at a given wave vector of the first

Brillouin zone are associated with the irreducible representations of these groups. As an example, the *K* point wave vector group is non-abelian with a two-dimensional irreducible representation, *E*, corresponding to a two-fold degenerate band. This is the case of the p_z-derived bands of graphene, which exhibit the famous Dirac points characterized by the crossing of two linear dispersive bands at the *K* points [4]:

$$E_{\pm}(\vec{q}) = \pm\hbar v_F \vec{q} \tag{1}$$

where $\vec{q} = \vec{k} - \vec{k}_K$ is the crystal momentum with respect to the *K* point and v_F the Fermi velocity.

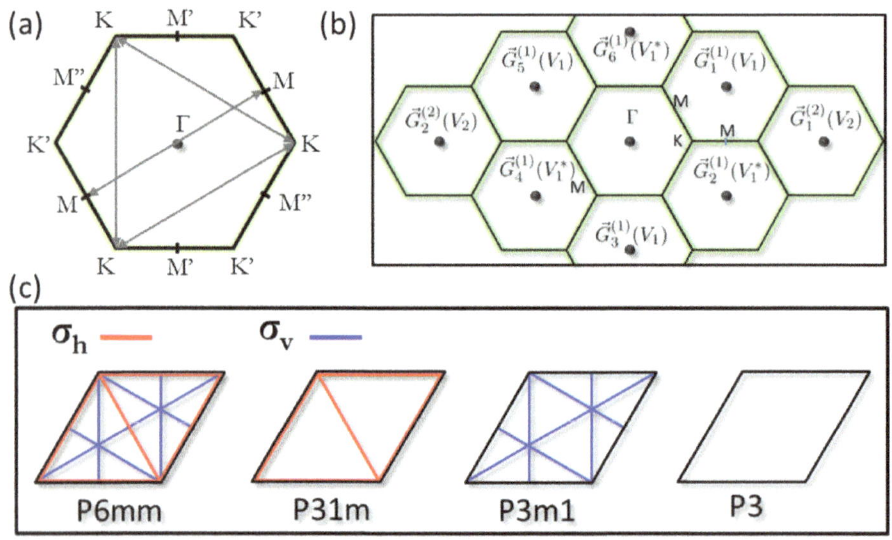

Figure 1: (**a**) The first Brillouin zone of a reciprocal space of 2D hexagonal lattices. The different *M* and *K* points are indicated (equivalent points are connected by a reciprocal space vector); (**b**) The unit cell with the symmetry mirrors for P6$_{mm}$, P31$_m$, P3$_m$1 and P3 space groups; (**c**) The reciprocal vectors in the first $(\vec{G}_n^{(1)})$ and second $(\vec{G}_n^{(2)})$ rings around Γ, as well as the corresponding Fourier component of the potential (see text).

Let us consider the breakdown of this symmetry by removing one of the mirror families. The removal of the σh family leads to the $P3m1$ (or C_{3v}^1) space group. The consequence is a change of the wave vector groups; at Γ, it is now C_{3v} or 3m, at M, C_{1h} or m, and at K, C_3 or three. As the K-vector group is now abelian, the essential degeneracy is left (the 2D-irreducible representation, E, of C_{3v} is decomposed in two 1D-representations of C_3), and one expects the band gap opening with the disappearance of the linear dispersion. Therefore, the Dirac points characterized by massless fermions for P6mm disappear, and the symmetry breaking leads to massive fermions and, then, to a gap. The dispersion relation close to the K points can be written:

$$E_{\pm}(\vec{q}) = \pm\sqrt{(\hbar v_F \vec{q})^2 + \Delta^2} \tag{2}$$

where $\Delta = m^* v_F^2$ is the mass term and 2Δ the gap width. This is what happens in the hexagonal boron nitride layer or epitaxial graphene, for example, on Ir(111) [8]. We will discuss in the following the triangular reconstruction of Ag/Cu(111), which exhibits this peculiar symmetry, and we will investigate the consequences on the Shockley surface state properties.

If the other mirror family (σ_v) is removed, the situation is completely different. The space group is then $P31m$ (or C_{3v}^2), and the wave vector groups are C_{3v} at Γ and C_{1h} at M, but remain C_{3v} at K. Therefore, no change is expected in the band degeneracy, and the band states corresponding to the two-dimensional E irreducible representation remain massless (linear dispersion with no gap), like in the high symmetry P6mm group. This behavior shows that the pertinent symmetry breaking is related to the symmetry of the wave vector group. For the lowest symmetric P3 group, both mirror families are removed, leading to the opening of a gap at K points.

The Ag/Cu(111) Reconstruction

As discussed above, the (111) surfaces of noble metals are known to exhibit surface states with a nearly free electron-like behavior (Shockley surface states [9]). Due to their surface localization, these Shockley bands are very sensitive to any surface modification. Deposition of atoms and formation of an ultra-thin epitaxial layer yield a change of the surface state energy and/or effective mass. For example, one monolayer of Ag epitaxially grown on Cu(111) leads to an energy shift of the surface state, but also to a reconstruction, due to the different atomic radius of Cu and Ag. As a consequence, a super-periodic cell ~(9.5 × 9.5) is observed, characterized by an hexagonal array of triangular dislocations [7,10]. The surface Brillouin zone of the reconstructed layer is small, and the parabolic dispersions of the Shockley state exhibit deviations close to the zone boundaries with the gap opening [11,12,13]. These gaps result from the Bragg mechanism at these high symmetry points and lines. For example, the *M* point of the first Brillouin zone is situated between the Γ point and one reciprocal vector, $\vec{G}_1^{(1)}$ (Figure 1b). In the nearly free electron framework, the band states at *M* are simply linear combinations of the incoming state, $|\vec{k}_M\rangle$, and the diffracted one, $|\vec{k}_M - \vec{G}_1^{(1)}\rangle$ (in the first approximation). The energy difference between these two band states $(|\vec{k}_M\rangle \pm |\vec{k}_M - \vec{G}_1^{(1)}\rangle)/\sqrt{2}$ is $2|V_{\vec{G}_1^{(1)}}|$, where $V_{\vec{G}_1^{(1)}} = V_1$ is the Fourier component of the surface potential, like in a simple 1D nearly-free-electron gas, discussed in solid state textbooks [14]. At the *K* point, the situation is more complicated. The *K* point is at the center of an equilateral triangle defined by Γ, $\vec{G}_1^{(1)}$ and $\vec{G}_2^{(1)}$, and three states composed from $|\vec{k}_K\rangle$, $|\vec{k}_K - \vec{G}_1^{(1)}\rangle$ and $|\vec{k}_K - \vec{G}_2^{(1)}\rangle$ are involved.

The widths of the gaps in the surface band structure give partial information about the potential associated with the reconstruction, but only occupied bands are accessible with ARPES, limiting the number of observable gaps. However, it has been shown that, by depositing K atoms, it is possible to shift the energy of the surface Shockley band towards higher binding energies [15,16]. We used this property in order to tune the energy gaps below the Fermi energy. These experimental gap widths allow one to determine several Fourier components and, then, to deduce the surface potential [17].

Symmetry and Potential Properties

We would like to discuss the link between the symmetry of the surface and the properties of the Fourier components of the potential, at least the first two ones. Firstly, we would like to consider the highly symmetric case corresponding to the P6mmP6mm space group. The presence of the six-fold axis induces that the Fourier components corresponding to the first six equivalent reciprocal vectors are all identical. In particular, one has:

$$V_{\vec{G}} = V_{-\vec{G}} \tag{3}$$

Moreover, it is possible to show that this component is purely real. Indeed, time reversal symmetry (T), which, for a spinless particle, is simply the complex conjugate operator, transforms the potential according to:

$$T(V(\vec{r})) = T\left(\sum V_{\vec{G}} \exp i\vec{G}\cdot\vec{r}\right) = \sum V^*_{\vec{G}} \exp -i\vec{G}\cdot\vec{r} = \sum V^*_{-\vec{G}} \exp i\vec{G}\cdot\vec{r} = V(\vec{r}) \tag{4}$$

since the potential is real. Then, one deduces that:

$$V_{\vec{G}} = V^*_{-\vec{G}} \tag{5}$$

which proves with Equation (3) that the Fourier components, $V_{\vec{G}}$, are real. A typical band structure in the nearly free electron model is reported in Figure 2a (center panel). We also show the mirror symmetries in the Brillouin zone and the potential used to compute the band structure. This potential was obtained from the Fourier components associated with the reciprocal vectors in the first two rings around Γ, as illustrated in Figure 1b. The dispersions exhibit a first gap at M and linear dispersive crossing bands at the K point. This illustrates the symmetry origin of this singular behavior, since it is encountered both in graphene (tight binding bands) and in Ag/Cu(111) (nearly free electron bands).

When we remove one mirror family, either σ_h or σ_v, two kinds of reciprocal vectors appear in the first ring around Γ: three equivalent ones associated with $\vec{G}_1^{(1)}$, the three equivalent other ones being associated with $\vec{G}_2^{(1)}$ (Figure 1b). For the P31m space group (Figure 2b), the remaining mirrors (σ_h) are along the ΓK directions and contain the six first reciprocal vectors in the second ring $(\vec{G}_n^{(2)}, n = 1, \ldots 6)$. Therefore, this mirror symmetry does not couple the two kinds of reciprocal vectors of this ring, and the corresponding Fourier components of the potential can be complex $(V_{\vec{G}_n^{(2)}} = V_2, n = 1, 3, 5$ and $V_{\vec{G}_n^{(2)}} = V_2^*, n = 2, 4, 6)$. On the other hand, the σh mirror couples the two sets of reciprocal vectors in the first ring $(\vec{G}_n^{(1)})$. Then, the first components (associated with the six reciprocal vectors of the first ring) are purely real: $V_{\vec{G}_n^{(1)}} = V_1, \forall n$. The contrary behavior occurs for the P3m1 space group (Figure 2c). The remaining mirrors (σv) are along the ΓM directions and contain the six first reciprocal vectors in the first ring and couple the two kinds of reciprocal vectors in the second ring. As a consequence, the Fourier components of the second ring are purely real $V_{\vec{G}_n^{(2)}} = V_2, \forall n$, ∀n, whereas the components belonging to the first ring can be complex. The band structure of Figure 2b illustrates the consequence on the electronic properties. As discussed above, the wave vector group at K remains C_{3v} for $P31_m$, and no gap opens; whereas it is only C_3 for $P3_m1$, and a gap appears with curvature in the band dispersion. Finally, if the two families of mirrors are removed (Figure 2d and the P3 space group), the first components (first ring) and the second ones (second ring) are both complex, the K-wave vector is C_3 and a gap is also exhibited at the K points.

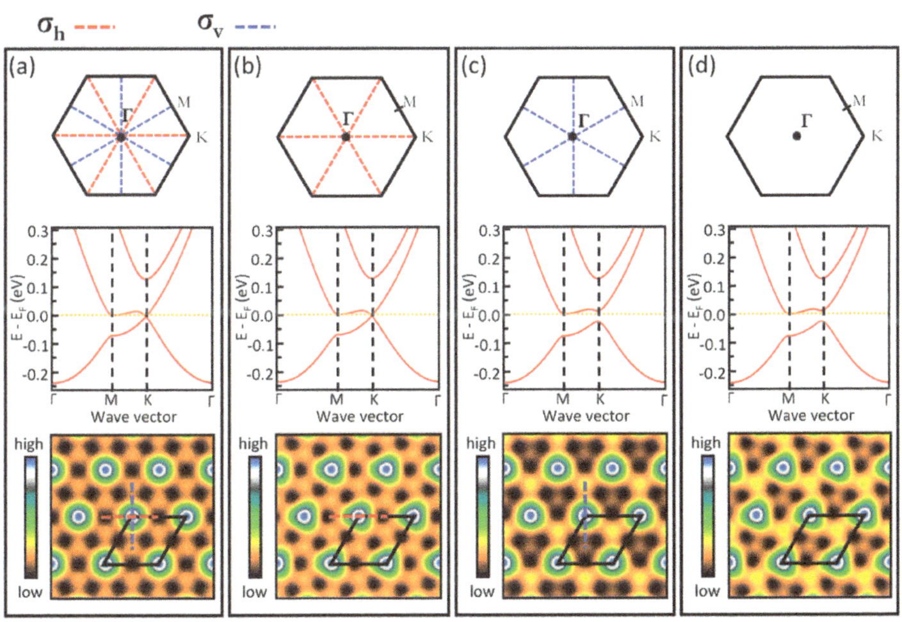

Figure 2: Brillouin zone with the mirror symmetries (top panel), schematic band structures in the nearly free electron model (center panel) and the corresponding direct space surface potential (bottom panel) for the different space groups: (a) P6mm; (b) P31m; (c) P3m1; and (d) P3. Comparison between these band structures shows that the σh mirrors are necessary to have a gapless Dirac point at K.

This symmetry analysis shows that the opening of a gap at the *K* point associated with massive fermions is simply related to the wave vector group at this high symmetry point. In the nearly free electron approach valid for Shockley surface states of a (111) noble metal surface, the symmetry imposes some rules for the Fourier components, in particular, the value of the imaginary parts of these components.

Electronic Properties of Ag/Cu(111)

Figure 3a shows an STM image of the Ag/Cu(111) surface. It exhibits the triangular lattice of dislocation loops, evidencing that only the σ_v mirrors are symmetry elements, leading to a P3m1 space group for this 2D crystal. Figure 3b reports the ARPES dispersion (the second derivative of the experimental signal) in the ΓM and ΓK directions for the Ag/Cu(111) surface and a K-doped surface. By doping the surface with K, it is possible to tune the surface state energy and, then, to measure the second energy gap at the M point, which is above the Fermi energy for the undoped surface. For the bare Ag/Cu(111) surface, a 80 meV gap is evidenced just below E_F , whereas the change in curvature suggests that a gap could exist at EF at the K point. This is corroborated by the results on the K-doped surface, which indicates the gap opening at K and a third band dispersing upward (close up of Figure 3b). Moreover, inspection of the reciprocal space (Figure 1a) shows that the Γ, the K of the first Brillouin zone and the M point in the second zone are aligned. Therefore, the experimental dispersion along the ΓK direction allows one to probe the first gap at K and the second gap at M. This is exactly what is measured in the experimental spectra of the K-doped surface (Figure 3b). From the gap widths, it is possible to estimate the Fourier components of the surface potential. Indeed, due to the nearly free electron behavior of the Shockley states, only a few plane waves are involved in the high symmetry points, mainly two for M points and three for K points. As a consequence, in a first approximation, the gaps at M and K only depend on the first Fourier component, since it is straightforward to show that:

$$E_{M2} - E_{M1} = 2|V_1| \quad \text{and} \quad E_{K2} - E_{K1} = 2\sqrt{3}Im|V_1| \tag{6}$$

leading to a determination of the real and imaginary parts of the V_1 component. We find $V_1 = (47 + i9.6)$ meV. The second component $V_2 = 15$ meV (purely real; see above) can be deduced for the gap appearing close to the Fermi energy for the doped surface at the M point. We would like to point out that, as the gap width is proportional to the imaginary part of V_1, a purely real Fourier component leads to a zero gap at K, corroborating our discussion above based on symmetry arguments.

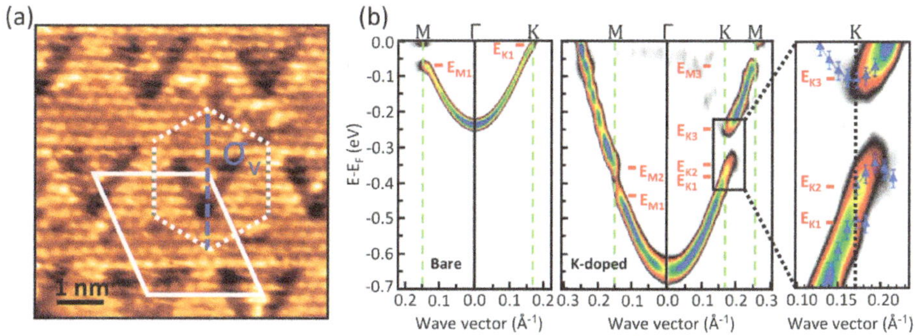

Figure 3: (**a**) Scanning tunneling microscopy (STM) image of the Ag/Cu(111) surface with the unit cell and the three-fold σ_v mirror. Solid and dotted lines represent the unit and Wigner–Seitz cells, respectively; (**b**) Second derivative angle-resolved photoemission spectroscopy (ARPES) signal representing the band dispersions measured on Ag/Cu(111) (left) and a K-doped surface (closeup illustrating the three bands close to the *K*point).

The $|V_1|$ value is corroborated by the intensity of the folded band. Indeed, although the band dispersions are periodic in the reciprocal space, the spectral weight is not periodic, and the folded band intensity in the second Brillouin zone is a function of the Fourier component [18] and can be simply calculated. Figure 4a shows the experimental intensity in the ΓMΓM direction for the main band and its replicate for the undoped surface. We also show the calculated ARPES intensity with $|V_1|$ experimentally deduced from the gaps. A very good agreement between experimental and calculated spectral intensities is obtained. With the V_1 and V_2 components we deduce from the gap widths at *M* and *K*, we can rebuild the potential associated with the Ag/Cu(111) reconstructed surface, as shown in Figure 4b. By solving numerically the Schrödinger equation with this potential, it is possible to compute the band structure in the high symmetry directions. The results are reported in Figure 4c and compared with the ARPES data (for the occupied states) and the density of states (occupied and unoccupied states) measured by scanning tunneling spectroscopy. A satisfactory agreement is found, demonstrating that this simple approach captures the essential features of the electronic properties.

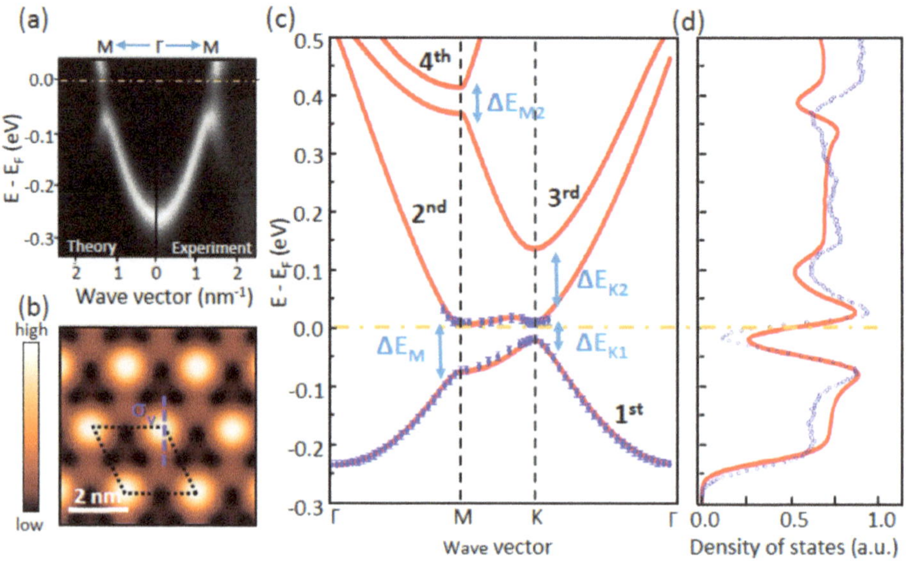

Figure 4: (**a**) Comparison between the experimental and calculated ARPES intensity for Ag/Cu(111) showing that both the energy gap width and the intensity of the folded band are reproduced. The spectra have been divided by the Fermi function in order to evidence the spectral weight in the 3kTkTenergy range above the Fermi energy; (**b**) The surface potential built from the V_1 and V_2 components; (**c**) The surface band structure (lines) calculated from the experimentally determined surface potential and experimental dispersions (symbols) up to the Fermi energy; (**d**) Experimental scanning tunneling spectrum (symbols) and calculated local density of states (lines) evidencing the energy gaps of the band structure.

CONCLUSIONS

In this paper, we show that the peculiar electronic properties usually encountered in hexagonal 2D lattices (*i.e.*, the existence of a gapless Dirac point at *K* points of the Brillouin zone with linear dispersion) result from the group of wave vectors at this specific high symmetry point. This behavior appears for states belonging to the two dimensional irreducible representation of the corresponding wave vector group. A symmetry breaking that modifies the wave vector group at the *K* point leads to the opening of an energy gap associated with finite mass Dirac electrons. More precisely, it is related to the breakdown of the mirror symmetry crossing the *K* point. This behavior is illustrated on the Shockley state of the Ag/Cu(111) reconstructed surface. In this system, the surface electronic properties can be described in the nearly fee electron model. From the measured gap widths, the first Fourier components can be obtained, and the surface potential can be built.

CONFLICTS OF INTEREST

The authors declare no conflict of interest.

REFERENCES

1. Anderson, P.W. More is different. *Science* **1972**, *177*, 393–397.
2. Higgs, P.W. Broken symmetries and the masses of gauge bosons. *Phys. Rev. Lett.* **1964**, *13*, 508–509.
3. Anderson, P.W. *Basic Notions of Condensed Matter Physics*; The Benjamin/ Cummings Publishing Company, Inc.: Menio Park, CA, USA, 1984.
4. Castro Neto, A.H.; Guinea, F.; Peres, N.M.R.; Novoselov, K.S.; Geim, A.K. The electronic properties of graphene. *Rev. Mod. Phys.* **2009**, *81*, 109–162.
5. Kogan, E.; Nazarov, V.U. Symmetry classification of energy bands in graphene. *Phys. Rev. B***2012**, *85*, 115418:1–115418:5.
6. Starodub, E.; Bostwick, A.; Moreschini, L.; Nie, S.; El Gabaly, F.; McCarty, K.F.; Rotenberg, E. In-plane orientation effects on the electronic structure, stability, and Raman scattering of monolayer graphene on Ir(111). *Phys. Rev. B* **2011**, *83*, 125428:1–125428:9.
7. Meunier, I.; Tréglia, G.; Gay, J.-M.; Aufray, B.; Legrand, B. Ag/Cu(111) structure revisited through an extended mechanism for stress relaxation. *Phys. Rev. B* **1999**, *59*, 10910–10917.
8. N'Diaye, A.T.; Bleikamp, S.; Feibelman, P.J.; Michely, T. Two-dimensional Ir cluster lattice on a graphene moiré on Ir(111). *Phys. Rev. Lett.* **2006**, *97*, 215501:1–215501:4.
9. Reinert, F.; Nicolay, G.; Schmidt, S.; Ehm, D.; Hüfner, S. Direct measurements of the L-gap surface states on the (111) face of noble metals by photoelectron spectroscopy. *Phys. Rev. B***2001**, *63*, 115415:1–115415:7.
10. Bendounan, A.; Cercellier, H.; Fagot-Revurat, Y.; Kierren, B.; Yurov, V.Y.; Malterre, D. Modification of Shockley states induced by surface reconstruction in epitaxial Ag films on Cu(111). *Phys. Rev. B* **2003**, *67*, 165412:1–165412:10.
11. Schiller, F.; Cordón, J.; Vyalikh, D.; Rubio, A.; Ortega, J.E. Fermi gap stabilization of an incommensurate two-dimensional superstructure. *Phys. Rev. Lett.* **2005**, *94*, 016103:1–016103:4.
12. Bendounan, A.; Forster, F.; Ziroff, J.; Schmitt, F.; Reinert, F. Influence of the reconstruction in Ag/Cu(111) on the surface electronic structure: Quantitative analysis of the induced band gap. *Phys. Rev. B* **2005**, *72*, 075407:1–075407:6.
13. Bendounan, A.; Forster, F.; Reinert, F.; Kierren, B.; Fagot-Revurat, Y.; Malterre, D. Comment on fermi gap stabilization of an incommensurate two-dimensional superstructure. *Phys. Rev. Lett.* **2006**, *96*.
14. Ashcroft, N.W.; Mermin, N.D. *Solid State Physics*; Cengage Learning: Stamford, CT, USA, 1976.
15. Abd El-Fattah, Z.M.; Matena, M.; Corso, M.; García de Abajo, F.J.; Schiller, F.; Ortega, J.E. Lifshitz transition across the Ag/Cu(111) superlattice band gap tuned by interface doping.*Phys. Rev. Lett.* **2011**, *107*, 066803:1–066803:5.

16. García de Abajo, F.J.; Cordón, J.; Corso, M.; Schiller, F.; Ortega, J.E. Fermi gap engineering by Au doping of the Ag/Cu(111) dislocation network. *Nanoscale* **2010**, *2*, 717–721.

17. Vasseur, G.; Fagot-Revurat, Y.; Kierren, B.; Sicot, M.; Malterre, D. Electronic surface potential from angle resolved photoemission. *Phys. Rev. B.* **2013**. in press.

18. Malterre, D.; Kierren, B.; Fagot-Revurat, Y.; Pons, S.; Tejeda, A.; Didiot, C.; Cercellier, H.; Bendounan, A. ARPES and STS investigation of Shockley states in thin metallic films and periodic nanostructures. *New J. Phys.* **2007**, *9*, 391:1–391:29.

INDEX